SURFACE PROPERTIES OF LAYERED STRUCTURES

Physics and Chemistry of Materials with Low-Dimensional Structures

VOLUME 16

The titles published in this series are listed at the end of this volume.

SURFACE PROPERTIES OF LAYERED STRUCTURES

Edited by

G. Benedek

Department of Physics,
University of Milan, Italy

SPRINGER-SCIENCE+BUSINESS MEDIA, B.V.

Library of Congress Cataloging-in-Publication Data

Surface properties of layered structures / edited by G. Benedek.
p. cm. -- (Physics and chemistry of materials with low-dimensional structures ; v. 16)
Includes bibliographical references and index.
ISBN 978-94-010-5188-0 ISBN 978-94-011-2684-7 (eBook)
DOI 10.1007/978-94-011-2684-7
1. Surfaces (Physics) 2. Layer structure (Solids) 3. Surface chemistry. I. Benedek, G. (Giorgio) II. Series.
QC173.4.S94S9636 1992
530.4'27--dc20 92-26745

ISBN 978-94-010-5188-0

Printed on acid-free paper

TABLE OF CONTENTS

Preface

Layered materials are characterized by a large anisotropy of either structural or compositional nature. Structural layered materials are strongly bound in two directions and loosely bound in the third one (the c axis). Compositional layered materials are stacks of thin slabs of different composition. Of course any polyatomic crystal may be viewed as a suitable periodic stack of atomic planes of different composition. We conveniently speak of a compositional layered material when the stack period along the c axis is appreciably larger than the unit cell size in the layer plane.

Structural layered materials do often exhibit compositional anisotropy (with the obvious exception of monolayered crystals such as graphite). On the contrary certain compositional layered materials, such as semiconductor and metal superlattices, have a small structural anisotropy, their physical and technical interest arising from subtler forms of anisotropy, e.g., transport properties or magnetic behaviour. Structural anisotropy normally implies a large anisotropy in both phonon dispersion and electronic band structure, as well as a fortiori in transport and magnetic properties. Thus in either forms of layered materials a number of physical properties may show a pronounced two-dimensional character, which is currently taken as the experimental signature of a layered crystal.

What is so interesting in the surface properties of layered structures? The recent experimental studies on the surface properties of layered crystals were motivated by two different thoughts. The first and simpler thought is that the structure and the dynamics of the surface layer should be very similar to those of the bulk layers due to the weakness of interlayer forces, and therefore the microscopy and spectroscopy with surface-sensitive probes actually provide an approximate information on bulk properties. Such possibility would permit to overcome the difficulty in using penetrating probes (for example neutrons) for layered structures which can hardly be grown as thick, stacking-fault-free monocrystals.

The second, apparently opposite argument is that the structure and the dynamics of the surface layer are intrinsically different from those of the bulk layers, as for any other crystal, but, due to the weak interaction between adjacent layers, such differences are strongly localized at the surface and indicative of a genuine two-dimensional behaviour. In this case the layered structures may be viewed as the physical support for a two-dimensional world, and their surface properties have their own intrinsic value. In this sense the layered structures may also work as excellent substrates for extrinsic two-dimensional systems, i.e., physically adsorbed phases. The weak interlayer coupling to the layered substrate ensures chemical inertness and structural perfection of the surface - two ideal conditions for physisorption studies.

The structural analysis of layered crystal surfaces by means of tunneling microscopies as well as the studies of the electronic and vibrational spectra, discussed in the first two parts of this book, will naturally have the reader face, and possibly choose one of the two points of view. Graphite is the simplest, paradigmatic example for a layered structure. Theory and experiment on the surface phonons of graphite agree that a negligible difference exists between surface and bulk phonon structures, so that surface sensitive probes like electron energy loss spectroscopy and helium atom scattering yield full information on the bulk phonon structure, otherwise inaccessible to neutrons. On the contrary inverse photoemission studies on graphite combined with electronic band structure calculations provide evidence for surface-localized excited states.

This book is organized in three parts. The first part is devoted to the crystallographic structure of layered crystal surfaces; the second part to their electronic and vibrational structures; the third part to the characterization and physical properties of adsorbed phases on layered crystal surfaces. R. Wiesendanger and D. Anselmetti present a general introduction to the surface crystallography of layered materials by means of scanning tunneling (STM) and atomic force (AFM) microscopies, and rewiev the highlights of STM and AFM studies on graphite, graphite intercalation compounds, transition metal chalcogenides, insulating layered compounds and high-T_C superconductors. The wide and intriguing class of transition metal di-chalcogenides supporting charge density waves deserves a deeper insight. This is given in the next chapter by R. V. Coleman, Z. Dai, W. W. McNairy, C. G. Slough and C. Wang, who report in great detail on the pioneering work accomplished at their Department on transition metal dichalcogenides and the microscopic characterization of charge density waves.

The dispersion of electronic states in layered structure have been thoroughly investigated below and above the Fermi level by means of direct and inverse angle-resolved photoemission, respectively, and previously reviewed for this volume series by Giorgio Margaritondo. These experiments basically inform us about the bulk electronic structure, except for graphite, where surface (excited) states have been observed and have stimulated a major theoretical effort: A.J. Freeman, A. Continenza, M. Posternak and S. Massidda review their fundamental work on the local density functional theory of the electronic surface and interlayer states in graphite, graphite intercalated compounds and hexagonal Boron nitride by means of the full potential linearized augmented plane wave (FLAPW) method. This chapter includes an extension to the electronic structure of AlB_2-like compounds, semiconductor superlattices and two-dimensional magnetism at iron-semiconductor interfaces to show all the potential of the FLAPW method for low-dimensional systems.

The experimental study of surface phonons on layered compounds by means of Helium atom scattering was originally intended to complement inelastic neutron scattering for bulk phonon spectroscopy, but an unexpected zoo of surface phonon anomalies has been found. This intriguing story concerning graphite, layered chalcogenides and high-T_C superconductors is told by J. Skofronick and J. P. Toennies. Rare gases do not just work as probes of surface phonons but can give a great variety of physically adsorbed phases on the surface of graphite. N. Shripton, Milton Cole, W. A. Steele and M. H. W. Chan review this fascinating subject combining surface science with low-temperature and statistical physics. There is another class of layered compounds - the transition metal dihalides - whose surfaces provide the appropriate substrate for very interesting studies on physisorbed phases. Unlike graphite these materials are ionic and the hosting surface cooperate to generate an interesting variety of adsorbed phases. Y. Lahrer has deeply analyzed these systems and will report on this subject in the closing chapter of this book.

Editing a book on the surface properties of layered compounds appeared to the Series and Volume Editors quite a challenging operation, the subject being still a narrow and rapidly evolving field. For this reason a few chapters required a fairly long incubation and substantial revision, so that an updating of the other chapters came into consideration. This was not needed, however: the broad perspective offered by each chapter in its own field, the fundamental and crucial aspects that the authors were able to enucleate and emphasize, even in those fields which are still open and controversial, will ensure the timeliness of this volume as it is. We are grateful to the authors for their effort. The volume editor acknowledges a Research Award from the Alexander von Humboldt Stiftung enjoyed at the Max-Planck-Institut für Strömungsforschung in Göttingen, Germany, where much of the editing work could be done.

The Editor

STM AND AFM STUDIES OF LAYERED MATERIALS: GENERAL

R. Wiesendanger and D. Anselmetti
University of Basel, Dept. of Physics,
Klingelbergstrasse 82, CH-4056 Basel,
Switzerland

1. Introduction

1.1. STM AND AFM FOR STUDYING SURFACE PROPERTIES ON THE ATOMIC SCALE

In 1982, G. Binnig, H. Rohrer, Ch. Gerber and E. Weibel [1-3] reported on the first successful vacuum tunneling experiments between a sample surface and an atomically sharp probing tip which can be raster scanned across the surface in atomic dimensions by means of piezoelectric transducers. A novel, powerful tool for studying local surface properties of conducting samples down to the atomic scale was born: the scanning tunneling microscope (STM). Since then, surface science has profited remarkably from the ability of STM to map the surface atomic and electronic structure directly in real space. For instance, several reconstructions particularly on semiconductor surfaces were solved; surface imperfections such as point defects, steps, dislocations and grain boundaries were characterized on the atomic scale; adsorbed and chemisorbed species and even the physical properties of supported clusters on flat substrates were studied locally by STM.

The principles of operating a STM are simple: a conducting sample and a sharp metal tip, which acts as a local probe, are brought within a distance of a few ångstroms, so that a significant overlap of the electronic wavefunctions occurs (fig. 1). After applying a bias voltage U_s (typically between 1 mV and 4 V), a tunneling current I_t (typically between 0.1 nA and 10 nA) flows from the occupied electronic states near the Fermi level E_F of one electrode into the unoccupied states of the other electrode. This current, which is of quantum mechanical origin, depends exponentially on the separation between the two electrodes. Therefore, a high spatial resolution of the order of 1 Å laterally and less than 0.01 Å vertically can be achieved, because a change in the distance of only 1 Å typically leads to a one order of magnitude change in the current. Varying the bias voltage U_s offers the possibility to select electronic states near E_F which can contribute to the tunneling

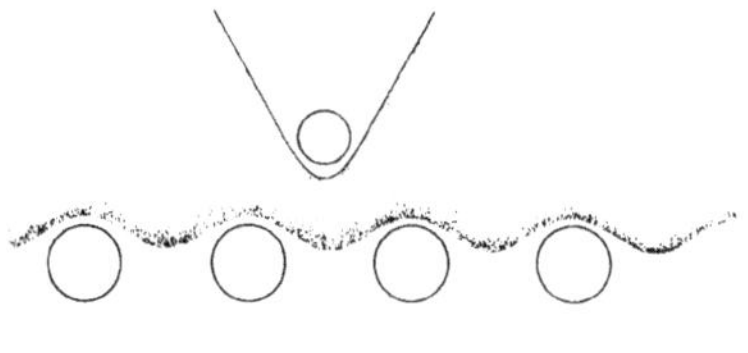

Figure 1: *In STM, a sharp metal tip is brought close to the surface of a conducting sample until a significant overlap of the electronic wavefunctions of both electrodes occurs.*

G. Benedek (ed.), Surface Properties of Layered Structures, 1–26.

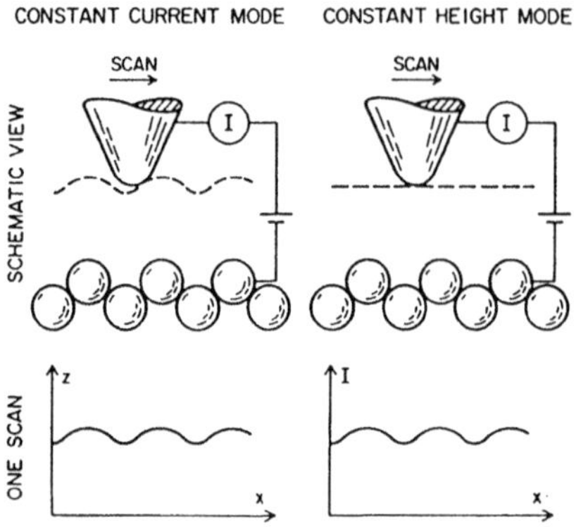

Figure 2: *Two possible modes of operating a STM. (Figure taken from P.K. Hansma and J. Tersoff [5].)*

current. Therefore, STM can also be used as a spectroscopic tool, yielding the spatially and energetically resolved electronic states on a sample surface. Additional information about the surface can also be extracted by measuring the mean potential barrier height separating the electronic states of the two electrodes.

A three-dimensional map of the surface topography is obtained by scanning the tip over the surface by means of a piezoelectric transducer. This can be done by using a variety of operation modes. The most commonly known operation mode is called "constant current mode": a feedback loop adjusts the tip to sample distance by applying an appropriate voltage to the piezoelectric transducer in such a way, that the tunneling current is held constant while scanning. This implies, to a first approximation, that the tip follows the surface contours at a constant distance s, if the tunneling barrier height is not changing as a function of location. By recording the voltage, which has to be applied to the piezoelectric element in order to keep the tunneling current constant, i.e. recording the height of the tip z(x,y) as a function of position, a "topographical image" can be obtained (fig. 2a). This constant current mode of operation, which was originally used by G. Binnig, H. Rohrer and coworkers, has the advantage that the height information of surface features can directly be obtained. On the other hand, the response time of the feedback loop limits the scan speed to about 100 Hz. Therefore, another mode of operation has been introduced which, however, is only applicable to atomically flat surfaces: the "constant height mode" [4]. In this mode, the tip is rapidly scanned at constant height over the surface while only the mean tunneling current is held constant by a feedback loop. The rapid variations in the tunneling current, which are recorded as a function of location, then contain the topographic information (fig. 2b). A significant advantage of this mode of operation is the faster scan rate which can be reached (well above 1 kHz). Therefore, possible image distortions such as thermal drift and piezoelectric hysteresis can be reduced, and dynamic processes on surfaces, occuring on a time scale of seconds, can be studied. However, extracting the topographic height information from the recorded variations of the tunneling current in the constant height mode of operation is not a straightforward task because the exact value of the local tunneling barrier height is often unknown. Furthermore, as already mentioned, the constant height mode of operation is only applicable to atomically flat surfaces; otherwise one may risk crashing the tip into a surface protrusion while scanning at high speed.

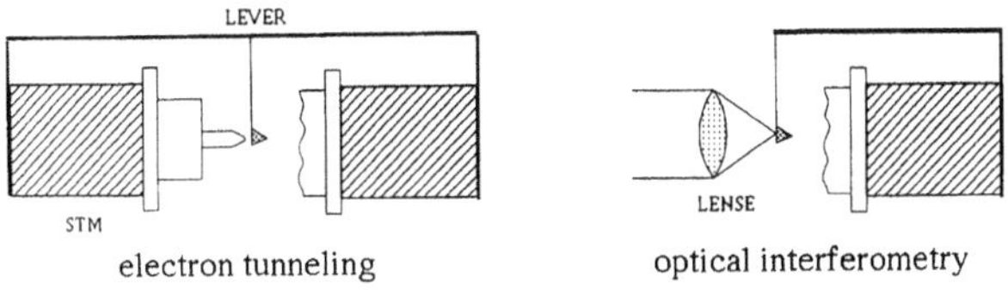

Figure 3: *In AFM, the sample surface is scanned relative to a sharp tip mounted on a cantilever which deflects according to the interaction forces between the tip and the sample surface. The small deflections of the cantilever can be monitored, for example, by a STM tip or by optical methods. (Courtesy of H. Heinzelmann and E. Meyer.)*

To get further information about the surface under investigation, for instance by using the STM as a spectroscopic tool or by mapping spatial variations of the local tunneling barrier height, a variety of modulation and feedback gating techniques has been introduced. Here, we refer the reader to review articles on STM [5-13] , since the detailed description of these additional operation techniques is beyond the scope of this introductory paragraph.

A significant drawback of the STM technique is its limitation to conducting samples. Therefore other types of interaction between a sharp probing tip and a sample surface have been considered after the invention of the STM leading to related "SXM" techniques, i.e. scanning microscopies with an arbitrary interaction mechanism X between the probing tip and the sample surface. Among these novel SXM techniques, scanning force microscopy (SFM) - also called atomic force microscopy (AFM) which was invented in 1986 by G. Binnig, C.F. Quate and Ch. Gerber [14] - has become the most important related scanning technique. Similar to a stylus profilometer, the sample is scanned against a sharp tip mounted on a cantilever-type spring. The cantilever deflects according to the force acting between the sample surface and the probing tip. These deflections can be monitored at high sensitivity either by a STM tip tunneling from the rear-side of the cantilever or by optical methods (fig. 3).

As in scanning tunneling microscopy, different modes of operation also exist in AFM. One can operate the AFM with constant cantilever deflection which is equivalent to the condition of constant force. This mode of operation is therefore called equiforce mode. Similar to the constant current mode in STM, the equiforce mode suffers from relatively low scan speeds, since a feedback loop is also required. A second mode of operation, the "variable deflection mode" has been introduced by AFM experimentalists in analogy to the constant height or current monitoring mode in STM. In this second mode of AFM operation, the cantilever deflection is allowed to change during scanning. Much faster scan rates can be achieved with this second mode with similar advantages and disadvantages compared to the constant height mode of operation in STM. For more details about the different modes of AFM operation, the reader is referred to reference [15].

The interaction forces in an AFM experiment are as small as 10^{-6}-10^{-10} N approaching the magnitude of interatomic forces, whereas the deflections of the cantilever are typically of the order of 100 Å with a variation of about 0.01-1 Å during scanning. Since the interaction force between the probing tip and the sample surface is present independent of the sample conductance, the AFM can be used for atomic scale surface investigations of both conductors and insulators. Additionally, physical phenomena related to force interaction, such as friction, tribocharging etc., can now be studied down to the atomic scale.

1.2. APPLICATION OF STM AND AFM TO LAYERED STRUCTURES

Materials with layered structure can be classified into three categories:

1. Layered materials which are built up by identical sheets. These sheets can be shifted relative to each other by a fraction of the in-plane lattice constant. For graphite, as an example, two types of layered structures exist: hexagonal graphite with an ABABAB... stacking sequence and rhombohedral graphite with an ABCABC... stacking sequence. Compound layered materials such as the nitrides (e.g. BN) also belong to this category.

2. Layered materials which are built up by a repetition of a sandwich of sheets. The individual sheets belonging to the sandwich are chemically different but still of simple structure. The transition metal dichalcogenides TX_2 (T = transition metal, e.g. Ti, Nb, Mo, Ta; X = chalcogenide: S, Se or Te), for instance, belong to this category. In this case, the sandwich of sheets consists of a top chalcogenide layer, a middle transition metal layer and a bottom chalcogenide layer.

3. Complex layered materials such as mica or the novel high-T_c oxides. The discovery of high-T_c superconductivity in materials with layered structure in 1986 by J.G. Bednorz and K.A. Müller [16] has undoubtedly led to renewed interest in layered materials as well as in materials with complex structure in general, which have hitherto been relatively neglected in solid state physics.

Besides these three categories mentioned above, there also exist intercalation compounds for all three categories. For the first and second category, the corresponding intercalation compounds have already been studied for quite a long time whereas the corresponding intercalation compounds for the third category are waiting to be explored.

One special property of layered materials has made them the most favourable samples for atomic resolution studies by scanning probe microscopies such as STM and AFM: they can easily be cleaved, leaving well defined surfaces which can be atomically flat over areas as large as several thousand ångstroms. This property is especially desirable for the constant height mode of operation in STM or the corresponding variable deflection mode of operation in AFM. In addition, many of the above mentioned layered materials exhibit a relatively inert surface, e.g. graphite or the transition metal dichalcogenides with a top S or Se layer. This allows for atomic resolution STM and AFM studies even in air, whereas other samples have to be prepared and characterized in ultrahigh vacuum or in an inert gas environment. Otherwise, a protective oxide layer would prohibit the investigation of the clean surface.

Although the layered materials are the most widely studied samples by STM and AFM, the experimental results raised more questions than they solved. This is far from being a drawback since the novel interesting questions originate from the fact that for the first time it has become possible to study the local aspects of surface properties down to the atomic scale by using STM and AFM. We will put special emphasis on these local aspects for the rest of this chapter. In the following second section, we will describe STM and AFM experiments on the most simplest layered material: graphite. However, the wealth of unexpected experimental findings for this apparently simple material will nicely demonstrate the challenges for understanding STM and AFM results on layered materials. The extension of STM investigations to graphite intercalation compounds with their modified electronic structure and the occurrence of superlattices will be presented in the third section. In the fourth section, we will report STM and AFM studies on different conducting and insulating layered materials belonging to the first and second category, such as the transition metal dichalcogenides and the insulators BN. Here, we will only shortly describe combined STM and AFM experiments at room temperature on transition metal dichalcogenides exhibiting charge density wave phenomena since a whole chapter by R.V. Coleman in this book is exclusively devoted to this topic. Finally, we will focus on STM and AFM investigations on complex layered materials such as mica and the novel high-T_c superconductors. Here, we will concentrate on the structural aspects as they can be studied at room temperature rather than the spectroscopic investigations which have to be performed at low temperatures. Concluding comments about STM and AFM studies of the surface properties of layered structures can be found at the end of this chapter.

2. STM and AFM Studies of the Surface Properties of Graphite

2.1. INTRODUCTION

Since the first report of atomic resolution STM studies of the highly oriented pyrolytic graphite (HOPG) surface by G. Binnig et al. [17], graphite has become the most widely studied sample by STM and AFM and is nowadays used as a calibration standard on the atomic scale for STM and STM-related microscopies performed in air. The easy sample preparation by pealing off a few carbon sheets with adhesive tape provides an atomically flat surface over thousands of square ångstroms. The inertness of the graphite surface against chemical reactions allows atomic resolution studies to be performed in air. Despite these advantages, the experimental STM and AFM results on the graphite surface have raised many questions; some of them are still unsolved or at least a topic of lively discussions. In the following, we will present some of the peculiarities of the experimental STM and AFM results on the graphite surface together with possible interpretations, as they have been suggested in the past.

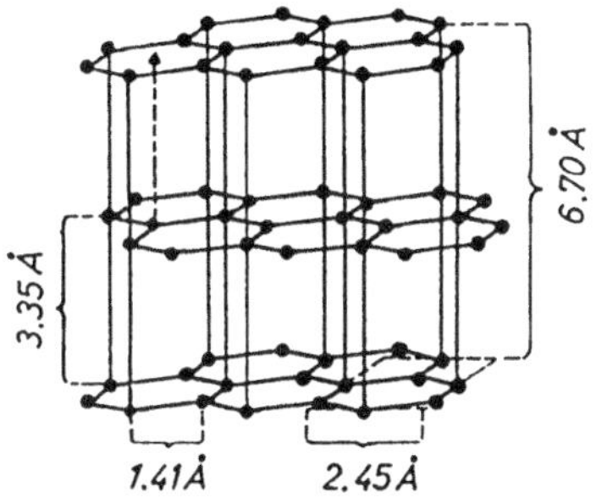

Figure 4: *Crystal structure of graphite. The lattice parameters are 2.46 Å (in-plane) and 6.7 Å (perpendicular to the layers).*

2.2. SITE ASYMMETRY AND ELECTRONIC STRUCTURE OF THE GRAPHITE SURFACE

In fig. 4 we present a schematic drawing of the atomic structure of hexagonal graphite with an ABABAB... stacking sequence of the carbon layers. The in-plane structure is defined by a honeycomb lattice with a nearest neighbour distance of 1.42 Å and a lattice constant of 2.46 Å. If one expects simply to see the atomic surface structure by STM, the honeycomb lattice should exclusively appear in STM images. This is definitely not the case. Instead, a hexagonal lattice with an in-plane lattice constant of 2.46 Å is observed in the majority of STM images of the graphite surface (fig. 5a), independent of the mode of STM operation. It seems that only every other carbon atom appears as a protrusion, which we will refer to as "carbon site asymmetry". This experimental result can not be explained by the surface atomic structure of graphite, which would be strongly contradictory to results from helium scattering experiments, but must have its origin in the surface electronic structure.

I.P. Batra et al. [18] and D. Tománek et al. [19, 20] have shown that the local electronic density of states (LDOS) near the Fermi level, which is probed by STM to a first approximation [21], is different for the two carbon atoms within the surface unit cell. The b-sites, which are the sites of those carbon atoms located above the center of a carbon hexagon in the next layer, have the highest LDOS and should therefore appear as a protrusion in STM images. The a-sites, which are the sites of those carbon atoms with a neighbour atom directly below in the next layer, only appear as saddle points whereas the centres of the hexagonal carbon rings (hollow sites) with the lowest LDOS appear as depressions.

Another interpretation of the experimentally observed hexagonal pattern was given by J. Tersoff [22]. He argued that for graphite as well as for other semiconductors and semimetals of low effective dimensionality, where the Fermi surface collapses to points at

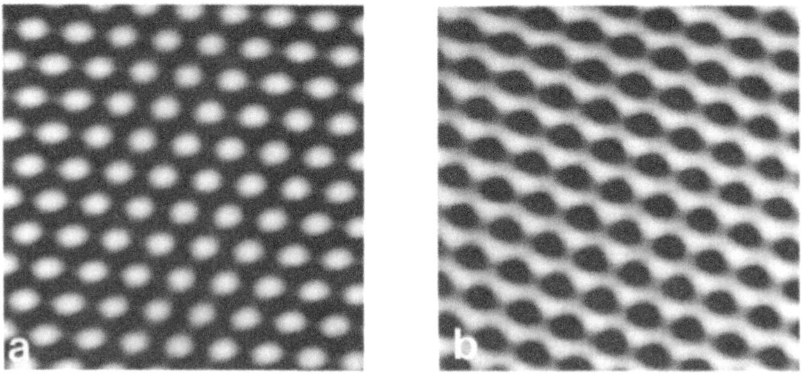

Figure 5: **(a)** *STM image of graphite (20×22 Å^2, I_t=2 nA, U_s=100 mV), which shows a hexagonal lattice with a lattice constant of 2.5 Å. In this image only every other carbon atom is visible, which is referred to as "carbon site asymmetry".* **(b)** *In this 22×22 Å^2 STM image of graphite, a honeycomb lattice structure is visible (I_t=1 nA, U_s=100 mV).*

the corners of the surface Brillouin zone, the STM image corresponds to an individual state. The nodal structure of this state then gives rise to the measured STM corrugation with the periodicity of the unit cell (2.46 Å), regardless of the underlying atomic structure, i.e. independent of the atomic positions within the unit cell.

Despite the different possibilities for interpretation, it has become clear that STM images of the graphite surface are dominated by the surface electronic structure rather than the surface atomic structure. The situation is, however, even more complicated, since in some STM images of the graphite surface, a honeycomb lattice (fig. 5b) or stripe-like features have been observed. The most likely explanation for these "anomalous" STM images has been given by H.A. Mizes et al. [23] under the assumption, that the surface is not imaged by a single atom tip but with a multiple tip. All experimentally observed "anomalous" STM images of the graphite surface could be simulated this way.

2.3. GIANT CORRUGATIONS AND ELASTIC RESPONSE OF THE GRAPHITE SURFACE

Besides the necessity of interpreting the observed in-plane lattice constant, another peculiarity can appear in STM experiments on the graphite surface: the giant corrugations. This means, that anomalous large amplitudes in STM traces of up to several ångstroms and even more can be measured, particularly in experiments performed in ambient air and with small tunneling gap resistances. A first explanation for this experimental observation has been given by J.M. Soler et al. [24] based on an elastic deformation of the graphite surface induced by interatomic forces between the probing tip and the sample surface, leading to a huge enhancement of the electronically based corrugation [22]. Another lecture was told to the STM community: interaction forces between the probing tip and the sample surface have to be taken into account especially at small tip to sample separations, i.e. at small tunneling gap resistances and for soft materials such as graphite or other layered materials. This is the essential point, regardless from the debate, how large the area is over which the interaction force between tip and sample surface is exerted.

H.J. Mamin et al. [25] showed that for STM experiments performed in air, the deformation of the graphite surface is mediated by a contamination layer and that the force necessary to deform the surface must be spread over several thousand square ångstroms. J.B. Pethica [26] also argued that the actual contact area between tip and sample surface may be more than one atom wide and emphasized, that imaging a simple periodic structure does not strictly require a tunnel beam of single-atom width, but rather a fluctuation in total conductivity between tip and sample which varies in registry with unit lattice shear. An indication for the existence of such an imaging mechanism is the observation by D.P.E. Smith et al. [27], that atomic resolution on the graphite surface can still be obtained with point contact between tip and sample surface where the potential barrier already has completely collapsed. It is obvious that the theoretical concepts developed for the STM with tip - sample separations of several ångstroms [21] break down at small tip to surface distances or even at point contact. New mechanisms for current flow may become dominant, possibly via tip induced localized states [28-30] .

In the meantime, the forces acting between the tip and the graphite surface in a STM experiment performed in air could be directly measured. A remarkable high value of about 10^{-6}N was obtained [31, 32]. The force interaction between the probing tip and the sample surface seems to be a key issue for understanding STM images of the graphite surface and suggests the application of another scanning probe microscopy: atomic force microscopy.

2.4. ATOMIC FORCE MICROSCOPY OF THE GRAPHITE SURFACE

In the history of atomic force microscopy, graphite was again an exceptional material since it was the first that could be imaged with atomic resolution [33]. The interpretation of AFM results on the graphite surface is however not easier than the interpretation of STM results.

AFM images of the graphite surface also preferentially reveal a hexagonal lattice with an in-plane lattice constant of 2.46 Å as shown in fig. 6. In contrast to the STM results where the observed site asymmetry could be explained by the dominance of electronic structure effects, the interpretation of the "every other atom images" in AFM experiments seems to be less obvious. In addition, the fact that atomic resolution can be achieved with repulsive forces in the range of 10^{-8} N, or even more, is difficult to understand since such forces should lead to a large elastic compressive deformation of the graphite surface. Assuming a monoatomic tip, such forces are expected to destroy the graphite surface because the strength of the in-plane interatomic bonding forces is only of the order of 10^{-8} N. Therefore, it has been proposed [34] that the tip is dragging a flake of graphite as it scans the surface so that the force per atom can be sufficiently small to cause no surface damage. As a consequence, not individual atoms on the graphite surface would be imaged by AFM, but rather repulsion maxima associated with misregistry of the graphite tip unit cell and the graphite surface unit cell. This model has also proven to be successful in explaining other "anomalous" AFM images of the graphite surface [34]. However, the described model is only applicable for layered materials such as graphite, where loose flakes can easily be picked up by the tip. It does not explain atomic resolution AFM images on non-layered materials such as NaCl [35] or LiF [36]. Therefore, the detailed imaging mechanism in atomic resolution AFM studies can still be regarded as unsolved. A key issue in understanding this imaging mechanism may be the occurrence of frictional forces on the atomic scale. C.M. Mate et al. [32] succeeded in the observation of atomic scale features in the frictional force acting on an AFM tip while sliding on the basal plane of a graphite surface. These atomic scale features show the periodicity of the graphite surface. Therefore this experiment can be regarded as a first step towards "atomic tribology".

Figure 6: *AFM image (39×53 Å^2) of graphite obtained with a loading force of about 10^{-8} N. (Courtesy of E. Meyer, see ref. [15].)*

2.5. IMAGING OF DEFECTS ON THE GRAPHITE SURFACE

Up to now, we have described STM and AFM experiments performed on the ideal graphite lattice and have discussed possible imaging mechanisms. However, the essential advantage of STM and related scanning techniques lies in the possibility of studying local surface properties such as defects (i.e. structural and chemical inhomogeneities down to the atomic scale), isolated adsorbed molecules or inhomogeneities in the distribution of charge. Imaging of defects on the atomic scale by STM and related techniques is not only of interest with regard to the characterization of these defects but can also contribute to a clarification of the imaging mechanism in scanning probe microscopies.

STM studies of isolated defects or adsorbates on the graphite surface turned out to be a particular exciting topic, because STM images of graphite are dominated by the surface electronic structure as already discussed. Long-range electronic perturbations have been observed near isolated adsorbed molecules on the graphite surface which perturb the charge density and give rise to periodic oscillations similar to Friedel oscillations [37]. These oscillations have a wavelength of $\sqrt{3}$ times that of the graphite lattice and the symmetry of the oscillations was shown to reflect the nature of the defect [37]. In fig. 7 we present an example of a STM image showing the electronic induced $(\sqrt{3}\times\sqrt{3})$ superlattice in the vicinity of an isolated structural defect, which can occasionally be found on a cleaved graphite surface.

The possibility of imaging local perturbations of the ideal graphite lattice and of the local charge density implies that the imaging mechanism can not always be based on a point contact between the STM tip and the sample surface with contact areas of several thousand square ångstroms. On the contrary, the "tunneling channel" has to be quite localized for explaining STM images such as that shown in fig. 7. Interestingly, a $(\sqrt{3}\times\sqrt{3})$ superlattice has also been observed locally in AFM experiments on a graphite surface with adsorbed species, which has been attributed to adsorbate-induced scattering of electron states in the graphite [38, 39]. Again, the observation of these highly localized perturbations seems to contradict the assumption of a macroscopic contact area between the probing tip and the sample surface.

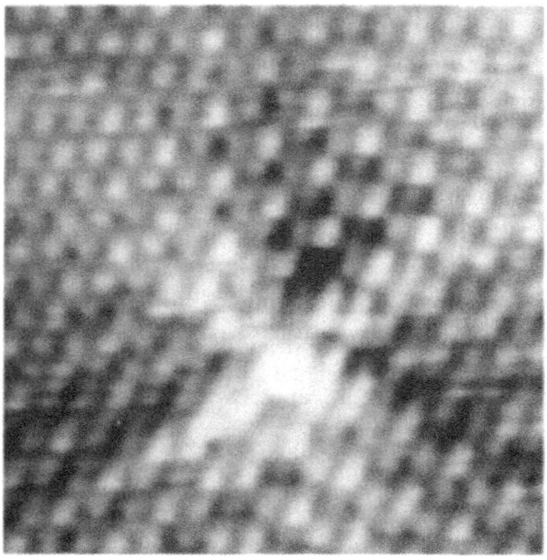

Figure 7: *In this 40×40 Å^2 STM image of graphite (I_t=1.2 nA, U_s=270 mV), an electronically induced $(\sqrt{3}\times\sqrt{3})$ superlattice appears in the vicinity of an isolated structural defect. The origin of this phenomenon is probably a multibonded chemisorbed molecule. This has been concluded from the superlattice symmetry and additional computer simulations.*

Another interesting STM study of defects on the graphite surface focussed on the observation of tilt boundaries [40]. Besides the determination of the structure and width of the grain boundary region, it was possible to observe Moiré patterns near grain boundaries when multiple tips, scanning over different grains, contributed to the image simultaneously. Such images gave support to the model, that multiple isolated tips, occasionally hundreds of ångstroms apart, can sometimes contribute to STM images.

In conclusion, STM and AFM studies of defects on the graphite surface can not only reveal the nature of these defects but also make an important contribution to the clarification of the imaging mechanism in STM and AFM which may vary depending on the experimental parameters, tip and sample surface preparation and environment.

3. STM and AFM Studies of Graphite Intercalation Compounds

3.1. INTRODUCTION

The extension of STM and AFM studies from graphite to graphite intercalation compounds (GIC) is of interest for several reasons: Firstly, it is possible to study the hitherto unknown surface structure of GIC's on the atomic scale by using STM. Secondly, since the STM images of the graphite surface were shown to be dominated by the surface electronic structure, the question arises, how the changes in the electronic structure due to the intercalant species, such as the local density of states near the Fermi level and the shape of the Fermi surface, influence the STM results. Thirdly, in both STM and AFM experiments on GIC's, the elastic response of the surface should change due to the presence of the intercalant.

Whenever a new type of material is studied with novel experimental techniques such as STM and AFM, it is desirable to start with the simplest prototype system. In the case of the GIC's, these are the alkali metal intercalated compounds, for which the bulk properties have already been studied for a long time. Indeed, the first atomic resolution STM investigations of GIC's [41-46] focussed on the alkali metal GIC's, and we will therefore restrict our present description of STM studies on GIC's to the alkali metal compounds.

3.2. EXPERIMENTAL STM RESULTS ON ALKALI METAL GIC'S

In contrast to graphite, alkali metal GIC's can not be studied in air because their surfaces are known to be highly reactive. On the other hand, the low vapour pressure of alkali metals prevents STM experiments in ultrahigh vacuum at room temperature. Successful STM studies of alkali metal GIC's could be performed in an inert gas environment. The microscope has to be operated for instance in a stainless steel glove box containing a high purity argon atmosphere and connected to a gas purification system which reduces impurities such as H_2O, O_2 and N_2 below 1 ppm. After in-situ cleaving of the sample, the freshly exposed GIC surface can be studied on the atomic scale by STM without deterioration of the image quality over a time period of up to several hours.

In the following description of experimental STM results on alkali metal GIC's, we will distinguish between the intercalation compounds of the light alkali metal lithium and the heavier alkali metals K, Rb and Cs, because the atomic scale structures as determined by STM seem to differ between Li-GIC's and the other alkali metal GIC's. In fig. 8, a typical topographic overview image of a stage 1 Li-GIC surface is shown. Besides atomically flat surface regions, highly defective areas are visible. In general, the density of defects on cleaved GIC surfaces is significantly higher than on cleaved pristine graphite surfaces.

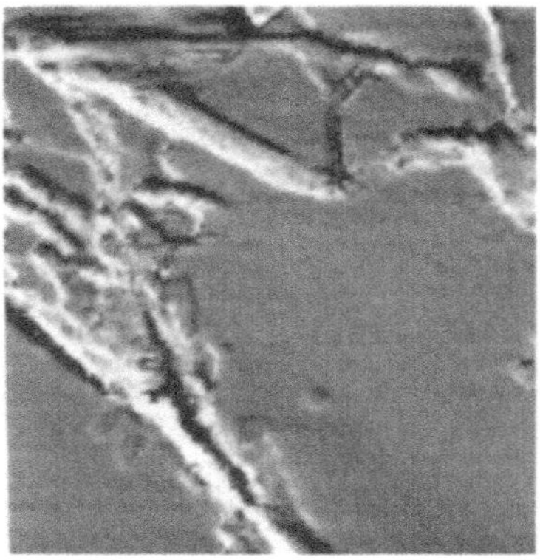

Figure 8: *In this figure a 2500×2500 Å^2 STM overview image of C_6Li is presented. It clearly shows flat surface regions, separated by steps and other defect structures (I_t=1 nA, U_s=-200 mV).*

After zooming into flat, defect-free surface regions, ordered superlattices can be observed on the atomic scale (fig. 9a-c). Interestingly, not only one type of superlattice structure but three different types of superlattices have been found [41, 42, 45] at different locations with in-plane lattice constants of about 3.5±0.2 Å (fig. 9a), 4.2±0.2 Å (fig. 9b) and 4.9±0.2 Å (fig. 9c). The lattice constants determined from other STM images of the stage 1 Li-GIC surface are all grouped around these three values as shown in the histogram of fig. 10. The in-plane lattice constant of 4.2 Å corresponds to a commensurate ($\sqrt{3}\times\sqrt{3}$) superlattice (fig. 11a) which is also present in bulk C_6Li. On the other hand, the in-plane lattice constant of 4.9 Å corresponds nicely to a commensurate (2×2) superlattice (fig. 11b) which is usually found in the bulk of stage 1 heavy alkali metal GIC's but can obviously also be observed at the surface of the light alkali metal Li-GIC. The third and smallest observed in-plane lattice constant of about 3.5 Å could be attributed to a an incommensurate superlattice. However, this can not definitely be decided as long as the underlying graphitic host lattice does not appear in STM images simultaneously with the superlattice structure. Unfortunately, the non-observation of the underlying graphitic host

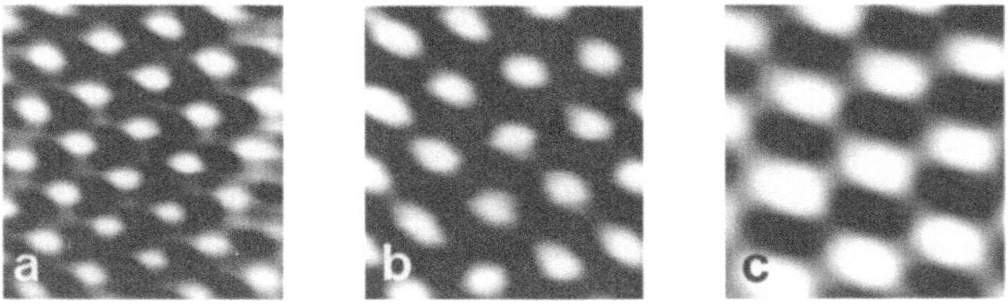

Figure 9: *Depending on the surface location, three different types of hexagonal superlattices have been found on the surface of C_6Li. The figures show 18×18 Å^2 STM images in a topview presentation. The lattice constants corresponding to the imaged hexagonal superlattices are 3.5±0.3 Å* (**a**), *4.2±0.2 Å* (**b**) *and 4.9±0.2 Å* (**c**). *The STM images were acquired in the "constant height mode" of operation, where the mean tunneling current and the sample bias voltage were set to I_t=1 nA and U_s=-100 mV respectively.*

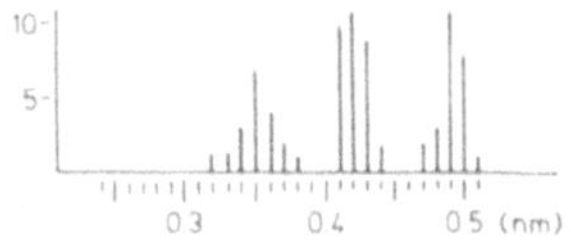

Figure 10: *This histogram shows the distribution of the observed in-plane lattice constants on the surface of C_6Li GIC's.*

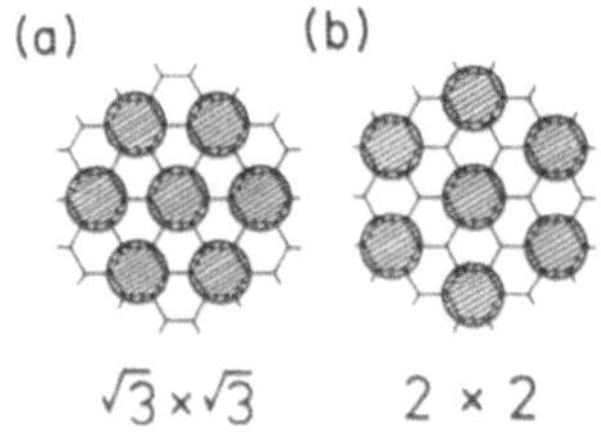

Figure 11: *Bulk structures of C_6M* **(a)** *and C_8M* **(b)** *graphite intercalation compounds, where M denotes an alkali metal.*

lattice in STM images of the Li-GIC surface seems to be typical in contrast to STM images of the heavier alkali metal GIC surfaces which will be described below. Despite the lack of simultaneous observation of the underlying graphitic host lattice, we believe that the observed in-plane lattice constant of about 3.5 Å can be attributed to an incommensurate superlattice of nearly close-packed lithium. The nearest neighbour distance in hexagonal lithium is 3.11 Å. However, electrostatic repulsion, originating from the charge transfer from the lithium to the graphite, is likely to increase this spacing. The larger spread of the observed in-plane lattice constants around the value of 3.5 Å, as seen in the histogram of fig. 10, may be explained by the incommensurability of the corresponding superlattice.

The experimental STM results obtained on the heavier alkali metal stage 1 GIC's of K, Rb and Cs are different from the results obtained on stage 1 Li-GIC's in several respects. First, it is comparatively easy to image both the superlattice structures and the underlying graphitic host lattice simultaneously. For instance, a commensurate (2×2) superlattice together with the underlying graphitic host lattice has been observed on the surfaces of the stage 1 compounds of K, Rb and Cs [43, 44, 46]. In fig. 12 we present a STM image showing this (2×2) superlattice structure on a stage 1 Rb-GIC surface. The

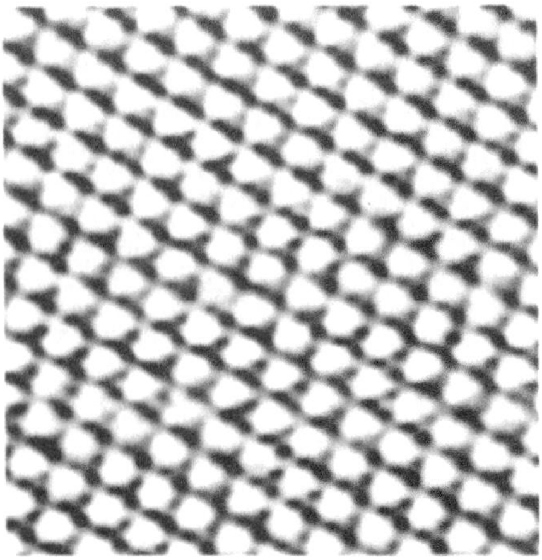

Figure 12: *The (2×2) superlattice with the underlying graphite host lattice is presented in this 50×60 Å^2 STM image of a C_8Rb GIC (I_t=3.4 nA, U_s=50 mV).*

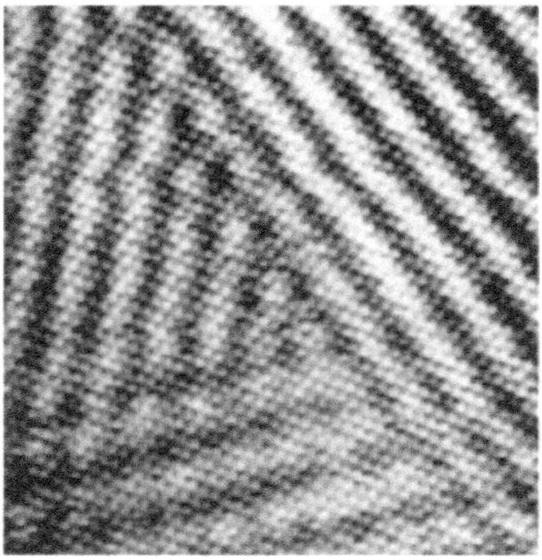

Figure 13: *This 100×90 Å^2 STM image of a C_8Rb GIC shows the boundary zone of three different ($\sqrt{3}\times 4$) domains at high resolution (I_t=2 nA, U_s=20 mV). The domains are inclined at 120^0 with respect to each other. A defect in the underlying graphite host lattice can not be seen.*

(2×2) superlattice is known to be present also in the bulk of these compounds. However, for the stage 1 K-GIC's one often finds only a graphitic surface structure without any superlattices superimposed [42, 45]. Likewise, a graphitic structure could be identified at the surfaces of stage 2 heavy alkali metal GIC's [44].

The second difference between experimental STM results on stage 1 Li-GIC and stage 1 heavy alkali metal GIC's is based on the possibility to observe non-hexagonal superlattices in stage 1 Rb- and Cs-GIC's [46]. In fig. 13 a STM image of a stage 1 Rb-GIC is presented, showing simultaneously the three possible rotational domains of a one-dimensional ($\sqrt{3}\times 4$) superlattice superimposed on the underlying graphitic host lattice. The domain boundaries are remarkably sharp. Similar non-hexagonal superlattices (($\sqrt{3}\times 2$) and ($\sqrt{3}\times\sqrt{13}$)) have already been identified in the bulk of unsaturated Cs-GIC's by using scanning transmission electron microscopy [47]. The character of the one-dimensional superlattice structures, as they appear in STM images of stage 1 heavy alkali metal GIC's, can however

Figure 14: *Mixed one-dimensional structures can also be detected on C_8M GIC's. This 65×60 Å^2 image shows chain-like structures corresponding to alternating ($\sqrt{3}\times\sqrt{13}$) and ($\sqrt{3}\times 4$) superstructures of the intercalated alkali metal atoms (I_t=2.6 nA, U_s=42 mV).*

Figure 15: *In this 2000×2000 Å^2 STM image we present a hexagonal superstructure with a lattice constant of ~300 Å on a C_6Li sample (I_t=1 nA, U_s=400 mV). The height of the structures is 20±3 Å with an average diameter of ~60 Å.*

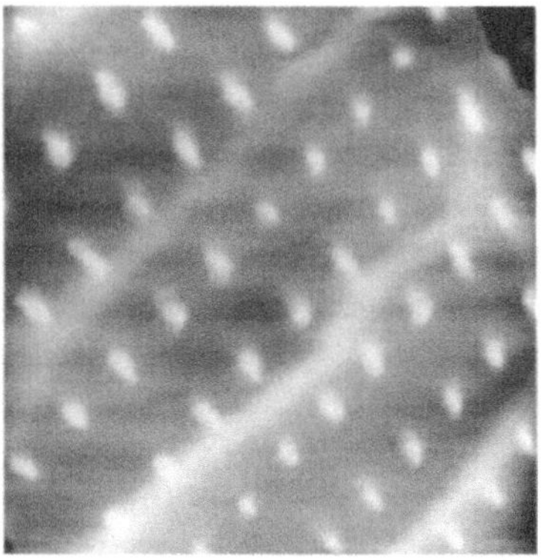

vary, e.g. ($\sqrt{3}\times4$) and ($\sqrt{3}\times\sqrt{13}$) superstructures can alternate as observed in the STM image presented in fig. 14.

Many hitherto unknown superstructures may additionally be identified on the surfaces of alkali metal GIC's and also other GIC's in the future by using STM. For instance, a remarkable superlattice on a scale of several hundred ångstroms has been found on the surface of cleaved stage 1 Li-GIC's (fig. 15). The origin of this superlattice is still unclear. However, this observation additionally emphasizes the great potential of STM for revealing novel surface structures on a local scale.

3.3. INTERPRETATION OF EXPERIMENTAL STM RESULTS AND COMPARISON WITH THEORETICAL PREDICTIONS

The discussion of the experimental STM results on pristine graphite has already demonstrated that the interpretation can be subtle and that the surface electronic structure can dominate the contrast observed in STM images rather than the surface atomic structure. In the case of graphite intercalation compounds, the question arises whether the maxima in STM traces can directly be attributed to the atomic positions of the intercalant or to electronic states introduced by the presence of the intercalant. A third interpretation is based on the possibility of a surface driven charge density wave (CDW) formation in C_8M (M = K, Rb Cs) compounds [48, 49], which has been favoured by Kelty and Lieber [43, 44] for explaining the (2×2) superlattice observed on the surface of C_8M compounds. However, the variety of hexagonal superlattice structures observed on the surface of stage 1 Li-GIC's as well as the variety of nonhexagonal one-dimensional superlattices observed on surfaces of C_8M compounds in addition to the (2×2) superlattice, make an interpretation solely based on the formation of CDW's unlikely. On the other hand, there is no doubt that similar to the case of graphite, electronic structure effects also play an essential role for understanding STM results on GIC's.

Unfortunately, all theoretical work about STM on GIC's reported so far is based on radical assumptions. Firstly, the superlattice structure present at the surface has been assumed to be fixed and the same as in the bulk [50] or has been omitted completely [51]. This is obviously not justified by the experimental STM results. Secondly, the top surface layer has always been assumed to be a carbon layer. However, from an experimental viewpoint, it seems obvious that after cleaving stage 1 GIC's, part of the intercalant will be left on top of the surface. Since for symmetry reasons, the intercalant is likely to be present on the surfaces of both parts of the cleaved sample, a rearrangement of the intercalant on top of the surface might occur, which is favoured compared to a rearrangement between the HOPG galleries. The nonhexagonal, one-dimensional $(\sqrt{3}\times 4)$ and $(\sqrt{3}\times\sqrt{13})$ superlattice structures observed on the surfaces of cleaved stage 1 Rb- and Cs-GIC's correspond to diluted systems compared to the C_8M stochiometry of the bulk and might be explained by this rearrangement of the alkali metal upon cleaving. However, we can not rule out the presence of a top carbon layer originating from an evaporation of the alkali metal at the surface. In this case, the observed one-dimensional superlattice structures might be due to a deformation of the topmost HOPG layer, induced by a linear arrangement of the alkali metal intercalant in the first HOPG gallery.

Finally, the distance, at which the STM tip probes the electronic structure of the GIC surface, might have an influence on the observed STM pattern in analogy to the anomalous distance dependence observed in STM images of a graphite surface with implanted gold atoms [52]. More theoretical and experimental work is needed for a definite interpretation of the remarkable STM results which have already been obtained and partially been presented in this section.

3.4. AFM EXPERIMENTS ON GIC'S

For the interpretation of experimental STM results on GIC's, it might be helpful to perform simultaneous STM and AFM experiments. In contrast to STM, which is sensitive to the electronic structure at the Fermi level, AFM is believed to probe the total charge density or the ion core repulsion. AFM might hopefully yield images, which reflect the surface topography more closely. For instance, different contrast in STM and AFM experiments on charge density wave systems is observed as will be reported in the next paragraph. Therefore, simultaneous STM and AFM measurements might allow to decide which superlattices observed on alkali metal GIC's can be attributed to charge density wave formation.

Furthermore, AFM experiments on GIC's should help to clarify the role of the elastic response of the surface to the proximity of the tip as a function of the applied loading force. It has already been suggested [53] that AFM should be a unique tool to determine the local surface rigidity and to measure the local surface layer distortion of graphite in the vicinity of intercalated species. Finally, AFM will also be useful for studying the interesting frictional properties of GIC's on a local scale.

4. STM and AFM Studies of Compound Layered Materials

4.1. INTRODUCTION

After describing STM and AFM experiments on graphite as one of the most simplest layered materials as well as on intercalation compounds of graphite, we now focus on STM and AFM studies of compound layered materials. Here, we may distinguish between layered materials where the top surface layer consists of only one element, such as for the transition metal dichalcogenides, and layered materials where the top surface layer is built up of at least two different atomic species such as for BN, mica or the novel high-T_c superconductors. In the first case, one may ask if the image contrast observed in STM and AFM images is mainly due to the top surface layer or if electronic states of the second atomic layer can dominate. In the second case, the question arises, whether all atomic species of the top surface layer contribute to the contrast in STM and AFM images or if one of the constituents is dominating. We also discuss in this fourth paragraph, how a distinction can be made between the atomic lattice and possible superlattices originating from collective transport phenomena such as charge density waves.

4.2. STM AND AFM STUDIES OF THE ATOMIC STRUCTURE OF TRANSITION METAL DICHALCOGENIDES: WHICH ATOMS ARE IMAGED ?

Transition metal dichalcogenides with layered structure are built up by a repetition of a three layer sandwich consisting of a top chalcogenide layer (S, Se or Te), a middle transition metal layer (Ti, V, Nb, Mo, Ta, W etc.) and a bottom chalcogenide layer (fig. 16). Very often, both the chalcogenide layers and the transition metal layer exhibit hexagonal symmetry with the same in-plane lattice constant (e.g. for MoS_2, TiS_2, TaS_2). Therefore, from the observation of a hexagonal pattern in STM images, it is not clear a priori, whether the contrast originates from electronic states of the top chalcogenide layer or from the d-states of the second transition metal layer. This problem dominated, for instance, the discussion of experimental STM results on MoS_2 [54-56] and $NbSe_2$ [57, 58].

On the other hand, there exist transition metal dichalcogenides such as WTe_2, where the transition metal layer does not exhibit the same hexagonal arrangement as the chalcogenide layers. In WTe_2, every two rows of W atoms pair to form zig-zag and slightly buckled chains. Tang et al. [59] have studied WTe_2 by STM. Their images showed a zig-zag structure, which they attributed to the W atoms. Therefore, Tang et al. concluded that the contrast in STM images of transition metal dichalcogenides is mainly determined by electronic states originating from the transition metal layer. However, T.R. Albrecht [39] has shown that the zig-zag structures observed in some STM images of WTe_2 are most likely due to multiple tip imaging which already explained anomalous STM images of the graphite surface. In conclusion, none of the STM images of transition metal dichalcogenides obtained so far contradict the assumption that STM is mainly sensitive to the electronic states of the top chalcogenide layer.

For finding a definite answer to the question, whether STM mainly probes the electronic states of the chalcogenide layer or the transition metal layer in transition metal dichalcogenides, one can think of simultaneous STM and AFM experiments: AFM is believed to respond almost exclusively to the topmost atomic layer, which would be the

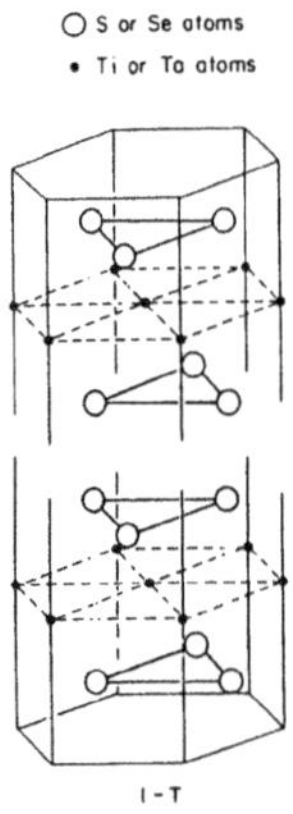

Figure 16: *The layered sandwich structure of transition metal dichalcogenides is presented in this figure. The sandwich structure consists of a top chalcogenide layer (S, Se or Te), a middle transition metal layer (Ti, V, Ta, ...) and a bottom chalcogenide layer.*

chalcogenide layer. Since the hexagonal transition metal layer is shifted relative to the hexagonal chalcogenide layer, one should only have to check, whether the maxima observed in STM traces are in-phase or out-of-phase with maxima in AFM traces. Such an experiment was performed by T.R. Albrecht [39]. Unfortunately, a definite answer from this experiment could not be obtained since the registry between the atomic lattice imaged by STM and the atomic lattice imaged by AFM shifted as the condition of the probing tip changed as a function of time. Therefore, similar to the case of graphite, the only conclusion that can be drawn by now is: both STM and AFM can resolve the unit cell periodicity of transition metal dichalcogenides. On the other hand, the assignment of the maxima observed in STM and AFM images to atomic positions within the unit cell is still left open.

4.3. COMPARATIVE STM AND AFM STUDIES OF CHARGE DENSITY WAVE SYSTEMS

Up to now, we have discussed STM and AFM investigations of the atomic structure of transition metal dichalcogenides. Some of these transition metal dichalcogenides are known to exhibit charge density waves (CDW). Coleman et al. [60] first showed that STM is sensitive to CDW formation which results from a Fermi surface instability. In the meantime, numerous CDW systems have been studied by STM from room temperature down to liquid helium temperature and many interesting new results were obtained by STM as a local probe [61]. Since a whole chapter in this book by R.V. Coleman is exclusively devoted to this topic, we will only focus on our comparative STM and AFM studies of charge density wave systems [62, 63], which demonstrate the different response in STM and AFM experiments.

1T-TaS_2 and 1T-$TaSe_2$ were chosen as samples exhibiting charge density waves even at room temperature which greatly facilitated the AFM studies. STM images of 1T-TaS_2 and 1T-$TaSe_2$ are dominated by the relatively strong charge density wave modulation, which corresponds to a measured corrugation of about 2-4 Å, compared to the simultaneously detected underlying atomic corrugation, which is less than 1 Å (fig. 17). This strong CDW modulation in 1T-TaS_2 and 1T-$TaSe_2$ may have the same origin as the electronically induced large corrugation on the graphite surface [22]. Another interesting aspect of STM

Figure 17: *This 140×140 Å^2 STM image of 1T-$TaSe_2$ has been obtained at 300 K (I_t=2.8 nA, U_s=-80 mV). The image is dominated by the commensurate ($\sqrt{13}\times\sqrt{13}$) CDW superlattice, which is superimposed on the atomic lattice. An isolated CDW defect at the left hand side can also be seen.*

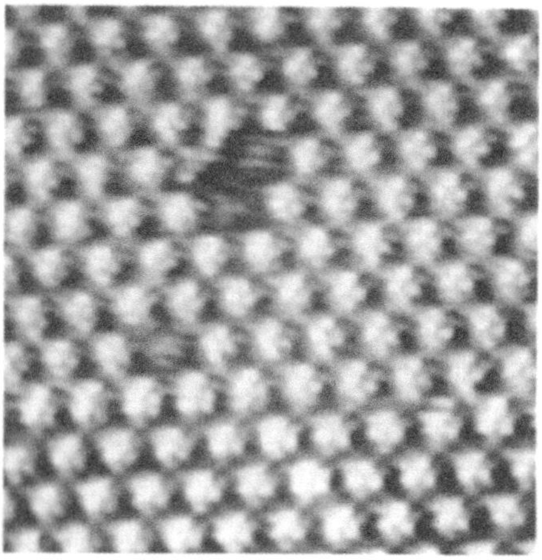

Figure 18: *Three missing CDW maxima have been imaged at the surface of a 1T-TaS_2 sample. The image shows a 120×120 Å^2 area in the commensurate phase at 77 K, where the charge density maxima and the underlying atomic lattice were imaged simultaneously. (I_t=3 nA, U_s=-224 mV, courtesy of S. Baumgartner).*

studies on CDW systems is the ability to observe point defects in the CDW superlattice such as missing charge maxima (fig. 18). Furthermore, since STM directly probes the amplitude of the CDW modulation, the expected reduction of the order parameter towards surface defects such as steps can be studied on a local scale. Finally, circular domains defined by a variation in the maximum CDW amplitude and forming a regular hexagonal superlattice on a nanometer scale could be studied by STM [64]. In fig. 19 we present a STM image of such a superlattice.

After the invention of the AFM, the question arose what AFM actually does respond to. The application of AFM to CDW systems might contribute to the solution of this problem since the AFM response to CDW's differs completely from the STM response. AFM images obtained on the same samples of 1T-TaS_2 and 1T-$TaSe_2$ with an applied loading force of 10^{-8}-10^{-7}N only show the atomic lattice (fig. 20). A sign of a CDW superlattice is neither seen in the real space AFM pictures nor in the two-dimensional Fourier transform patterns.

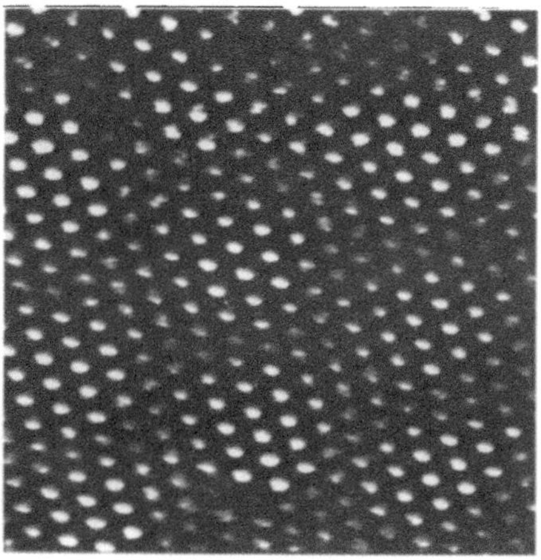

Figure 19: *Domains of CDW's were imaged at 300 K on 1T-TaS_2, indicating the existence of the domain-like discommensurate phase. The domains are hexagonally arranged with a period of ~70 Å. (200×200 Å^2, I_t=6 nA, U_s=-19 mV, courtesy of S. Baumgartner).*

Figure 20: *AFM image (67×86 Å^2) of 1T-TaS_2 obtained with a loading force of about 10^{-8} N. (Courtesy of E. Meyer, see ref. [62, 63].)*

The different response of STM and AFM to CDW's might not seem surprising at first sight. STM is known to be sensitive to the electronic density of states near the Fermi level which is affected by the CDW formation. On the other hand, AFM is believed to respond to the total charge density at the sample surface or to the ion core repulsion. Since only a small fraction of the total number of electrons is condensed in the CDW state and the small periodic lattice distortion associated with the CDW formation is beyond the lateral resolution capabilities of present scanning probe microscopies, it might be concluded that AFM should be less sensitive to the CDW modulation compared to STM. However, helium scattering experiments on 1T-TaS_2 and 1T-$TaSe_2$ [65, 66] have revealed a significant distortion of the top surface layer due to the CDW formation which should also be detectable by AFM. Therefore, the nonobservation of CDW's in an AFM experiment is most likely the result of the probing mechanism itself. This means, that either the applied loading force is too large, leading to a pressure induced suppression of the CDW state, or the frictional forces influence the CDW state with an applied loading force of 10^{-8}–10^{-7} N. Recent AFM experiments performed with a loading force below 10^{-8} N indeed show a CDW superlattice superimposed on the atomic lattice [67].

The different response of STM and AFM to the CDW state is an interesting fact. It might allow, for instance, to decide which superlattices observed on the surfaces of stage 1 alkali metal GIC's can indeed be attributed to CDW formation and which not.

4.4. AFM STUDIES OF INSULATING COMPOUND LAYERED MATERIALS

We now focus on insulating compound layered materials which can only be studied on an atomic scale by means of atomic force microscopy. Boron nitride (BN) was the first insulator that could be imaged with atomic resolution [39, 68]. In fig. 21a the crystal structure of BN is shown. In contrast to graphite and the transition metal dichalcogenides, the top surface layer is built up of two different atomic species whereas the stacking sequence ABABAB... is similar to hexagonal graphite. The AFM image of BN [39], which is presented in fig. 21b, shows a triangular lattice instead of a honeycomb lattice indicating that only one type of atoms appears as maxima. For BN, the inequivalence of the two lattice sites is given by the two different atomic species in the top surface layer. However, determining which site, either the boron or the nitrogen site, appears as a bright spot

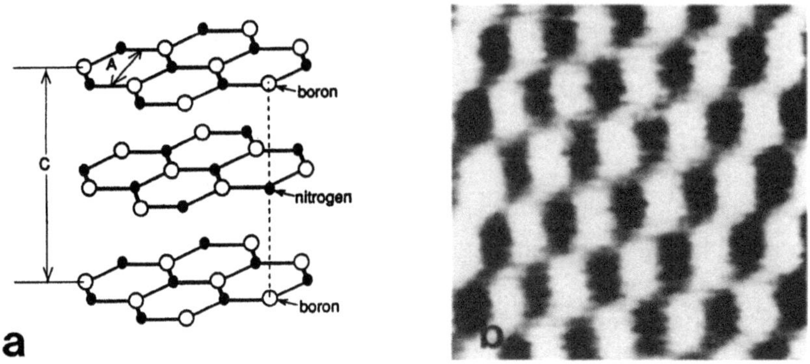

Figure 21: **(a)** *Crystal structure of boron nitride. The lattice parameters are A=2.504 Å and C=6.66 Å.* **(b)** *AFM image of boron nitride. Only three bright spots per hexagon appear with a spacing of 2.5 Å. (Courtesy of T.R. Albrecht, see ref. [68, 69]).*

is difficult. By making several assumptions, T.R. Albrecht [39] showed that the applied force in AFM experiments should compress the boron surface atoms to a larger extent compared to the nitrogen surface atoms. Therefore the maxima in AFM images of the BN surface were attributed to the nitrogen sites. However, it should be emphasized, that as long as the contrast mechanism in AFM is not fully understood, such conclusions have still to be regarded with great care.

AFM experiments on complex layered materials such as mica (fig. 22) also revealed atomic-scale features [71], as shown in fig. 23. Individual atoms could not be resolved so far on the mica surface. Instead, the surface unit cell periodicity of 5.2 Å appears in AFM images. The same periodicity is also observed in the frictional force between the mica surface and the probing tip [72] indicating that these frictional forces may play an important role in the imaging mechanism in AFM experiments.

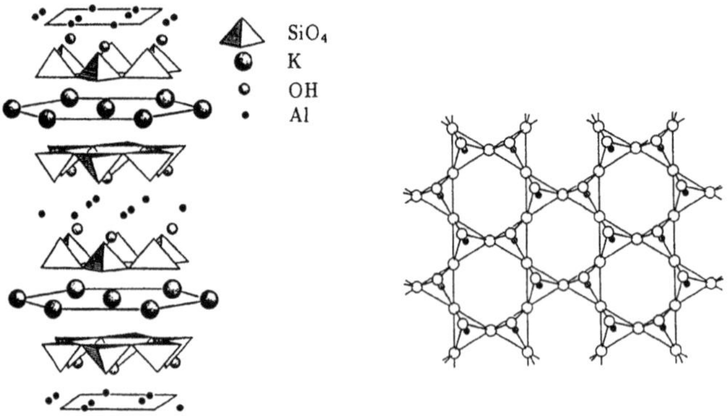

Figure 22: *Crystal structure of mica. The in-plane lattice constant is 5.2 Å. (Figure from ref. [70]).*

Figure 23: *AFM image of mica. The surface unit cell periodicity of 5.2 Å is clearly resolved. (Courtesy of H. Heinzelmann).*

4.5. STM AND AFM STUDIES OF HIGH-T_C SUPERCONDUCTORS

Since the discovery of high-T_c superconductivity in layered materials of complex structure [16], almost all experimental techniques have been applied to these materials in order to investigate the mechanism for superconductivity. There are two fields where STM and - to a lesser extent - AFM have contributed to the present knowledge about high-T_c superconductors: First, STM has been used as a spectroscopic tool at low temperatures to determine the superconducting energy gap on a local scale, which turned out to be advantageous, particularly for sintered materials, where the grain size hardly exceeds several microns. These investigations will not further be discussed. Instead, we will focus on the second application of STM (and also AFM) to high-T_c superconductors: the investigation of the surface atomic scale structure performed at room temperature.

The Bi-compounds (Bi-Sr-Ca-Cu-O) are particularly suited for atomic resolution STM studies because fresh surfaces of these compounds can easily be obtained by cleaving the samples such as for graphite and other layered materials [73-76] . The cleavage process most likely leaves a BiO plane on top of the surface. Atomic resolution STM images, which were obtained under ultrahigh vacuum conditions [74, 76], show both the atomic lattice and an incommensurate superstructure consisting of a sinusoidal modulation with a periodicity of nine to ten unit cells (fig. 24). This superstructure has also been observed by transmission electron microscopy, and its origin has recently been explained [77]. The protrusions in STM images of cleaved Bi-compounds are attributed to Bi atoms whereas the oxygen atoms are believed to remain invisible. Spectroscopic STM measurements [75, 76] have revealed a very small electronic density of states at the Fermi level, implying that the surface BiO layer is not metallic.

For other high-T_c superconducting materials such as the (123)-compounds, for which fresh surfaces cannot be prepared as easily as for the Bi-compounds, the question arises, whether the crystal structure is maintained up to the topmost atomic layer. STM and particularly AFM measurements on single crystals of (123)-compounds [78, 79] have shown that the height of surface steps is always a multiple of the c-axis lattice constant indicating that the crystal structure indeed persists right up to the surface. The lack of a suitable

Figure 24: *STM image (150×200 Å²) of a (Bi-Sr-Ca-Cu-O) high T_c superconductor taken at 150 mV bias and 0.2 nA tunneling current. The bright protrusions are attributed to the Bi sites. A superstucture consisting of a sinusoidal modulation along the a-axis is visible. (Courtesy of M.D. Kirk, see ref. [74]).*

surface cleaning procedure for the (123)-compounds, however, has prevented atomic resolution STM studies so far.

It is a dream of high-T_c superconductivity research to correlate structural investigations with superconducting properties on a local scale. STM might contribute to the fulfillment of this dream in two ways: firstly, by simultaneously mapping the topography and the distribution of the superconducting energy gap [80, 81] and secondly, by simultaneously observing the topography and the Abrikosov flux lattice [82]. The latter observation might also be possible by means of magnetic force microscopy [83, 84] probing the Meissner effect on a local scale.

5. Concluding Comments

The focus of this chapter has been the application of novel local experimental techniques, scanning tunneling microscopy and atomic force microscopy, to materials with layered structure. The experimental results obtained within the short time since the invention of these scanning probe microscopies already demonstrate their great potential for studying local surface properties which are hardly accessible by any other surface analytical technique. Among them are defects within the atomic lattice or within the superlattice of charge density waves, novel superlattice structures on the surfaces of alkali metal graphite intercalation compounds and sinusoidal modulations on the surfaces of high-T_c superconductors. Besides structural investigations down to the atomic scale, STM allows to determine the energy and spatially resolved electronic density of states of metallic and semiconducting materials. AFM offers for the first time the possibility to study the atomic surface structure of insulators and to investigate the microscopic origin of frictional forces and of pressure dependent physical quantities, both being particularly important for materials with layered structure. On the other hand, it has also become clear in the course of this chapter, that the interpretation of STM and AFM results on layered materials is still in its infancy, even for materials with simple structure such as graphite. However, the application of STM and AFM to layered materials has proved to be instructive in many ways, e.g. for the realization of the influence of forces in STM experiments or the importance of frictional forces in AFM experiments. Therefore, one can hope that past and future STM and AFM studies of materials with layered structure will finally lead to both an improved conception of the STM and AFM technique itself as well as a microscopic comprehension of the surface properties of layered materials in general.

ACKNOWLEDGEMENTS

We would like to thank Prof. Dr. H.-J. Güntherodt for his encouragement and stimulation as well as all members of the STM/AFM group at the University of Basel for lively discussions. We are also grateful to Prof. Dr. F. Lévy for providing various samples with layered structure, Prof. Dr. G. Güntherodt and Prof. Dr. E. Kaldis for offering single crystals of high-T_c superconductors, Dr. A.W. Moore for providing HOPG and V. Geiser for preparing intercalation compounds. We also thank T.R. Albrecht, S. Baumgartner, H. Heinzelmann, M.D. Kirk and E. Meyer for kindly providing some of the STM and AFM images of layered materials presented in this chapter. Finally, we thank Prof. Dr. D. Tománek and Prof. Dr. K. Lüders for their continuous interest in this work and for stimulating discussions. Financial support from the Swiss National Science Foundation is gratefully acknowledged.

REFERENCES

[1] G. Binnig, H. Rohrer, Ch. Gerber, and E. Weibel. *Phys. Rev. Lett.* **49**, 57 (1982).
[2] G. Binnig, H. Rohrer, Ch. Gerber, and E. Weibel. *Appl. Phys. Lett.* **40**, 178 (1982).
[3] G. Binnig, H. Rohrer, Ch. Gerber, and E. Weibel. *Physica B* **109 & 110**, 2075 (1982).
[4] A. Bryant, D.P.E. Smith, and C.F. Quate. *Appl. Phys. Lett.* **48**, 832 (1986).
[5] P.K. Hansma and J. Tersoff. *J. Appl. Phys.* **61**, R1 (1987).
[6] G. Binnig and H. Rohrer. *IBM J. Res. Dev.* **30**, 355 (1986).
[7] C.F Quate. *Physics Today* **August**, 26 (1986).
[8] J.A. Golovchenko. *Science* **232**, 48 (1986).
[9] R.J. Behm and W. Hösler. In R. Vanselow and R. Hove, editors, *Chemistry and Physics of Solid Surfaces VI*, page 361, Springer Series in Surface Science, Springer, Berlin, (1986).
[10] T.E. Feuchtwang and P.H. Cutler. *Physica Scripta* **35**, 132 (1987).
[11] G. Binnig and H. Rohrer. *Rev. Mod. Phys.* **59**, 615 (1987).
[12] Y. Kuk and P.J. Silverman. *Rev. Sci. Instr.* **60**, 165 (1989).
[13] J.E. Demuth, U. Koehler, and R.J. Hamers. *J. Microsc.* **152**, 299 (1988).
[14] G. Binnig, C. F. Quate, and Ch. Gerber. *Phys. Rev. Lett.* **56**, 930 (1986).
[15] E. Meyer, H. Heinzelmann, P. Grütter, Th. Jung, Th. Weisskopf, H.R. Hidber, R. Lapka, H. Rudin, and H.-J. Güntherodt. *J. Microsc.* **152**, 269 (1988).
[16] J. G. Bednorz and K. A. Müller. *Z. Phys. B - Condensed Matter* **64**, 189 (1986).
[17] G. Binnig, H. Fuchs, Ch. Gerber, H. Rohrer, E. Stoll, and E. Tosatti. *Europhys. Lett.* **1**, 31 (1986).
[18] I.P. Batra, N. García, H. Rohrer, H. Salemink, and S. Ciraci. *Surf. Sci.* **181**, 126 (1987).
[19] D. Tománek, S.G. Louie, H.J. Mamin, D.W. Abraham, R.E. Thomson, E. Ganz, and J. Clarke. *Phys. Rev. B* **35**, 7790 (1987).
[20] D. Tománek and S.G. Louie. *Phys. Rev. B* **37**, 8327 (1988).
[21] J. Tersoff and D.R. Hamann. *Phys. Rev. B* **31**, 805 (1985).

[22] J. Tersoff. *Phys. Rev. Lett.* **57**, 440 (1986).

[23] H.A. Mizes, S.-I. Park, and W.A. Harrison. *Phys. Rev. B* **36**, 4491 (1987).

[24] J.M. Soler, A.M. Baro, N. García, and H. Rohrer. *Phys. Rev. Lett.* **57**, 444 (1986).

[25] H.J. Mamin, E. Ganz, D.W. Abraham, R.E. Thomson, and J. Clarke. *Phys. Rev. B* **34**, 9015 (1986).

[26] J. B. Pethica. *Phys. Rev. Lett.* **57**, 3235 (1986).

[27] D.P.E. Smith, G. Binnig, and C.F. Quate. *Appl. Phys. Lett.* **49**, 1166 (1986).

[28] S. Ciraci and I.P. Batra. *Phys. Rev. B* **36**, 6194 (1987).

[29] E. Tekman and S. Ciraci. *Physica Scripta* **38**, 486 (1988).

[30] E. Tekman and S. Ciraci. *Phys. Rev. B* 40, 10286 (1989).

[31] S. L. Tang, J. Bokor, and R. H. Storz. *Appl. Phys. Lett.* **52**, 188 (1988).

[32] C. M. Mate, R. Erlandsson, G. M. McClelland, and S. Chiang. *Surf. Sci.* **208**, 473 (1989).

[33] G. Binnig, Ch. Gerber, E. Stoll, T.R. Albrecht, and C.F. Quate. *Europhys. Lett.* **3**, 1281 (1987).

[34] F.F. Abraham and I.P. Batra. *Surf. Sci. Lett.* **209**, L125 (1989).

[35] G. Meyer and N.M. Amer. *Appl. Phys. Lett.* **56**, 2100 (1990).

[36] E. Meyer, H. Heinzelmann, H. Rudin, and H.-J. Güntherodt. *Z. Phys. B - Condensed Matter* **79**, 3 (1990).

[37] H.A. Mizes and J.S. Foster. *Science* **244**, 559 (1989).

[38] T. R. Albrecht, M. M. Dovek, C. A. Lang, P. Grütter, C. F. Quate, S. W. J. Kuan, C. W. Frank, and R. F. W. Pease. *J. Appl. Phys.* **64**, 1178 (1988).

[39] T. R. Albrecht. PhD thesis, Stanford University, (1989).

[40] T.R. Albrecht, H.A. Mizes, J. Nogami, Sang-il Park, and C.F. Quate. *Appl. Phys. Lett.* **52**, 362 (1988).

[41] D. Anselmetti, R. Wiesendanger, and H.-J. Güntherodt. *Phys. Rev. B* **39**, 11135 (1989).

[42] D. Anselmetti, R. Wiesendanger, V. Geiser, H.R. Hidber, and H.-J. Güntherodt. *J. Microsc.* **152**, 509 (1988).

[43] S.P. Kelty and C.M. Lieber. *J. Phys. Chem.* **93**, 5983 (1989).

[44] S.P. Kelty and C.M. Lieber. *Phys. Rev. B* **40**, 5856 (1989).

[45] R. Wiesendanger, D. Anselmetti, V. Geiser, H.R. Hidber, and H.-J. Güntherodt. *Synth. Met.* **34**, 175 (1989).

[46] D. Anselmetti, V. Geiser, G. Overney, R. Wiesendanger, and H.-J. Güntherodt. *Phys. Rev. B* **42(2)**, in press (1990).

[47] D.M. Hwang, N.W. Parker, M. Utlaut, and A.V. Crewe. *Phys. Rev. B* **27**, 1458 (1983).

[48] D. Marchand, C. Fretigny, N. Lecomte, M. Laguës, and J.E. Fischer. *Synth. Met.* **23**, 165 (1988).

[49] M. Laguës, J.E. Fischer, D. Marchand, , and C. Fretigny. *Solid State Comm.* **67**, 1011 (1988).

[50] A. Selloni, C.D. Chen, and E. Tosatti. *Physica Scripta* **38**, 297 (1988).

[51] X. Qin and G. Kirczenow. *Phys. Rev. B* **39**, 6245 (1989).

[52] A. Bryant, D.P.E. Smith, G. Binnig, W.A. Harrison, and C.F. Quate. *Appl. Phys. Lett.* **49**, 936 (1986).

[53] D. Tománek, G. Overney, H. Miyazaki, S.D. Mahanti, and H.-J. Güntherodt. *Phys. Rev. Lett.* **63**, 876 (1989).
[54] G.W. Stupian and M.S. Leung. *Appl. Phys. Lett.* **51**, 1560 (1987).
[55] M. Weimer, J. Kramar, C. Bai, and J.D. Baldeschwieler. *Phys. Rev. B* **37**, 4292 (1988).
[56] D. Sarid, T.D. Henson, N.R. Armstrong, and L.S. Bell. *Appl. Phys. Lett.* **52**, 2252 (1988).
[57] H. Tokumoto, H. Bando, W. Mizutani, M. Okano, M. Ono, H. Murakami, S. Okayama, Y. Ono, K. Watanabe, S. Wakiyama, F. Sakai, K. Endo, and K. Kajimura. *Jap. J. Appl. Phys.* **25**, L621 (1986).
[58] D.C. Dahn, M.O. Watanabe, B.L. Blackford, and M.H. Jericho. *J. Appl. Phys.* **63**, 315 (1988).
[59] S.L. Tang, R.V. Kasowski, and B.P. Parkinson. *Phys. Rev. B* **39**, 9987 (1989).
[60] R.V. Coleman, B. Drake, P.K. Hansma, and G. Slough. *Phys. Rev. Lett.* **55**, 394 (1985).
[61] R.V. Coleman, B. Giambattista, P.K. Hansma, A. Johnson, W.W. McNairy, and C.G. Slough. *Adv. Phys.* **37**, 559 (1988).
[62] E. Meyer, D. Anselmetti, R. Wiesendanger, H.-J. Güntherodt, F. Lévy, and H. Berger. *Europhys. Lett.* **9**, 695 (1989).
[63] E. Meyer, R. Wiesendanger, D. Anselmetti, H.R. Hidber, and H.-J. Güntherodt. *J. Vac. Sci. Technol. A* **8**, 495 (1990).
[64] X.L. Wu and C.M. Lieber. *Science* **243**, 1703 (1989).
[65] G. Boato, P. Cantini, and R. Coella. *Phys. Rev. Lett.* **42**, 1635 (1979).
[66] P. Cantini, G. Boato, and R. Coella. *Physica B* **99**, 59 (1980).
[67] E. Meyer, private communication.
[68] T.R. Albrecht and C.F. Quate. *J. Appl. Phys.* **62**, 2599 (1987).
[69] T.R. Albrecht. PhD thesis, Stanford University, (1989).
[70] K.-H. Hellwege. *Einführung in die Festkörperphysik*. Springer-Verlag, Berlin, (1976).
[71] B. Drake, C. B. Prater, A. L. Weisenhorn, S. A. C. Gould, T. R. Albrecht, C. F. Quate, H. G. Hansma D. S. Cannell, and P. K. Hansma. *Science* **243**, 1586 (1989).
[72] R. Erlandsson, G. Hadzioannou, C. M. Mate, G. M. McClelland, and S. Chiang. *J. Chem. Phys.* **89**, 5190 (1988).
[73] M.D. Kirk, C.B. Eom, B. Oh, S.R. Spielman, M.R. Beasley, A. Kapitulnik, T.H. Geballe, and C.F. Quate. *Appl. Phys. Lett.* **52**, 2071 (1988).
[74] M.D. Kirk, J. Nogami, A.A. Baski, D.B. Mitzi, A. Kapitulnik, T.H. Geballe, and C.F. Quate. *Science* **242**, 1673 (1988).
[75] M. Tanaka, T. Takahashi, H. Katayama-Yoshida, S. Yamazaki, M. Fujinami, Y. Okabe, W. Mizutani, M. Ono, and K. Kajimura. *Nature* **339**, 691 (1989).
[76] C.K. Shih, R.M. Feenstra, J.R. Kirtley, and G.V. Chandrashekhar. *Phys. Rev. B* **40**, 2682 (1989).
[77] Y. Le Page, W.R. McKinnon, J.-M. Tarascon, and P. Barboux. *Phys. Rev. B* **40**, 6810 (1989).
[78] H. Heinzelmann, D. Anselmetti, R. Wiesendanger, H.-J. Güntherodt, E. Kaldis, and E. Wisard. *Appl. Phys. Lett.* **53**, 2447 (1988).
[79] H. Heinzelmann, D. Anselmetti, R. Wiesendanger, H.R. Hidber, H.-J. Güntherodt, M. Düggelin, R. Guggenheim, H. Schmidt, and G. Güntherodt. *J. Microsc.* **152**, 399 (1988).

[80] A.L. de Lozanne, S.A. Elrod, and C.F. Quate. *Phys. Rev. Lett.* **54**, 2433 (1985).

[81] J.R. Kirtley, S.I. Raider, R.M. Feenstra, and A.P. Fein. *Appl. Phys. Lett.* **50**, 1607 (1987).

[82] H.F. Hess, R.B. Robinson, R.C. Dynes, J.M. Valles, Jr., and J.V. Waszcak. *Phys. Rev. Lett.* **62**, 214 (1989).

[83] J. J. Sáenz, N. García, P. Grütter, E. Meyer, H. Heinzelmann, R. Wiesendanger, L. Rosenthaler, H.-R. Hidber, and H.-J. Güntherodt. *J. Appl. Phys.* **62**, 4293 (1987).

[84] Y. Martin and H. K. Wickramasinghe. *Appl. Phys. Lett.* **50**, 1455 (1987).

STM & AFM of Layered Transition Metal Compounds

by

R. V. Coleman, Zhenxi Dai, W. W. McNairy,
C. G. Slough and Chen Wang

Department of Physics
University of Virginia
Charlottesville, VA 22901

Abstract. We have used scanning tunneling microscopes (STMs) operating at 4.2, 77 and 300 K, and atomic force microscopes (AFMs) operating at 300 K, to image charge-density waves (CDWs) in layered structure transition metal dichalcogenides. These materials grow in a variety of phases depending on the growth temperature and the stability of a given phase in a given temperature range. The STM and AFM results reported in this review cover data obtained on the 1T, 2H and 4Hb phases of representative dichalcogenides. Experimental results are presented for the complete range of CDW amplitudes and structures observed in these materials. The spectroscopic mode of the STM has been used to study the energy gap structure associated with the CDWs, and a systematic correlation to the strength of the CDWs has been made. The ability of the STM to detect defects and CDW domain structures is also presented.

G. Benedek (ed.), Surface Properties of Layered Structures, 27–95.

1. Introduction

The layer structure dichalcogenides are assembled from close-packed layers of metal atoms sandwiched between two close-packed layers of chalcogen atoms. These sandwiches are in turn stacked to form the various phases which are determined by the sandwich type and the number of sandwiches per unit cell. A van der Waals gap exists between each sandwich forming the crystal. These crystals undergo a variety of electronic phase transitions including charge-density wave (CDW), metal-insulator, magnetic, and superconducting transitions. The CDW transitions have received the greatest degree of experimental and theoretical study using many techniques and the CDW structure has been well established in many of the compounds and crystal types. The scanning tunneling microscope (STM) and the atomic force microscope (AFM) have proved to be excellent instruments for detecting the real space CDW structure as well as the atomic structure of these compounds.

2. Charge-Density Wave Formation

Charge-density wave (CDW) formation can directly modify the conduction electron wave functions and the local density of states (LDOS) at the Fermi energy. The CDW transition is driven by the electron-phonon interaction and results in the formation of a gap which obliterates a fraction, or in some cases all, of the high temperature Fermi Surface (FS). The conduction electron standing wave formed in this process usually has a wavelength close to a multiple of the atomic lattice, but can be either commensurate or incommensurate with the underlying lattice. The amplitude and charge transfer associated with the CDW is found to vary over a wide range depending on the specific material or phase. This in turn can be related to geometric features of the FS which favor nesting of FS sheets for which the electron-hole interaction at $2k_F$ can drive the FS instability. CDW formation is therefore found more often in quasi-one or two dimensional metals where the band structure favors the formation of a FS with parallel sheets (planes or cylinders). The onset of a CDW and the standing wave charge density is accompanied by a periodic lattice distortion since ion motion is required to screen the electronic modulation induced by the CDW. The presence of the CDW can therefore be detected by neutron, electron, or x-ray diffraction techniques and, of course, by electrical transport or susceptibility measurements which are directly sensitive to changes in the FS. These techniques average over macroscopic volumes of the crystal and thus measure bulk properties of the crystal. The STM, on the other hand, looks directly at the local conduction electron density or the LDOS at E_F and can detect subtle modifications of electronic structure on an atomic scale at the surface of the crystal.

The transition metal dichalcogenides represent a series of compounds exhibiting a wide range of CDW behavior and can be grown as single crystals with excellent surfaces for STM studies. The CDWs in these compounds were first observed using STM by Coleman et al.[1] The formula unit is MX_2 and the crystals grow in the form of close-packed three layer sandwiches separated by a van der Waals gap between each sandwich. The sandwiches form with either trigonal prismatic or octahedral coordination between the transition metal and chalcogen atoms leading to a variety of phases or polytypes exhibiting CDW transitions at widely different temperatures.

The CDW formation in the layer structure dichalcogenides has been studied extensively by Wilson et al.[2] using electron and neutron diffraction, and the CDW wavevectors and superlattice formation have been worked out in detail. The strongest CDW amplitudes are observed in the 1T phases (octahedral coordination) of the Ta based layer structure dichalcogenides, with CDW transitions exhibiting onset temperatures well above room temperature. The 2H phases (trigonal prismatic coordination) show substantially weaker CDW amplitudes and exhibit CDW onset temperatures below room temperature. The 4Hb phases (mixed coordination) show a combination of CDW amplitudes and onset temperatures which maintain a correspondence to the CDWs observed in the pure coordination phases of the same materials.[1] 1T-$TiSe_2$ shows a fairly strong CDW onset at 202 K while 1T-TiS_2 shows no CDW transition down the lowest temperatures. Both of these latter compounds have been studied by STM and results on several other representative 1T layer compounds will also be presented.

The CDW transitions occur at onset temperatures which vary from above 600 to ~20 K. In addition, incommensurate to commensurate transitions can occur well below the onset temperature for some compounds, while others remain incommensurate at all temperatures. The STM studies to be reviewed here have been carried out with microscopes operating at 300, 77 and 4.2 K. Most of the compounds discussed have been successfully studied at both temperatures. In all, ten different compounds have been studied and they are listed along with the CDW transition temperatures and structure in Table I. They cover a complete range of CDW structures observed in these materials and allow a systematic comparison of STM response to the various types of CDW structures. In Section 6 we compare the observed STM response to the calculated and measured electronic structure and FS.

For the strong CDW structures we observe an enhanced STM response, and discuss possible mechanisms related to the electronic structure.

TABLE I

Compound	T_0(K)	T_1(K)	CDW superlattice	Lattice constants (Å)
1T-$TaSe_2$	>600	473	$a=b=\sqrt{13}\,a_0$, $c'=13c_0$	$a_0=3.477$, $c_0=6.263$
1T-TaS_2	>600	150*	$a=b=\sqrt{13}\,a_0$	$a_0=3.346$, $c_0=5.859$
1T-$TiSe_2$	202	202	$a=b=2a_0$, $c=2c_0$	$a_0=3.542$, $c_0=6.015$
1T-TiS_2	—	—	no CDW	$a_0=3.405$, $c_0=5.699$
1T-VSe_2	≈110	≈80	$a=b=4a_0$, $c\approx 3c_0$	$a_0=3.35$, $c_0=6.10$
2H-$NbSe_2$	≈33	—	$a=b\approx 3a_0$	$a_0=3.449$, $c_0=12.562$
2H-$TaSe_2$	122.3	≈90	$a=b=3a_0$	$a_0=3.434$, $c_0=12.692$
2H-TaS_2	75.3	—	$a=b\approx 3a_0$	$a_0=3.315$, $c_0=12.104$
4Hb-TaS_2 O TP	≤600 22	315 —	$a=b=\sqrt{13}\,a_0$ $a=b\approx 3a_0$	$a_0=3.332$, $c_0=23.62$
4Hb-$TaSe_2$ O TP	600 ≈75	410 —	$a=b=\sqrt{13}\,a_0$ $a=b\approx 3a_0$	$a_0=3.46$, $c_0=25.188$

Transition temperatures and CDW superlattice wavelengths of selected transition metal dichalcogenides. T_0 is the CDW onset temperature and T_1 is the temperature at which the CDW becomes commensurate. *The CDW in 1T-TaS_2 becomes commensurate at 150 K.

3. Experimental Techniques

The AFMs and STMs used in these investigations range from homebuilt to commercial models. Some of the homebuilt STMs are designed to operate in a cryogenic liquid environment. Operating with the microscope submersed in a cryogenic liquid has produced a high degree of thermal stability and has greatly aided in the detection of the many subtle electronic effects observed. The cryogenic STMs are all based on a similar design described below. A tube microscope design,[3] also described below, is the basis for the AFMs and some of the room temperature STMs.

The STMs used in most of these experiments follow a simple basic design using an x-y translator cut from a single block of Channel 5400 piezoelectric material (Channel Industries, Santa Barbara, CA) and a semicircular disk bimorph made of Channel 5800 piezoelectric material. The tunneling tip is mounted in the x-y translator and the crystal is mounted on the bimorph which provides the z-deflection.

The x-y translator design is shown in Fig. 1(a) and has two triangular openings cut through it to define the x and y arms. A collet, made from a 1-72 brass screw, is mounted at the junction of the arms and holds the scanning tip which is made from a $Pt_{0.8}\,Ir_{0.2}$ wire 0.5 mm in diameter. The two front spacers and the adjustable rear foot support the bimorph, which is clamped in place over the x-y translator with a BeCu spring. The assembled piezoelectric drives from a side view are shown in Fig. 1(b). The coarse adjust is accomplished with the rear foot pusher which is driven by an electric motor through a worm gear drive assembly as shown in Fig. 1(b). When tunneling current is detected, the motor shuts off and the bimorph maintains the tunneling gap through feedback to a logarithmic integrating error amplifier.

The x motion is generated by applying a voltage so that the electric field is in the direction of poling for one arm and opposite to the direction of poling for the other x arm. Thus, one arm expands and one contracts, moving the collet and tip to one side relative to the spacers, while maintaining the total distance between the spacers at a constant value. The motion is on the order of 0.33 Å/V at 4.2 K.

The y motion is generated by applying a voltage to the perpendicular arm which moves the collet and tip by bending the x translation arms. The lengths of the electrodes on the arms are arranged to give approximately the same calibration for both x and y motion. Final adjustment to obtain uniform x and y motion is made electronically.

The semicircular bimorph forming the z-translator is 1 cm in radius. The calibration of the z-motion is 3.6 Å/V at 4.2 K. The top and bottom bimorph surfaces are silver plated to form the electrodes and the sample is mounted on the front edge of the bimorph using an insulating pad and an appropriate glue or varnish.

The entire assembly is suspended with a thin 0.188 inch diameter stainless steel support tube and hangs just below the end of a long stainless steel tube 1.250 inches in diameter. This can be lowered into a liquid helium storage dewar which is supported on inflated inner tubes for vibration isolation.

A tube microscope can be modified to operate as an AFM.[4] The sample is now mounted to the single piezoelectric tube. A diamond or Si_3N_4 tip[5] mounted on the end of a Si_3N_4 cantilever is used as the AFM tip. A laser beam is reflected off

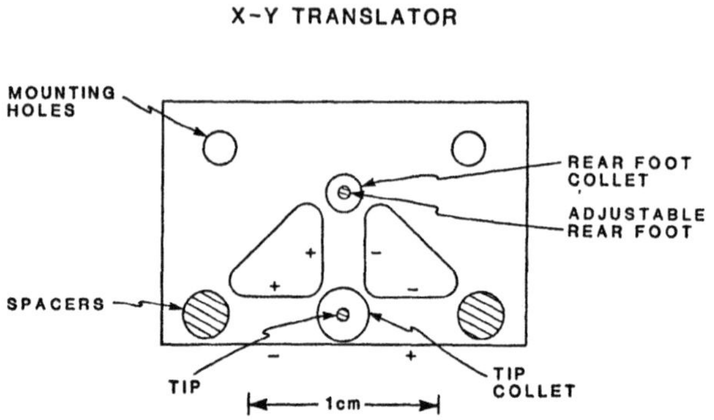

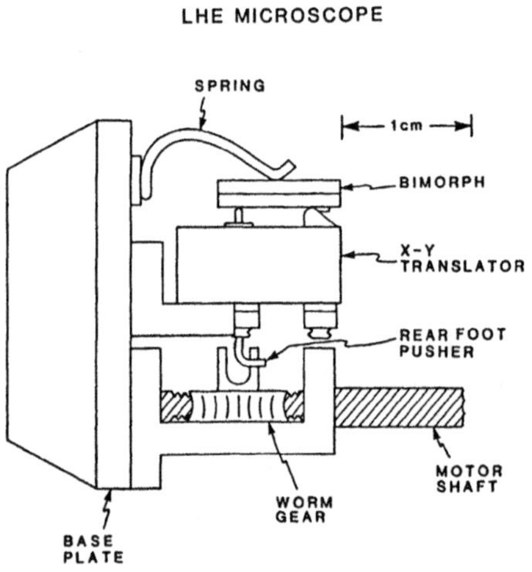

Fig. 1. (a) The X-Y translator used in the cryogenic STM. The plus and minus signs indicate the direction of poling on the arms of the translator. The bimorph rests on the two front spacers and on the adjustable rear foot. (From Reference 18.)

(b) Side view of the cryogenic STM. The tip-sample gap is controlled by advancing or retracting the rear foot pusher using the worm gear assembly, thereby raising or lowering the rear foot of the bimorph. The bimorph then pivots on the front feet, thus changing the tip-sample distance. (From Reference 18.)

a mirror mounted to the back of the cantilever and is detected by a spot detector which is able to sense cantilever deflections of less than 0.1 nm. Figure 2 shows a schematic of this setup. The same idea is used in the commercial microscope built by Digital Instruments of Santa Barbara California.[6]

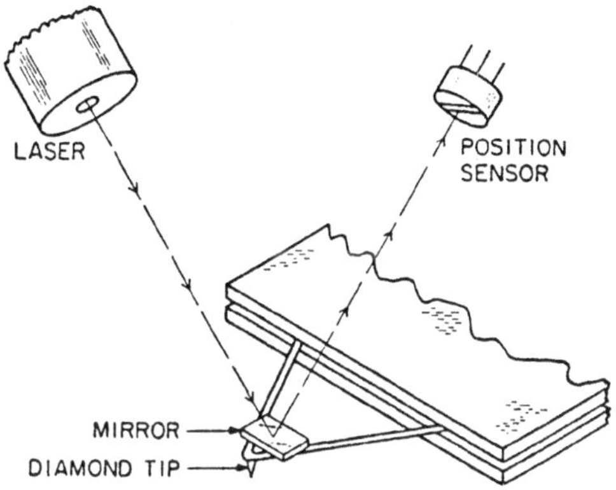

Fig. 2. The cantilever deflection detection schematic for the AFM. The sample is mounted to the single piezoelectric tube. A diamond or Si_3N_4 tip is mounted on the end of the Si_3N_4 cantilever. (From Reference 4.)

All images produced by the cryogenic STMs were taken in the constant current mode. The tunneling currents were typically in the range of 1-10 nA and the tunneling bias voltages were in the range of 10-30 mV. The tunneling current between the tip and the sample is detected across a series resistor. A PAR 118 preamplifier then amplifies the signal for processing by a homebuilt logarithmic integrating error amplifier. The error signal is sent to an IBM PC-AT computer which is used to assemble the image. A 12 bit analogue-to-digital converter (ADC) with a range of ± 10 V is used to digitize the error signal. This provides a nominal resolution of 0.02 Å at 4.2 K for the LHe microscope and 0.09 Å at 77 K for the LN_2 microscope. The computer also provides the voltages which drive the X-Y translator. Signal averaging can be accomplished through the computer acquisition program by varying the number of samples taken at each data point. Typically 50 to 100 samples are taken.

Analogue drives and a storage oscilloscope are used to search for a good scan area before hooking up the more sensitive computer drive. The microscopes are vibrationally isolated from building vibrations by either rubber tires or elastic cords. The resonant frequency of the system is ~ 0.5 Hz.

The transition metal dichalcogenides are grown by the method of iodine vapor transport. Stoichiometric quantities of powder were sealed in quartz tubes and pre-reacted by sintering at 900 °C. The sintered powders were then placed in another quartz tube with iodine gas, and the tube was then sealed. The exact temperatures and temperature gradients differ for each compound. The Ta-based octahedral and 4Hb crystals must be quenched from high temperatures (~900 °C for the 1T phase, ~700 °C for the 4Hb phase), whereas the Ta-based trigonal prismatic, and Ti and V crystals are slowly cooled from temperatures around 700 °C. High quality crystals resulted from these growth procedures, as confirmed by residual resistance ratios ($\rho_{300}/\rho_{4.2}$) of 60 to 300, and the observation of large magneto-quantum oscillations in the 2H and 4Hb compounds. Reference 7 contains a more detailed analysis of many properties of the transition metal dichalcogenides.

4. STM and AFM of 1T Phase Layer Structure Crystals

The crystal structure of the 1T phase exhibits trigonal symmetry and consists of close-packed three-atom-layer sandwiches with one sandwich per unit cell. Octahedral coordination exists between the metal atoms in the center layer and the chalcogen atoms in the top and bottom layers. The 1T phases of interest are in general semi-metals and exhibit large conductivity decreases at the CDW transitions, but remain metallic at low temperature. Results of STM studies on 1T-$TaSe_2$, 1T-TaS_2, 1T-$TiSe_2$, 1T-TiS_2, and 1T-VSe_2, are reviewed below.

The group V materials have received intense experimental and theoretical study and include the compounds TaS_2, $TaSe_2$, VSe_2 and $NbSe_2$. 1T-$TaSe_2$ and 1T-TaS_2 can be stabilized at low temperature by quenching the crystals from 900 °C to room temperature. 1T-VSe_2 is the equilibrium phase at room temperature and crystals can be cooled slowly to room temperature. $NbSe_2$ does not exhibit a 1T phase.

Band structure calculations indicate that these compounds are all good d band metals in the undistorted phase. Below the CDW transition the FS topology is modified and the conductivity can show varying degrees of change depending on the amount of FS gaping and the strength of the CDW.

In the case of 1T-$TaSe_2$ and 1T-TaS_2 the CDW is extremely strong and large changes in the FS topology and the DOS at the Fermi level are expected to occur. This is confirmed by the observation of substantial decreases in the electrical conductivity at the CDW transitions. The STM results to be discussed below also confirm this conclusion by the observation of a very large amplitude modulation at the CDW wavelength.

The CDWs forming in 1T-$TaSe_2$ and 1T-TaS_2 are similar in the low temperature commensurate phase and can be described by a six-pointed star-of-David cluster of thirteen Ta atoms as shown in Fig. 3. Charge is transferred toward the center atom in each of these thirteen atom clusters resulting in a superlattice as outlined by the large hexagon in Fig. 3. The commensurate superlattice vector is $\sqrt{13}\,a_0$ in length. 1T-$TaSe_2$ remains commensurate up to 473 K, while 1T-TaS_2 makes transitions to a number of different nearly commensurate phases before reaching a final incommensurate phase above 350 K. This difference between 1T-$TaSe_2$ and 1T-TaS_2 is clearly evident in both STM and AFM studies at room temperature and will be summarized below.

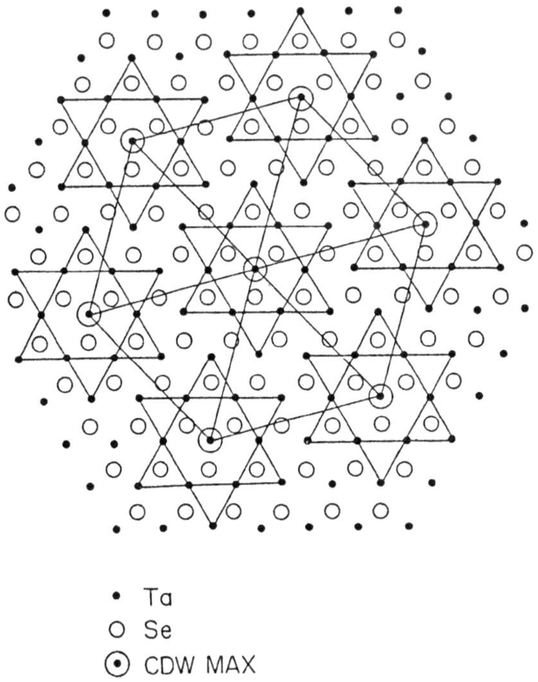

Fig. 3. CDW supercell and star-of-David thirteen Ta atom clusters in 1T-$TaSe_2$ and 1T-TaS_2 layer compounds. The supercell connects the charge hills at the center of each thirteen-atom cluster and is rotated from the underlying atomic lattice by 13.9°. The commensurate superlattice vector is $\sqrt{13}\,a_0$ in length. (From Reference 18.)

In contrast to the 1T phase compounds of Ta, 1T-VSe_2 exhibits a relatively weak CDW. Conductivity changes are small and the observed STM modulations at the CDW wavelength are small. The formation of the CDW is accompanied by a PLD on the order of 0.1 to 0.2 Å in the Ta layer and smaller distortions of the surface Se or S atoms normal to the surface. This will produce a change in the LDOS at the position of the tip, but the total z-deflection due to the CDW is more than an order of magnitude larger than the PLD. This suggests that additional mechanisms are involved in the STM response to the CDW structure.

4.1 STM OF 1T-TaS_2 AT 4.2 AND 77 K

In the commensurate CDW phase of 1T-TaS_2 at low temperatures, the $\sqrt{13}\,\mathbf{a}_0 \times \sqrt{13}\,\mathbf{a}_0$ unit cell contains three types of atom labeled a, b, and c in Fig. 4. Smith et al.[8] have developed a band structure model to describe the effects of the CDW formation and to interpret experimental results from angle-resolved photoemission spectroscopy. Their analysis of the CDW phase shows a collapse of the Ta d band into three subband manifolds separated by gaps and gives the total number of occupied electrons per atom on each type of atom. The calculated numbers are $n_a = 1.455$, $n_b = 1.311$ and $n_c = 0.611$. The calculation indicates that the CDWs have a fairly uniform amplitude within the seven atom cluster centered

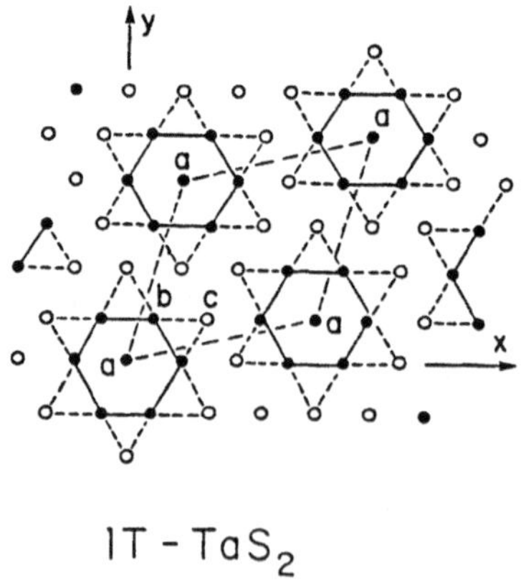

Fig. 4. Ta atom pattern contributing to the CDW charge transfer in 1T-TaS_2. The thirteen Ta atoms centered on the CDW maixima form a six-pointed star-shaped cluster. The electronic charge maximum occurs on the a-type atoms. Charge is transferred from the c-type atoms toward the b-type and a-type atoms. (From Reference 8.)

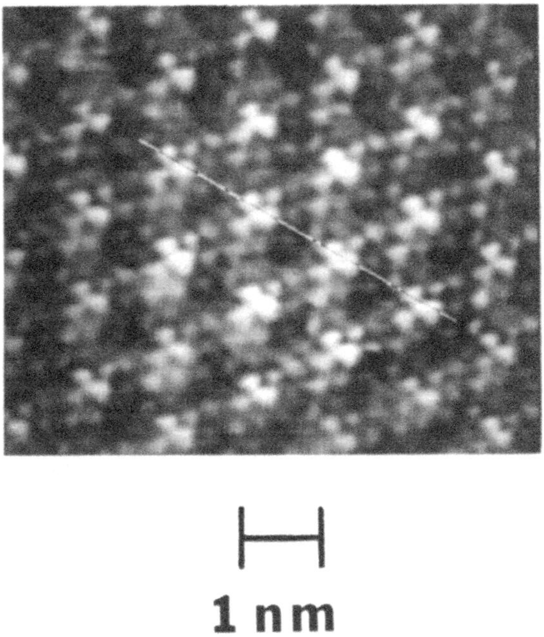

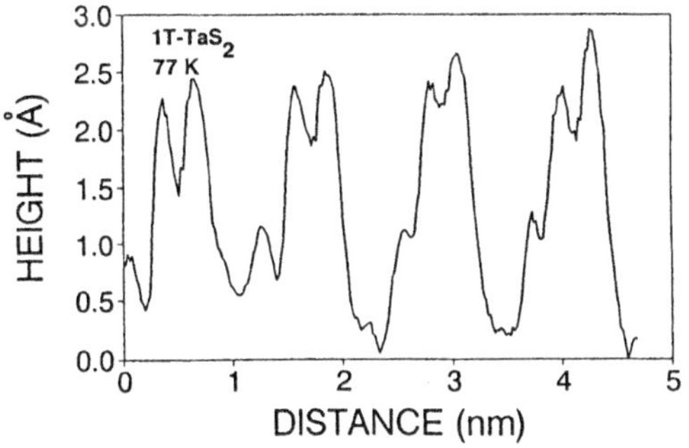

Fig. 5. (a) Grey scale STM image of atoms and CDWs on the surface of a 1T-TaS_2 crystal at 77 K. (I = 2.2 nA, V = 2 mV). This shows the $\sqrt{13}\,\mathbf{a}_0 \times \sqrt{13}\,\mathbf{a}_0$ CDW superlattice which is rotated 13.9° from the atomic lattice.

(b) Profile of z-deflection along the track indicated by the black and white line in (a). The total z-deflection is ~ 3 Å and is dominated by the CDW modulation at a wavelength of $\sqrt{13}\,a_0$. The atomic modulation contributes ~ 0.5 Å to total the z-deflection of ~ 3 Å. (From Reference 18.)

on the a-type atom, while the amplitude decreases strongly in the interstitial regions near the c-type Ta atoms.

A STM scan of $1T\text{-}TaS_2$ at 77 K is shown in Fig. 5(a) and a profile recorded along the track indicated by the black and white line is shown in Fig. 5(b). The total z-deflection is ~ 3 Å and is dominated by the CDW modulation at a wavelength of $\sqrt{13}\ a_0$. The atomic modulation contributes ~ 0.5 Å to the z-deflection. These observations confirm that the charge transfer associated with the CDW must be very large, although the total z-deflection indicates that some enhancement mechanism may be amplifying the STM response. The charge transfer calculated by Smith et al.[8] is extremely large, however the modulation of the conduction electron wavefunction above the surface would not be expected to extend to 3 Å. Nevertheless, the STM results indicate that the charge transfer and the accompanying FS modification are extremely large in $1T\text{-}TaS_2$, and a similar result is found for $1T\text{-}TaSe_2$ as described in Section 4.3.

The resistance versus temperature curve measured for $1T\text{-}TaS_2$ is shown in Fig. 6 and exhibits a complicated structure associated with the various CDW subphases which form in specific ranges of temperature. Two distinct incommensurate phases have been identified in $1T\text{-}TaS_2$ which can be characterized as nearly commensurate (NC). The NC $1T_2$ phase occurs on cooling between 350 and ~ 190 K, but on warming exists only between ~ 280 and 350 K. The T phase

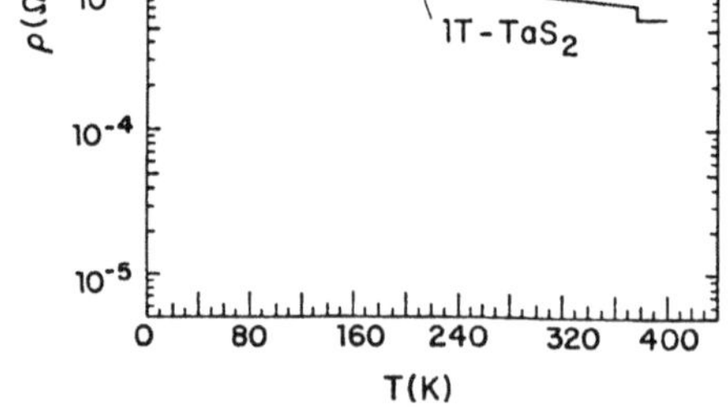

Fig. 6. Resistivity versus temperature for $1T\text{-}TaS_2$ measured parallel to the layers. $1T\text{-}TaS_2$ shows a small discontinuity at 350 K where rotation of the CDW superlattice begins. This is followed by a large hysteretic discontinuity centered around 200 K which corresponds to a lock-in of the CDW into a commensurate phase. (From Reference 45.)

exists upon warming between ~ 220 and 280 K, and indicates a breaking of three fold symmetry of the wavevector positions observed in the $1T_2$ phase. Above 350 K the final incommensurate phase is formed with the CDW superlattice aligned along the atomic lattice. The NC phase is characterized by higher harmonics of the fundamental CDW wavevectors in electron and x-ray diffraction experiments, thereby suggesting that the CDW superlattice is forming some type of two dimensional domain structure. This structure of the NC phase existing at room temperature has been studied extensively by STM and will be reviewed below.

At 4.2 K and below the commensurate CDW phase of 1T-TaS_2 exhibits a drive toward a metal-insulator transition. This may be associated with the CDW modification of the band structure in which a very narrow band forms leading to a possible Mott-Anderson transition. STM images at 4.2 K are very hard to obtain and give very poor definition. This behavior may be associated with the random onset of the metal-insulator transition and a low density-of-states at the Fermi level.

4.2 STM OF 1T-TaS_2 AT 300 K

STM scans of 1T-TaS_2 at room temperature show a two dimensional structure due to a long range modulation of the CDW amplitude. The period of this long range modulation is ~ 6 CDW wavelengths and corresponds to an amplitude variation of ~ 46% as a percentage of the maximum CDW amplitude. Wu and Lieber[9] were the first to report the observation of a two-dimensional domain-like structure which can be readily observed by adjusting the look up table (LUT) so that the amplitude variation is exaggerated by the black and white contrast as shown in Fig. 7. A profile through the centers of several domains in the original gray scale image along the track shown in Fig. 7 is shown in Fig. 8. The profile demonstrates the continuous modulation of the CDW amplitude which follows an approximately sinusoidal variation. A nonsinusoidal profile due to higher harmonics can be expected, and a limited nonsinusoidal behavior consistent with a wide discommensuration is indicated in the averaged profiles in Fig. 9. However, present experimental error does not allow a precise calculation of the higher harmonic component. The STM scan shown in Fig. 7 is completely dominated by the CDW amplitude with the atom modulation an order of magnitude smaller than the CDW amplitude. Any beat structure or interference modulation by the atomic structure is negligible. When the atomic modulation is appreciable the beat structure makes it difficult to separate the true CDW amplitude modulation, although this interference effect does not account for the major part of the observed long range amplitude modulation observed in such scans. The interference or beat structure occurs between the CDW modulation and the modulation produced by the surface S atoms. These two modulations are out of phase so that any amplitude modulation of the CDW due to interference will be

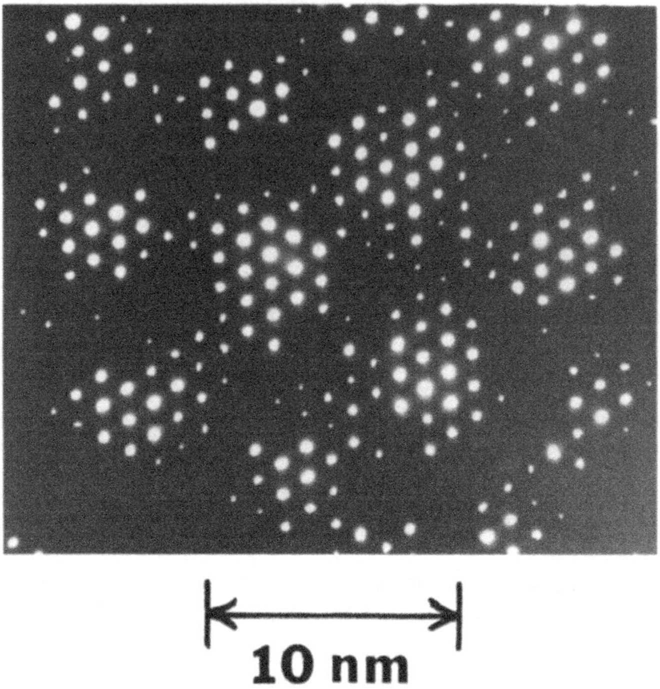

Fig. 7. Grey scale STM image of $1T\text{-}TaS_2$ taken at room temperature with the LUT adusted to emphasize the two-dimensional domain-like structure resulting from a variation of the CDW amplitude (I= 2.2 nA, V= 25 mV). (From Reference 19.)

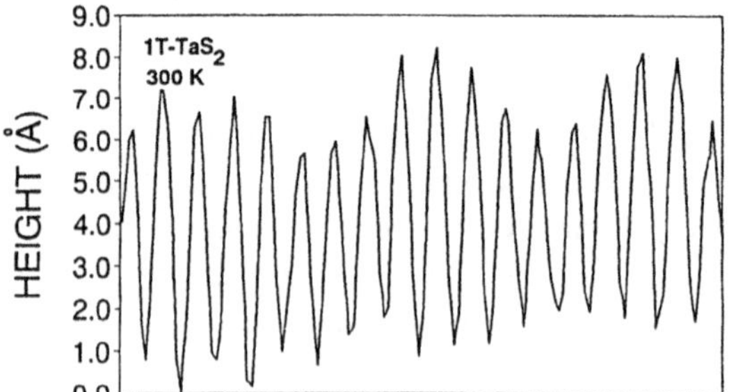

Fig. 8. Profile of z-deflection measured along the black and white track aligned along a row of CDW maxima as shown in Fig. 7. The CDW amplitude undergoes a continuous ~ sinusoidal modulation with a period of ~ 6 CDW wavelengths. (From Reference 19.)

small. In fact, such interference modulation, although repeating at ~ 6 CDW wavelengths, reaches a minimum at the center of the CDW domain where the CDW maximum is in phase with the Ta atom in the metal layer below the surface. Therefore any two dimensional domain-like structure created by the interference would not be in phase with the true CDW amplitude modulation structure which dominates the data shown here. These effects were considered by Coleman et al.[10] in discussion of measurements of amplitude modulation in 1T-TaS_2 at 300 K. The presence of beat structure in STM scans with appreciable atomic modulation superimposed on the CDW modulation makes it difficult to determine reliable phase information from graphical constructions of the real space data. A small superimposed atomic modulation will generate apparent localized phase slips due to interference which do not reflect a true phase slip between the CDW and atomic lattices. However, the amplitude modulation introduced by beating will remain extremely small in scans where the CDW amplitude is large and dominant. The observed amplitude of the CDW modulation also scales with the enhancement of the CDW amplitude and not with the atomic amplitude. Asymmetries in the CDW domain amplitude due to the out-of-phase interference component are also negligible.

Nakanishi and Shiba[11] originally predicted the existence of a two-dimensional domain-like structure in 1T-TaS_2 based on a free energy model calculation. This model was used to calculate domain structures, and predicted specific phase and amplitude variations of the CDWs based on the existence of higher harmonics indicated by diffraction and photoemission data. The domains would consist of more nearly commensurate regions in the center of the domain with phase shifts and amplitude variations across the domain. The model is consistent with wide discommensurations in which the CDW amplitude modulation would follow a continuous variation with the maximum amplitude occuring at the center of the domain. The amplitude modulation would deviate from sinusoidal due to the presence of higher harmonics and a nonuniform phase variation.

Fig. 9 shows a profile of CDW maxima obtained by averaging the profiles across four separate domains in a given STM scan of 1T-TaS_2. The solid line in Fig. 9(a) represents the variation of the maxima of one CDW component in the Nakanishi and Shiba[11] model as derived from a calculation by Burk et al.[12] The absolute CDW amplitude in the calculated curve has been scaled to match the experimentally observed maximum CDW amplitude measured from maximum to minimum at the center of the domain. This maximum CDW amplitude is measured from the zero of Fig. 9 and not from the minimum observed in the profile. A profile through a row of CDW maxima is offset from the corresponding absolute CDW minima by ~ a_0. The solid line represents the predicted variation of the CDW maxima along one of the CDW component directions. The experimentally observed

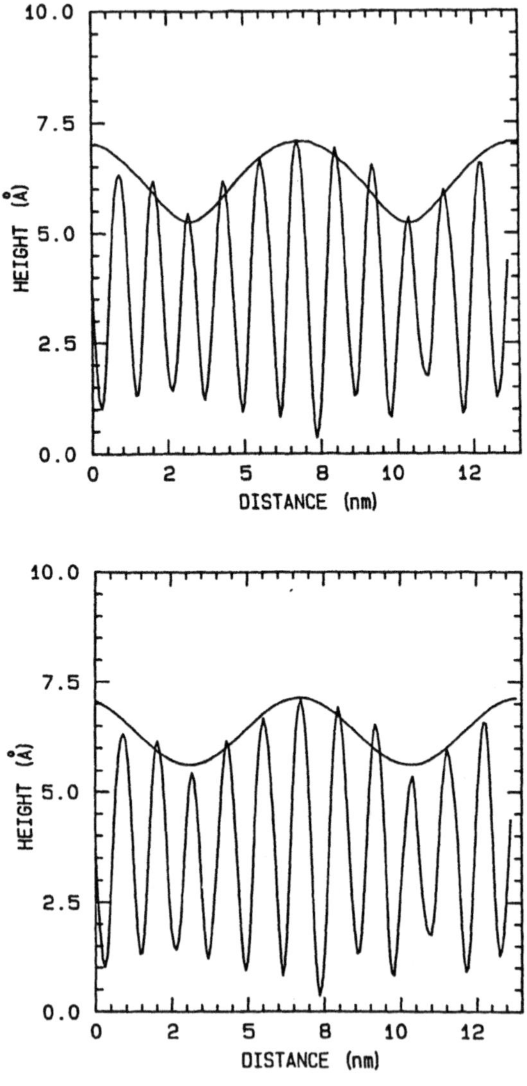

Fig. 9. (a) Profile of CDW maxima obtained by averaging profiles across 4 separate domains in the original grey scale image from which Fig. 7 was generated. The solid line represents the amplitude variation of the maxima of one CDW component of the Nakanishi and Shiba[11] domain model as calculated by Burk et al.[12] The calculated solid curve has been scaled to match the experimentally observed maximum CDW amplitude in the center of the domain. The experimentally observed relative amplitude variation is 0.54 and the calculated variation is 0.63.
(b) Experimentally observed profile of the CDW amplitude compared to a sine wave (solid line). At the center of the domain boundary the amplitude is consistently less than that required to match a sine wave. (From Reference 10.)

variation is extremely close to the calculated profile from the Nakanishi and Shiba model which contains wide discommensurations (domain boundaries). Fig. 9(b) shows a sine wave fit to the profile of the CDW maxima as indicated by the solid line. The profile calculated from the Nakanishi and Shiba model is a slightly better fit than the sine wave. The experimental profile indicates a depressed CDW amplitude at the minimum of the domain boundary consistent with the presence of higher harmonics. However, higher experimental accuracy would be required to deduce quantitative data on the higher harmonic content. The observed amplitude variation is certainly consistent with the Nakanishi and Shiba[11] model. Commensurate domains with narrow discommensurations at the boundaries would require a much stronger nonsinusoidal amplitude modulation than is observed. However, the amplitude modulation data alone do not confirm the complete domain model since no local phase variation information can be accurately determined from the real space data. The local phase variation would change over a number of CDW wavelengths and measurements of CDW maxima relative to atomic positions using graphical methods cannot be used to accurately determine the local variations in phase, angle or wavelength of the CDW. However, substantial regions at the commensurate angle of 13.9° do not appear to be present. The domains and CDW amplitude modulation result from local variations in charge transfer, CDW wavevector, and phase across the domain. The observation of a significant modulation of the CDW amplitude requires that the local CDW charge transfer, wave vector and PLD change nonuniformly across the domain. The detailed phase variation has to be measured by other techniques. Burk et al.[12] have recently obtained wide area STM scans in the constant height mode which enables them to resolve satellites in the Fourier tranforms. By using the location and intensity of these satellites a fit to the Nakanishi and Shiba[11] model can be made which determines both CDW amplitude and phase as calculated parameters. The results are in approximate agreement with the original Nakanishi and Shiba[11] model fit.

The STM and AFM are clearly the only experimental techniques which can verify this structure in local detail, but the complete analysis of the STM and AFM results is still under development. The continuous variation of amplitude as shown in Figs. 8 and 9 suggests a wide discommensuration with a phase variation extending over a number of CDW wavelengths. A localized discommensuration is not observed, although the STM does not give the precise phase or angle of the CDW with respect to the atomic lattice. The alignment of domain centers is observed by STM to make an angle of ~ 5.6° with respect to the CDW superlattice. The domains in the STM scans show a degree of irregularity when comparing the centers of adjacent domains and possible variations of the domain superstructure have been discussed by Wilson.[13] The present STM observations indicate an average angle of ~ 12° between the CDW superlattice and the atomic lattice. If commensurate regions exist near the center of the domains, they cannot extend beyond one or two CDW wavelengths, and the phase shift will occur over fairly wide discommensurations comprising at least 50% of the domain area. Local variations in domain size, alignment and amplitude are also observed. Some of this may be connected with defects although point defects produce very localized modifications of the CDW amplitude.[14]

The model of Nakanishi and Shiba[11] with the phase shifts spread across three or four CDW wavelengths can certainly be used to fit the STM amplitude data within experimental error. This detailed structure of 1T-TaS_2 at room temperature has also been examined by AFM and further discussion will be included under Section 4.4. A measurement of the precise phase variation can be made from high resolution Fourier transforms obtained from large area STM or AFM scans covering many domains.[12]

Initial studies of 1T-TaS_2 at 225 K have been carried out by Thomson et al.[15] after warming the sample from 143 K. The data was obtained with the STM operating in the constant height mode so that profiles and amplitudes could not be measured. However, apparent domain boundaries were detected and the geometry suggested long narrow domains with a width of 65-70 Å. This striped domain structure was oriented at ~ 26° to the CDW superlattice and within each domain the pattern of atoms and CDWs suggested commensurate regions. More detailed study of the T phase by both STM and AFM will be required before a complete analysis can be made.

4.2.1 *Defects and CDW Domain Structure in 1T-TaS_2 at 300 K.* The domain structure associated with the amplitude and phase modulation of the CDWs in the nearly commensurate phase might be quite sensitive to defects. In the commensurate phase at lower temperatures defects in the CDW structure are observed to be localized over one or two CDW wavelengths and examples will be discussed in Section 5. Similar localized defects have been observed in the nearly commensurate phase. These interrupt the continuous amplitude modulation of the CDW domain, but do not extend over more than one or two CDW wavelengths. The overall domain structure remains quite stable.

Examples of the amplitude profiles recorded through CDW defects in 1T-TaS_2 at 300 K are shown in Fig. 10. In Fig. 10(a) one CDW maxima has been entirely quenched at one location while in Fig. 10(b) one CDW maximum has been partially quenched. In both cases the amplitudes of the adjacent CDW maxima have been either enhanced or decreased, but after a few wavelengths the regular amplitude modulation of the domain is restored. The disruption of the charge transfer associated with the CDW formation remains localized and behaves in essentially the same way as observed in the commensurate phase. These localized defects in the

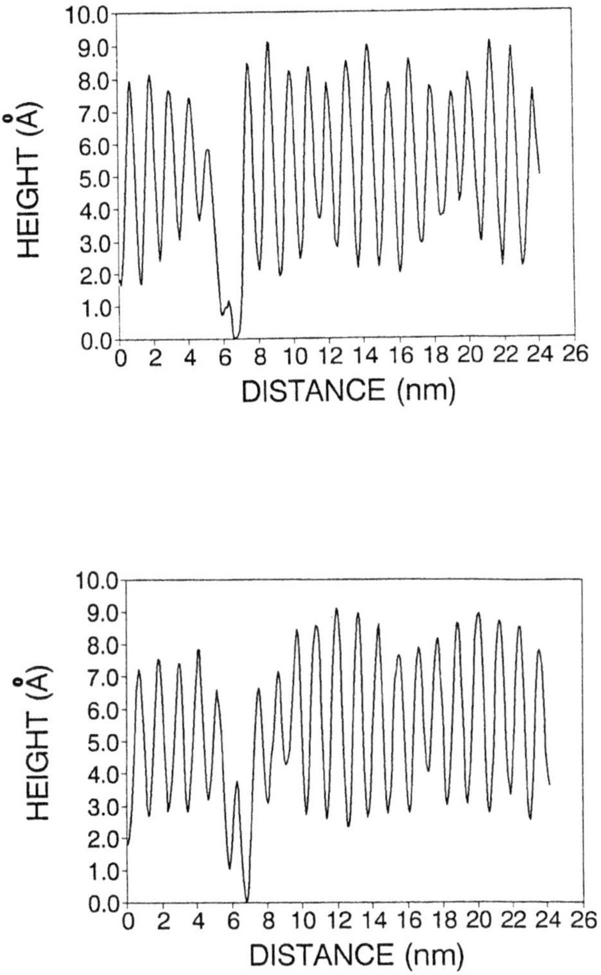

Fig. 10. (a) Profile of z-deflection through a CDW defect in which the charge maximum has been entirely quenched.

(b) Profile of z-deflection across a CDW defect in which the charge maximum has been decreased but not removed. The defects are well localized and do not perturb the more distant CDW pattern. (From Reference 14.)

CDW superlattice may be created by either vacancies or substitutional impurities in the Ta layer which interrupt the charge transfer within the local 13 atom clusters of Ta atoms. These defects do not destabilize the domain structure and would have to exist in much greater density before the domain structure is destroyed.

4.3 STM OF 1T-$TaSe_2$

A STM scan of 1T-$TaSe_2$ at 77 K recorded by computer and presented as contours of constant z-deflection is shown in Fig. 11. The CDW charge maxima appear as white plateaus centered on a $\sqrt{13}\,\mathbf{a_0} \times \sqrt{13}\,\mathbf{a_0}$ hexagonal lattice with an experimental spacing of 12.4 ± 0.2 Å, again in agreement with the expected CDW superlattice spacing. The contours of constant LDOS at the Fermi level show no evidence of modulation by the surface layer of Se atoms and the interstitial positions between the charge maxima show three deep minima and three saddle points symmetrically located relative to the maxima of the CDW superlattice. The entire pattern is therefore dominated by the CDW although the deep minima and saddle

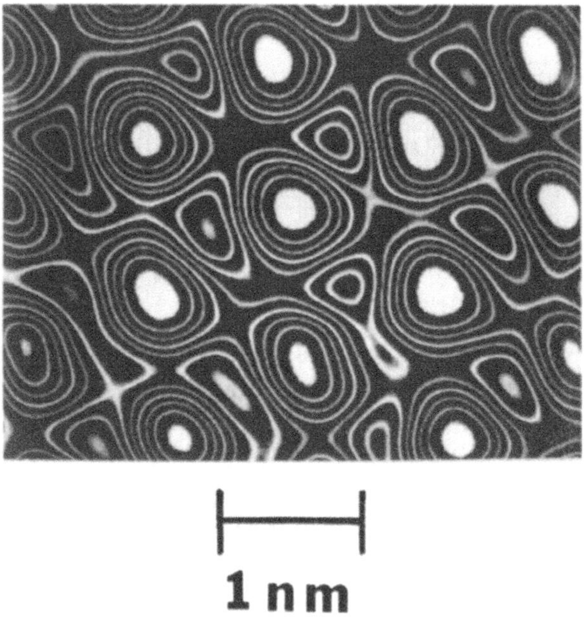

Fig. 11. A STM image of 1T-$TaSe_2$ taken at 77 K using a contour LUT which shows contours of constant z-deflection corresponding to contours of constant LDOS. The white plateaus indicate the CDW maxima centered on a $\sqrt{13}\,\mathbf{a_0} \times \sqrt{13}\,\mathbf{a_0}$ hexagonal superlattice with an experimental spacing of 12.4 ± 0.2 Å, in good agreement with the value of 12.54 Å measured by electron diffraction. (From Reference 18.)

points must reflect the inequivalent surface Se atom arrangements as well as the CDW modulations to the LDOS at the positions. As shown in Fig. 12 the saddle points are centered over a Se atom while the deep minima occur at positions between surface Se atoms which also have no Ta atom located in the layer below. The three pairs of interstitial positions should not be quite equivalent due to the 13.9° rotation of the CDW superlattice provided the surface Se atoms contribute to the modulation of the conduction electron density. In general the charge density contours show a small inequivalence at these interstitial positions, but it is close to the limit of resolution for the present STM operation. The calibrated z-deflection from the deep minima to the maxima in the image of Fig. 11 is very large with a value of 2.4 ± 0.2 Å.

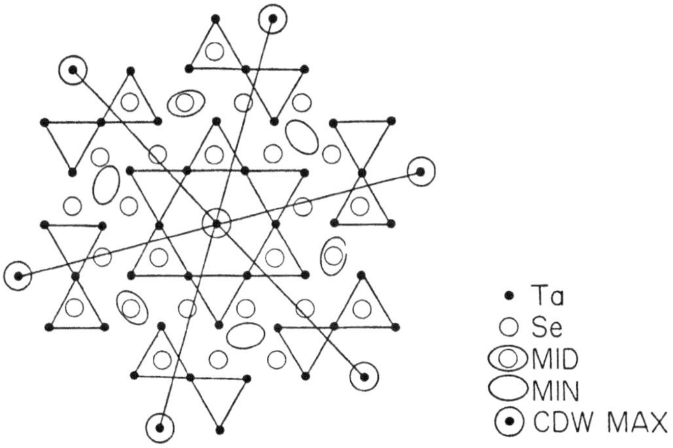

Fig. 12. Diagram showing positions of deep minima and saddle points of the CDW relative to the 1T-$TaSe_2$ atomic lattice. Deep minima occur between groups of three surface Se atoms which have no Ta atom in the layer below and saddle points over Se atoms. The ellipses indicate the location of minima and saddle points. (From Reference 18.)

1T-$TaSe_2$ in the CDW phase at room temperature is a semi-metal with a resistivity of 2 x 10^{-3} Ω cm parallel to the layers. The resistivity shows a monotonic decrease below room temperature, reaching a value of 10^{-4} Ω cm at 4.2 K. This increase in conductivity does not appear to have any major effect on the STM images obtained at 4.2 K. A computer scan recorded from 1T-$TaSe_2$ at 4.2 K presented as contours of constant z-deflection is shown in Fig. 13. The pattern is essentially the same as that observed at 77 K in Fig. 11 and again shows only the $\sqrt{13}\,\mathbf{a}_0 \times \sqrt{13}\,\mathbf{a}_0$ CDW pattern with no evidence of atomic modulation. The average maximum to minimum z-deflection observed at 4.2 K is 3.5 ± 1.4 Å and

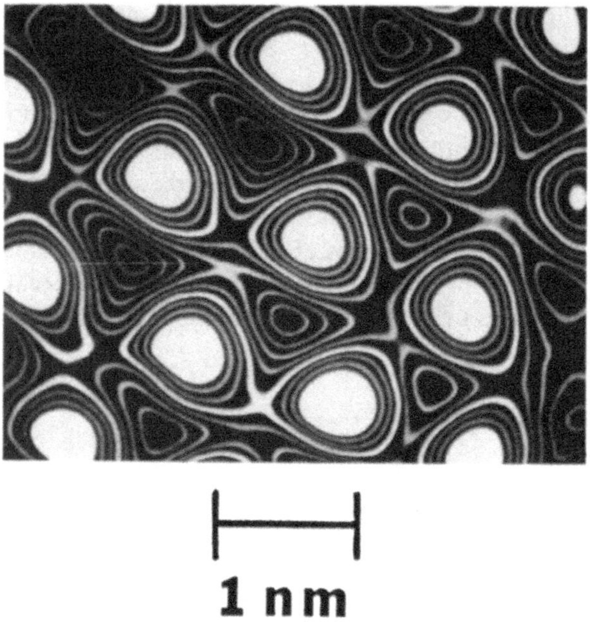

Fig. 13. A STM image of CDWs at the surface of a $1T\text{-}TaSe_2$ crystal at 4.2 K. The image is a plot of the contours of constant z-deflection. The white plateaus occur at the CDW charge maxima (I = 2.2 nA, V = 30 mV). (From Reference 37.)

although constant during a given run, it varies in the range 2.5 to 5 Å for different crystals or cleaved surfaces. The deflection observed at 4.2 K is on average larger than that observed at 77 K, but at both temperatures the deflection is anomalously large relative to the expected spacial modulation of the LDOS from an extended FS.

The relative strength of the atomic modulation and the degree of atomic resolution are extremely sensitive to the STM response. Even when the atomic modulation is clearly resolved it represents only a small fraction of the total z-deflection, ~ 0.2 Å out of a total of ~ 2 Å. Fig. 14 shows a gray scale scan and a profile recorded at 4.2 K. The atoms are well-resolved in Fig. 14(a) but the CDW amplitude completely dominates the profile as shown in Fig. 14(b).

The FS modifications of the $1T\text{-}TaSe_2$ band structure have not been studied in detail and at present no estimates of the charge transfer have been developed. The STM response to the $1T\text{-}TaSe_2$ CDW is similar in magnitude to that observed for $1T\text{-}TaS_2$, except that the atomic modulation is often absent in the $1T\text{-}TaSe_2$ scans at both 77 and 4.2 K. The CDW in the 1T phases of TaS_2 and $TaSe_2$ both form commensurate $\sqrt{13}\,\mathbf{a}_0 \times \sqrt{13}\,\mathbf{a}_0$ superlattices at low temperature. The presence of

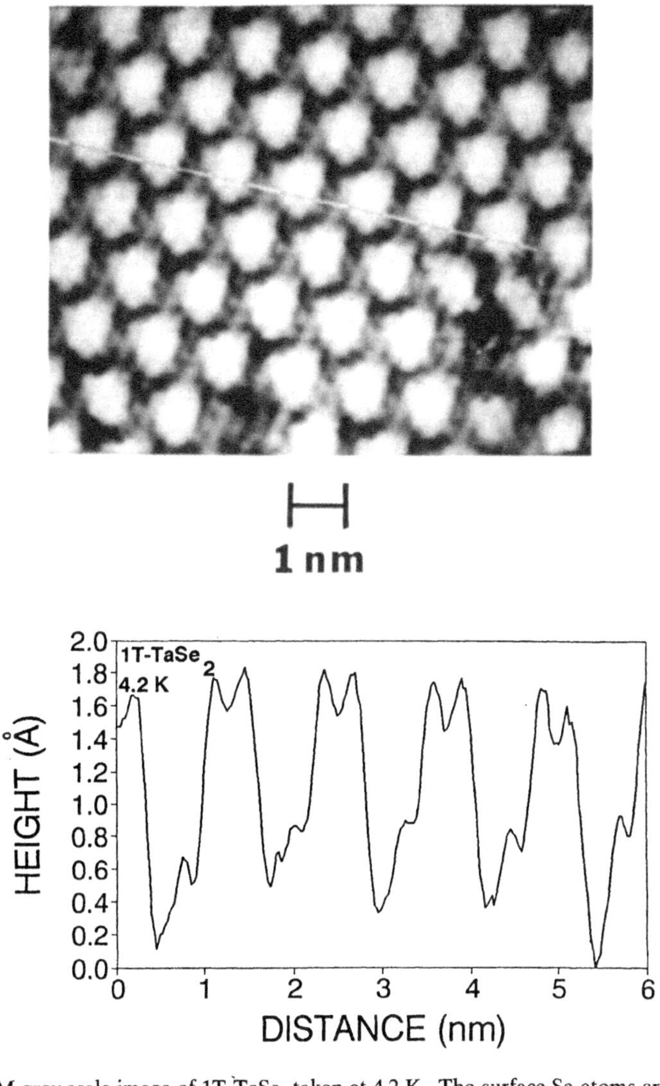

Fig. 14. (a) STM grey scale image of $1T\text{-}TaSe_2$ taken at 4.2 K. The surface Se atoms are resolved and appear superimposed on the CDW modulation of much larger amplitude. White areas with a weak superimposed atomic modulation are centered on the CDW maxmima. (From Reference 53.)
(b) Profile of z-deflection taken along the black and white track shown in (a). The total z-deflection of ~ 2.0 Å is dominated by the CDW modulation with the atoms contributing ~ 0.3 Å.

a weak atomic modulation in the STM scans of $1T\text{-}TaS_2$ and the substantially weaker atomic modulation in many STM scans of $1T\text{-}TaSe_2$ must reflect a subtle difference in the FS and electronic structure. The low temperature resistivity certainly reflects this difference although no information is presently available on the detailed differences in the two Fermi surfaces that result from the band folding.

The high-temperature Fermi surfaces are different based on the band calculations of Woolley and Wexler.[16] The 1T selenide has a pancake-shaped region around Γ while the sulphide does not. The calculated **k** dependence for the selenide is greater than that in the sulphide and this may enhance the importance of a nesting vector component parallel to the **c** axis. This FS difference is not due to an enhanced intersandwich interaction but to the accidental fact that the Fermi level of $1T\text{-}TaSe_2$ falls in the middle of the approximately t_{2g} (the g subscript refers to gerӓde) triplet band whose splitting due to the trigonal distortion is k dependent.

These differences in the high temperature FS can certainly propagate into a difference of FS topology in the CDW superlattice state and consequently produce a difference in the CDW modulation of the LDOS. The STM response suggests such a difference, but no detailed model has been worked out as yet. Differences in the CDW induced FS topology can be reflected in the spatial modulation of the LDOS through a mechanism proposed by Tersoff.[17] Examples of calculated profiles for CDW systems can be found in Ref. 18.

4.4 AFM OF $1T\text{-}TaS_2$ AT 300 K

The atomic force microscope has also been used to study the CDW modulated domain-like structure existing in $1T\text{-}TaS_2$ at room temperature.[19] The AFM detects the total charge density rather than the DOS at the Fermi level and this results in a substantially different amplitude response to the CDW superlattice than is observed with the STM. In the case of the STM, the CDW modulation was observed to dominate the z-deflection with the atomic modulation contributing only a small fraction of the total modulation. In the case of the AFM scans, the CDW and atomic modulations make equal contributions and the total z-deflection observed with the AFM is substantially less than that observed with the STM. The equal amplitudes of the CDW and atomic modulations make the AFM scans more suited to carrying out Fourier transform analysis of the relative orientations of the CDW and atomic lattice.

A typical AFM scan of $1T\text{-}TaS_2$ at 300 K is shown in Fig. 15. The surface S atoms are very well-resolved and the CDW modulation produces enhanced seven atom clusters with variable enhancement amplitudes that repeat over distances of ~ 6 CDW wavelengths. The seven atom clusters near the CDW maxima show a continuous variation in the amplitude and assymmetry of the individual atom

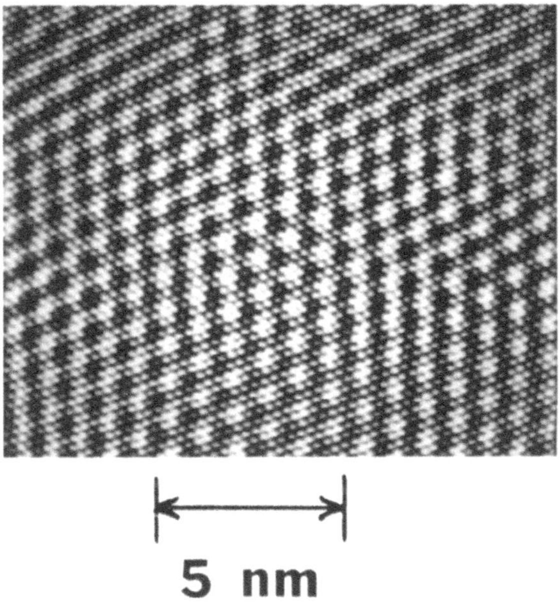

Fig. 15. Grey scale AFM image of $1T\text{-}TaS_2$ taken at room temperature. The incommensurate CDW produces a clearly visible modulated structure. The atoms and CDWs contribute approximately equal amplitudes to the AFM image (constant force mode with force adjusted to ~ 10^{-8} newtons). (From Reference 19.)

enhancements. This indicates an incommensurate CDW structure which appears to show a nearly continuous amplitude modulation similar to that observed in the STM scans. Profiles across one of the domains are shown in Fig. 16. Both the maxima and minima of the CDW show a continuous amplitude variation, thereby giving rise to the conclusion that the overall CDW amplitude reaches a maximum at the center of the domain and a minimum at the boundary. Abrupt amplitude changes indicating the existence of localized discommensurations are not observed. However, the precise magnitude of the CDW modulation cannot be easily corrected for the beat structure effects introduced when the atomic and CDW amplitudes are of equal magnitude. The total amplitude modulation due to beat effects and true CDW amplitude modulation will then be comparable. The AFM profiles are consistent with the more accurate STM determinations of the CDW amplitude modulation, but the AFM results alone cannot be used for an accurate determination of the absolute CDW amplitude. The discommensurate structure extends over a number of CDW periods and the precise phase shifts cannot be determined from either the STM or

AFM scans. The data is consistent with the wide discommensuration model of Nakanishi and Shiba[11] provided that the phase shifts are spread over a number of CDW periods. Regions of completely commensurate CDWs would be small and exist only at the center of the domain, if they exist at all.

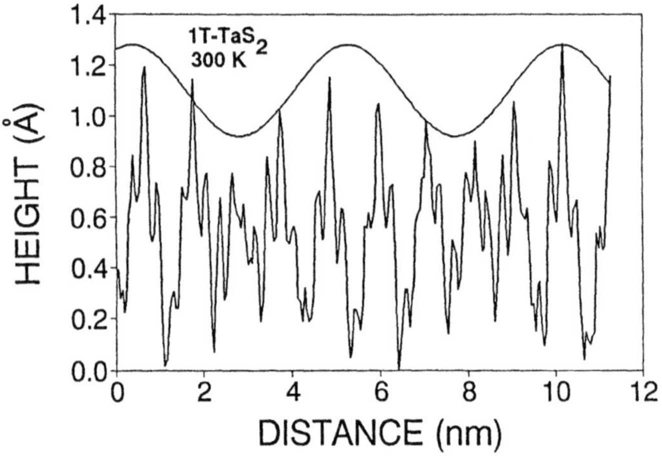

Fig. 16. Profile of the z-deflection along a row of CDW maxima in Fig. 15. It shows a continuous amplitude variation which approximately follows the sine wave shown by the solid curve.

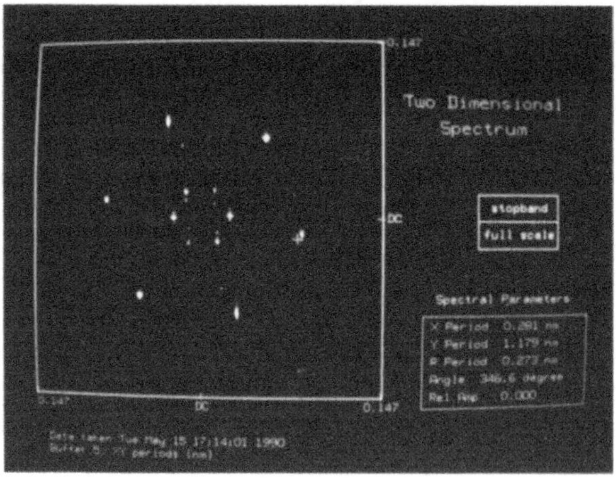

Fig. 17. Fourier transform of the original data set from which a subset was used for the AFM image shown in Fig. 15. The two hexagonal sets of spots represent the CDW superlattice and the atomic lattice. The relative angle of rotation measured using pairs of spots is 11.8° ± 0.4°, close to the average incommensurate value of 11.9°. (From Reference 19.)

The Fourier transform (FT) of the full 27nm x 27nm scan of 1T-TaS_2, from which a smaller area was shown in Fig. 15, is shown in Fig. 17. It gives sufficient resolution to show that the angle between the CDW superlattice and the atomic lattice is 11.8° ± 0.4°, and that the first order FT spot is therefore at the incommensurate position. No second order satellite spots can be sufficiently resolved in this FT for any measurement of position or strength of higher order coefficients. Even with higher resolution these are probably not enough domains in this scan to make a clear analysis of second order structure. A much larger area scan with good definition will probably be required to bring out satellites due to the CDW modulation structure. Calculations of exact phase variations within the CDW domain could then be attempted.

4.5 AFM OF 1T-$TaSe_2$ AT 300 K.

The AFM scans of 1T-$TaSe_2$ at 300 K show a uniform pattern of atoms and CDWs. An example is shown in Fig. 18. The uniformity of this pattern indicates a constant phase relation between the atoms and the CDW maxima. Enhanced seven atom clusters look identical and are spaced on a superlattice of wavelength $\sqrt{13}\ a_0$ as expected. The total z-deflection is on the order of 1 Å as observed in the AFM scans of 1T-TaS_2. The atomic and CDW amplitudes make equal contributions of ~ 0.5 Å each. The profiles of z-deflection show a uniform amplitude over many CDW wavelengths in agreement with the profiles expected for a uniformly commensurate superlattice.

The AFM scans of 1T-$TaSe_2$ form an excellent data set for a Fourier transform analysis of the relation between the atomic lattice and the CDW superlattice. A Fourier transform of a 13 nm x 13 nm AFM scan of 1T-$TaSe_2$ at 300 K is shown in Fig. 19. Using the three pairs of spots, the measured angle between the atomic lattice and the CDW superlattice is 14.2° ± 0.5°, thereby indicating good agreement with the expected commensurate angle of 13.9°. More details can be found in Reference 20.

4.6 STM OF 1T-VSe_2

1T-VSe_2 is a good d band metal and band structure calculations carried out by Myron[21] and Woolley and Wexler[16] predict high temperature FS cross sections which are close to those calculated for 1T-$TaSe_2$. A significant feature of the similarity is the occupied electron pocket in the center of the zone at the Γ point, unlike the FS found for 1T-TaS_2, where these states are unoccupied. The FS in 1T-VSe_2 also shows a greater k_z dependence than in the 1T sulphides.

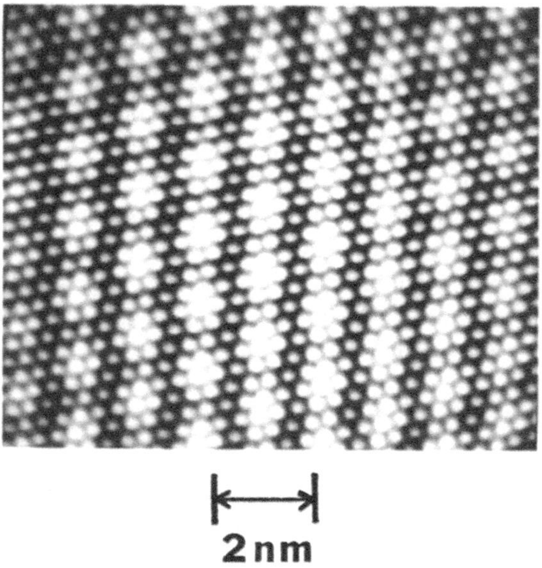

Fig. 18. Grey scale AFM image of 1T-$TaSe_2$ taken at room temperature. A uniform pattern of atoms and CDWs which maintain a constant relative phase indicates a commensurate structure as observed in the STM images (constant force mode with force equal to $\sim 10^{-8}$ newton). (From Reference 19.)

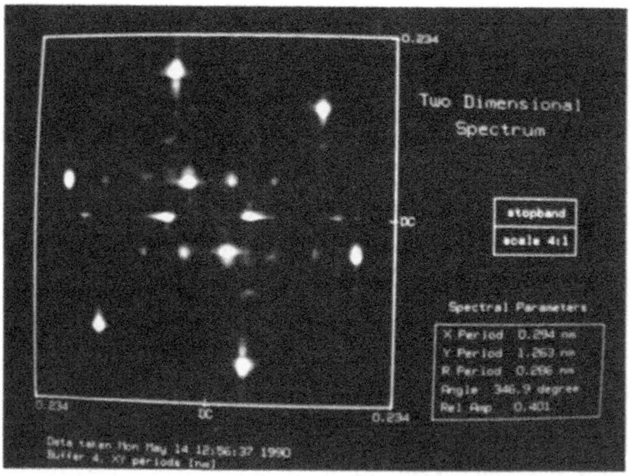

Fig. 19. Fourier transform of a 13nm x 13nm AFM scan of 1T-$TaSe_2$ taken at 300 K. The two hexagonal sets of spots represent the CDW superlattice and the atomic lattice. The relative angle of rotation between the CDW superlattice and atomic lattice is 14.2° ± 0.5° as measured using the three pairs of spots. This value is in good agreement with the expected commensurate angle of 13.9°. (From Reference 20.)

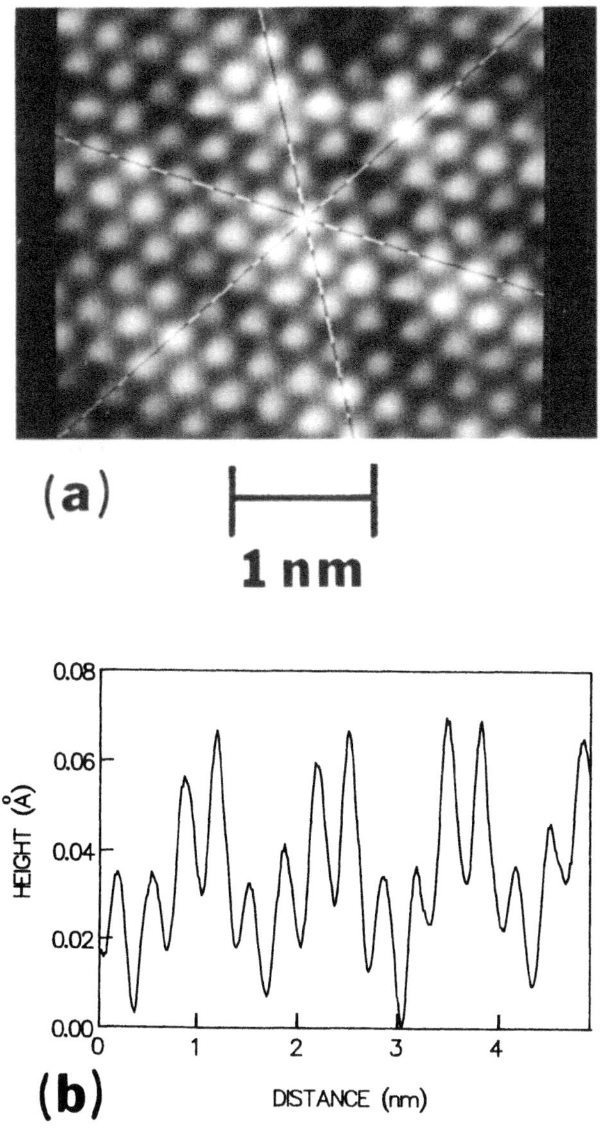

Fig. 20. (a) Grey scale image of a STM scan on 1T-VSe_2 recorded at 4.2 K. The weak $4\mathbf{a}_0 \times 4\mathbf{a}_0$ CDW superlattice is superimposed upon the pattern of surface Se atoms (I= 2.2 nA, V= 25 mV). (b) Profile of z-deflection along the line shown in (a). In the STM profiles of 1T-VSe_2 the atomic modulation and the CDW modulation are comparable. (From Reference 53.)

In spite of this similarity between the high temperature FSs of $1T\text{-}TaS_2$ and $1T\text{-}VSe_2$, the CDW formation is entirely different. The **q** vectors in the metal-metal plane of $1T\text{-}VSe_2$ are $\mathbf{a}_0^*/4$ rather than $\mathbf{a}_0^*/\sqrt{13}$ as found in $1T\text{-}TaSe_2$. The STM scans of $1T\text{-}VSe_2$ at both 77 and 4.2 K show a $4\mathbf{a}_0 \times 4\mathbf{a}_0$ superlattice within the layer plane. A typical grey scale image recorded at 4.2 K is shown in Fig. 20(a) and exhibits a profile in which the atomic modulation dominates as shown in Fig. 20(b). In this respect, the STM response to the CDW is completely different than observed for $1T\text{-}TaSe_2$ or $1T\text{-}TaS_2$. The total z-deflection can vary in the range 0.08 to 4 Å, but in all cases the CDW modulation is less than or at most equal to the atomic modulation. This major difference in CDW formation and STM response is probably related to the presence of a strong p/d hybridization in $1T\text{-}VSe_2$, as compared to $1T\text{-}TaSe_2$. In the latter the d-based band is separated by a large gap (~ 0.1 Ry) from the chalcogen p based valence band. In $1T\text{-}VSe_2$ a significant overlap of the d and p bands occurs and can lead to a more complicated pattern of FS instabilities.

In addition the STM scans can show widely different amplitudes for different components of the triple **q** CDW. This appears to reflect a competition between a

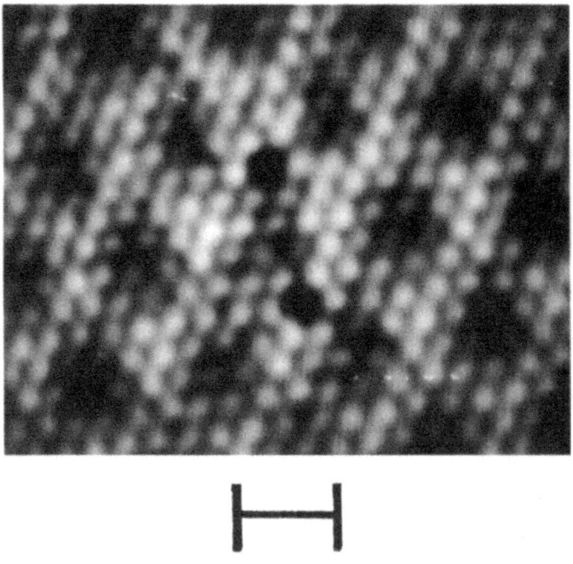

Fig. 21. Grey scale STM image of $1T\text{-}VSe_2$ recorded at 4.2 K. The pattern shows CDW maxima separated by rather diffuse minima in a hexagonal pattern instead of alternating deep minima and saddle points. This pattern suggests a double-q rather than a triple-q CDW structure (I= 2.2 nA, V= 25 mV). (From Reference 53.)

double and triple **q** structure. Fig. 21 shows a grey scale image of a STM scan on 1T-VSe_2 at 4.2 K in which the pattern shows a hexagonal array of diffuse minima rather than the alternating deep minima and saddle points characteristic of a 3**q** structure. This is due to the presence of large amplitude CDW modulations in only two out of the three equivalent directions.

Diffraction studies by a number of authors including Williams,[22] Tsutsumi et al.,[23] Van Landuyt et al.,[24] and Fung et al.[25] have suggested a variety of CDW superlattice structures which form upon cooling the crystal below the initial incommensurate CDW phase forming at 140 K. Recent electron and x-ray diffraction studies by Tsutsumi,[26] Yoshida and Motizuki,[27] and Eaglesham et al.[28] generally agree on a low temperature CDW superlattice with a commensurate component of $4\mathbf{a}_0$ parallel to the layer and an incommensurate component of $\sim 3\mathbf{c}_0$ perpendicular to the layer. Eaglesham et al.[28] observed a 3q to 2q transition below 80 K, and the STM scans at 77 and 4.2 K tend to confirm this possibility. Above 80 K a number of different possible CDW structures have been reported involving either domain structures of the 3q or 2q variety or different **q** vectors. These observations may be a function of the specific sample or cooling history and have not yet been completely resolved.

4.7 STM OF 1T-$TiSe_2$

The materials in the group IV transition metal compounds are more covalently bonded than the group V compounds and tend to be semiconductors rather than metals. Many experiments and calculations suggest that 1T-$TiSe_2$ is an exception with a small indirect overlap between a Se-based s/p band near Γ and a Ti-based d band near L. Small pockets of holes and electrons result from this indirect overlap and a second order phase transition is observed at 202 K in electron diffraction experiments.[29,30] This is identified with the formation of a $2\mathbf{a}_0 \times 2\mathbf{a}_0 \times 2\mathbf{c}_0$ CDW superlattice. The CDW forms in the commensurate state, and it has been suggested that the CDW involves an electron-hole interaction in which a zone-boundary phonon is driven soft. Angular resolved photoemission data, electron and x-ray diffraction data, magnetic susceptibility data, resistivity data and infrared reflectivity data by Woo et al.[29] and by Brown[31] generally support models based on band overlap. However, Stoffel et al.[32] have used photoelectron spectroscopy data in an analysis which concludes that $TiSe_2$ is a very narrow band semiconductor in which the Se 4p and Ti 3d bands have a strong interaction leading to CDW formation. The STM results show a CDW with medium intensity, but it involves substantially less charge transfer than occurs in the CDWs of the group V 1T phases of $TaSe_2$ and TaS_2.

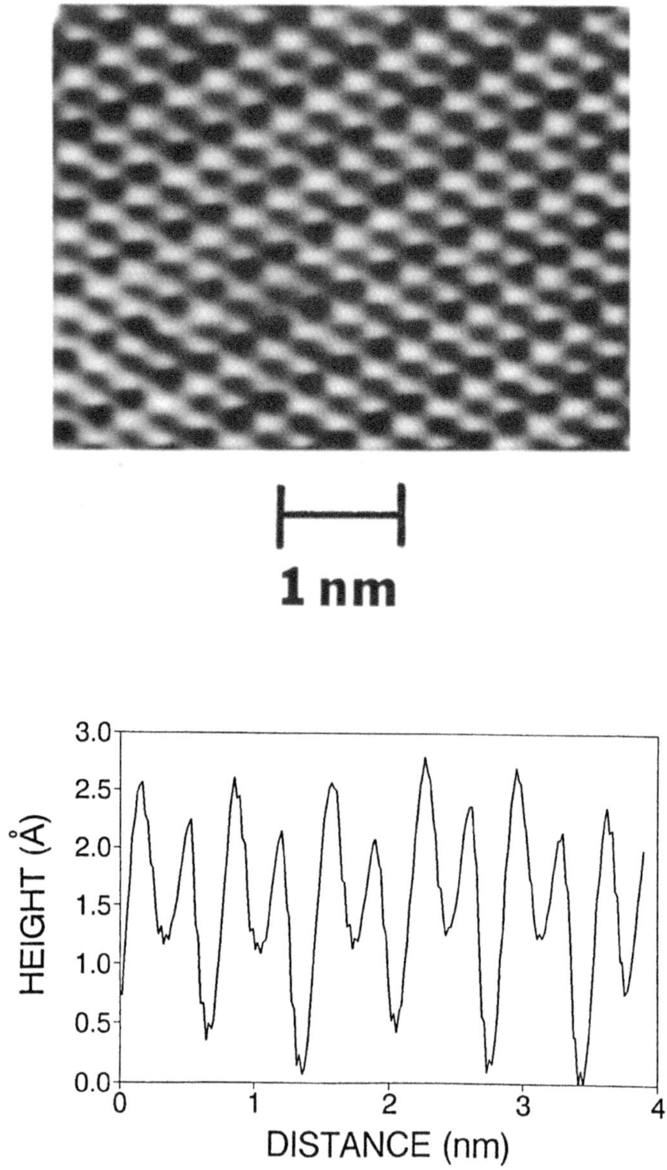

Fig. 22. (a) STM grey scale image of 1T-$TiSe_2$ at 77 K showing the CDW and atomic modulations (I = 8.6 nA, V = 50 mV).
(b) Profile of z-deflection along the line shown in (a). The $2a_0$ modulation due to the CDW is clearly resolved with the atoms and CDWs contributing equal amplitudes. (From Reference 18.)

A STM scan of 1T-$TiSe_2$ recorded at 77 K is shown in Fig. 22(a) and shows a well-resolved pattern of surface Se atoms. The CDW superlattice modulation is also detected as an enhancement of z-deflection on every second row of atoms. A profile along one of the atomic rows is shown in Fig. 22(b) and indicates that the atomic and CDW modulations generate approximately equal z-deflections of ~ 1 Å each. The CDW maxima show a phase shift relative to the seven atom clusters of surface Se atoms surrounding the maxima. As shown in the STM scan of Fig. 23 taken at 77 K, the center Se atom of the cluster is enhanced, but the overall intensity is shifted off center to form an asymmetrical cluster of 3 surface Se atoms. This would place the CDW maxima close to the Ti atom positions, but the exact charge distribution will depend on the relative roles of the Ti d band and Se s/p valence band in the CDW formation and charge transfer.

A band Jahn-Teller mechanism has been developed by Motizuki et al.[33] and extended by Suzuki et al.[34] This model includes a wave vector and mode dependent electron-lattice interaction as well as a FS nesting, both of which contribute to CDW formation. This model explains the stability of the triple **q** CDW state as well as many of the experimental results. Nevertheless, considerable controversy on the exact electronic structure exists, with some models and interpretation[29,31] confirming the indirect band overlap while other results[32] suggest a narrow band semiconductor.

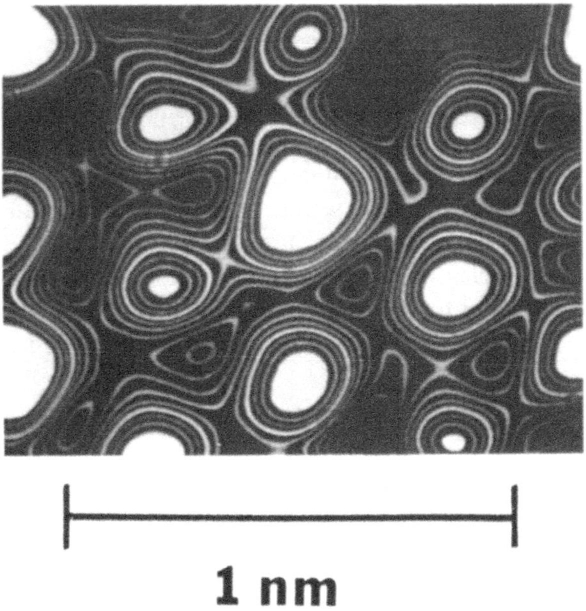

Fig. 23. Magnified contour plot of a STM scan showing the asymmetry in 1T-$TiSe_2$ at 77 K introduced by the phase difference between the atomic and CDW modulations. (From Reference 18.)

4.8 STM OF 1T-TiS_2

There is general agreement that 1T-TiS_2 is a small band gap semiconductor with a gap on the order of 0.2 to 0.3 eV. A STM scan of 1T-TiS_2 at 4.2 K is shown in Fig. 24(a) and shows well-defined atomic structure. Many of the STM scans show a long range modulation in z-deflection, but the individual atomic modulation amplitudes remain fairly constant at ~ 2 Å. This modulation can be fairly regular with a period of 7-8 atomic spacings, but it can also show random variations suggesting surface undulation rather than intrinsic modulation of the electronic structure. Fig. 24(b) shows a profile along a row of atoms and the long range undulation is clearly detected as a height variation of ~ 1 Å.

No evidence of CDW formation is observed, but the STM response to the atomic structure is very similar to that observed in 1T-$TiSe_2$. The atomic modulation in both cases is 1 to 2 Å and suggests some enhancement of z-deflection independent of the presence or absence of a CDW. The 2H phase layer compounds, which are good metals with weak CDWs, show much smaller z-deflections in the STM scans.

5. CDW Defects

The STM has the potential to detect a variety of defects on an atomic scale and examples have been reported for surface atom vacancies[1,18] and surface adatoms.[35] In the present studies of the 1T-phase crystals the surface atom patterns have appeared to be perfect in most cases where the surface atoms are well resolved. However, in all three 1T-phase crystals studied, localized defects in the CDW pattern have been observed. These have a common characteristic in that they often appear as a deficit in the LDOS which would have been associated with the CDW modulation. This produces a strong local minimum in the z deflection which often appears as a "black hole" in the gray-scale images. However, in some cases a localized enhancement of the CDW maxima is also observed on surrounding CDW maxima. In 1T-VSe_2 a Se atom vacancy may also be associated with the defect.

5.1 1T-$TaSe_2$

An example of 1T-$TaSe_2$ at 4.2 K is shown in Fig. 25(a). This gray-scale scan resolves only the CDW modulation and shows several such defects localized at the positions of the CDW maxima. Adjacent CDW maxima appear to be slightly enhanced, but in general the surrounding CDW maxima are remarkably unperturbed. The atomic modulation was not detected as confirmed in the higher magnification

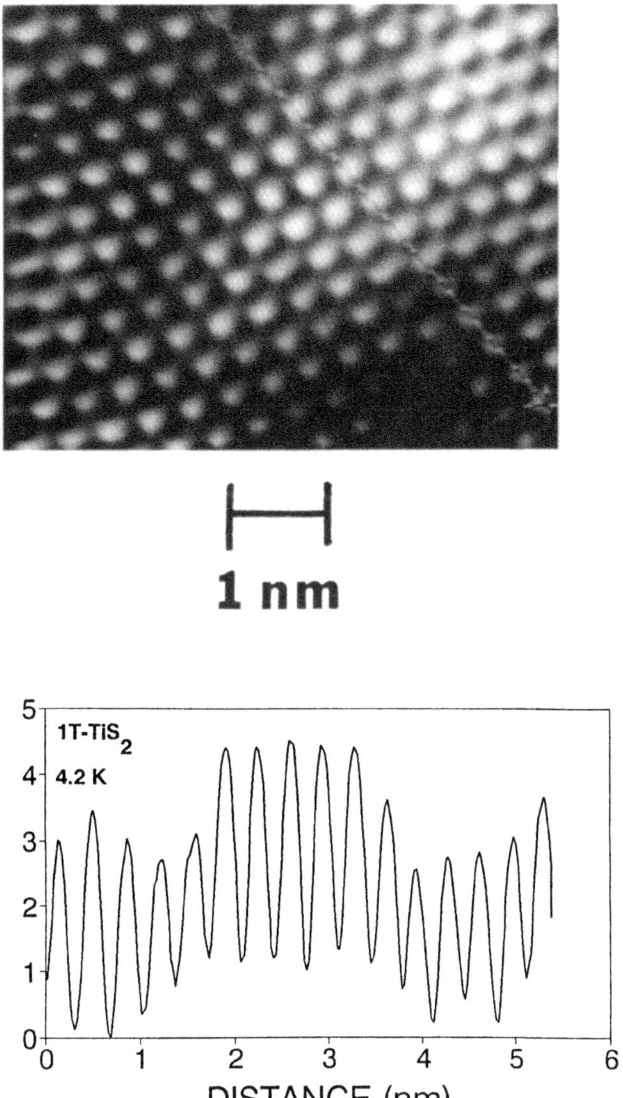

Fig. 24. (a) STM grey scale image of $1T\text{-}TiS_2$ at 4.2 K showing only an atomic modulation (I = 2.2 nA, V = 25 mV).
(b) Profile of z-deflection along the line shown in (a). Note the long range modulation of the surface atomic structure which suggests a possible two-dimensional supercell structure of the surface atoms. (From Reference 18.)

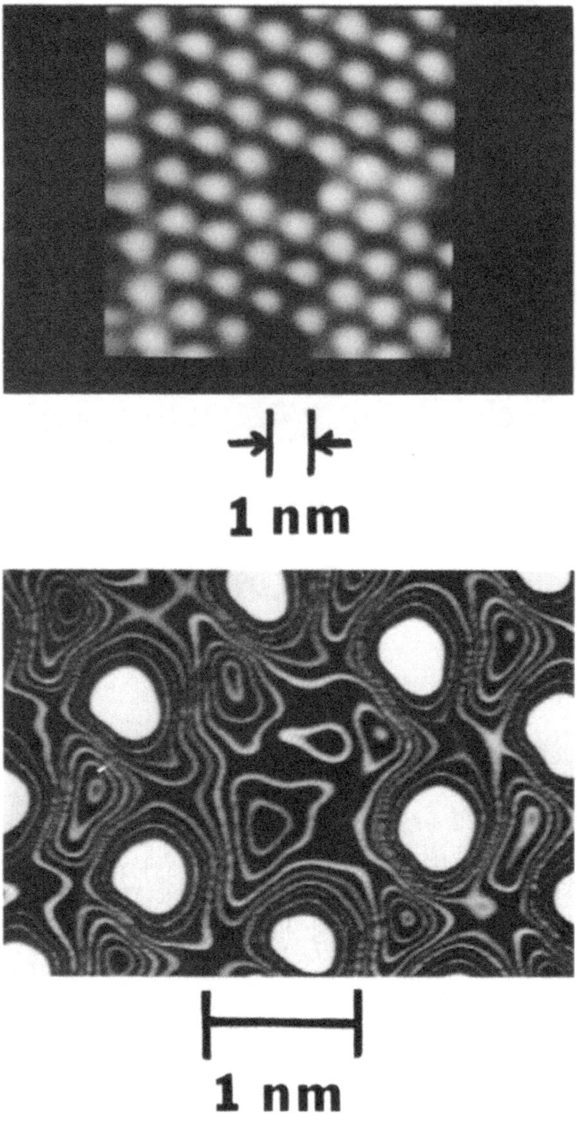

Fig. 25. STM scans on 1T-$TaSe_2$ at 4.2 K showing the presence of defects in the CDW superlattice. (a) Grey scale image showing missing CDW maxima. Pattern is totally dominated by the CDW superlattice.

(b) Contour plot of the STM scan shown in (a). Contours confirm the dominance of the CDW superlattice with the region of the missing CDW maximum showing a minimum and relatively uniform z deflection in the central region. (From Reference 18.)

contour scan of Fig. 25(b). The maxima, deep minima, and saddle points reflect only the CDW pattern with a spacing of 12. 6 ± 0.3 Å = $\sqrt{13}\, a_0$.

5.2 1T-TaS_2

In the case of 1T-TaS_2 similar defects corresponding to missing CDW maxima have been observed at 77 K as shown to the left in Fig. 26(a). In this case the surface atom pattern is also well resolved and no evidence of missing surface atoms can be seen near the missing CDW maximum. Figure 26(b) shows a profile through the defect and modulation by the surface atoms is uninterrupted while the CDW enhancement is absent. The adjacent CDW maxima again show a slight perturbation in height, as was observed for the 1T-$TaSe_2$ defects, but the major defect in the CDW pattern remains remarkably localized.

A second example of defects in the CDW superlattice of 1T-TaS_2 is shown in the STM scan of Fig. 27. The CDW superlattice is clearly resolved with little or no contribution from the atomic modulation. The image was recorded in the commensurate phase at 153 K and gives a very large maximum z deflection in the range of ~ 4 Å. The defects in the CDW superlattice pattern are observed in a region of the crystal surface showing an irregular pattern of CDW maxima suggesting some type of disorder in the three-atom layer sandwich supporting the CDW. Associated with these local domains are two distinct types of defects involving a symmetric cluster of seven CDW maxima.

In the lower section of the scan two defects are observed in which the center CDW maximum is missing and the six surrounding maxima are enhanced. As shown in the profile of Fig. 28(a) this enhancement of the CDW maxima is on the order of ~ 1 Å relative to the surrounding CDW maxima and must represent a strong perturbation of the charge transfer and LDOS in the 13-atom clusters contributing to each of these CDW maxima.

In addition to the two defects discussed above at least three others are characterized by a "black hole" indicating a missing CDW maximum, but these do not exhibit the strongly enhanced ring of surrounding CDW maxima. Figure 28(b) shows the profile recorded along a track through one of these defects. The CDW maxima surrounding the defect contribute approximately the same z deflection as observed for other nearby CDW maxima. However, the missing CDW maximum shows a deeper minimum resulting in an overall minimum to maximum z deflection of ~ 4 Å, the same as observed in the defects discussed in Fig. 25.

The cause of the CDW defects shown above would seem to be in the Ta layer of atoms since the defects represent rather drastic perturbations of the LDOS associated with the Ta d electrons participating in the CDW. Either substitutional impurity atoms or vacancies in the cluster of 13 Ta atoms associated with each

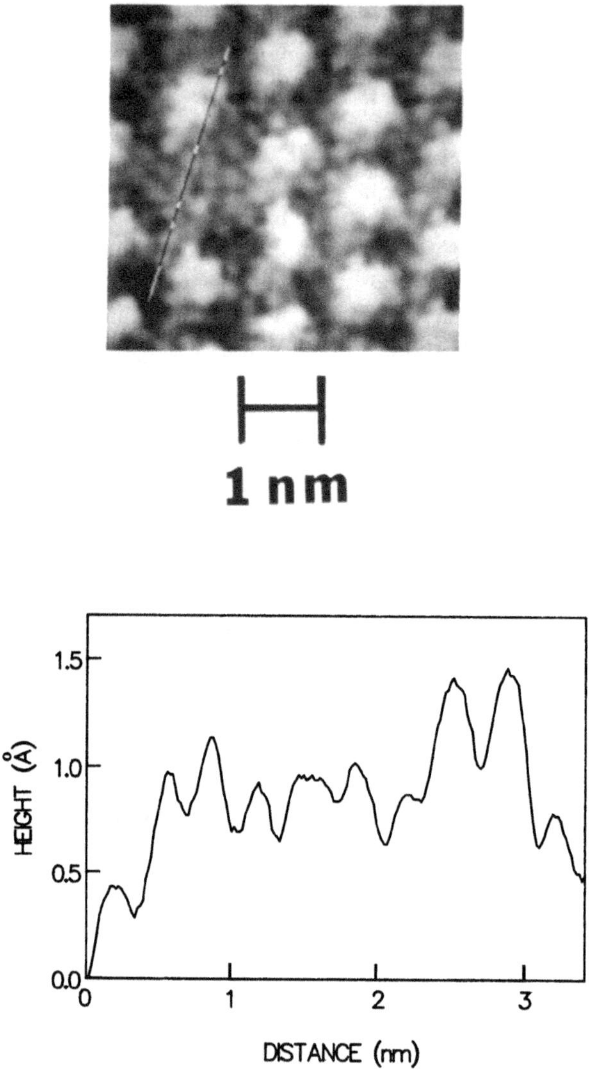

Fig. 26. (a) Grey scale image of a STM scan on 1T-TaS_2 at 77 K showing a missing CDW maximum. In this case the surface S atoms are resolved within the region of the missing maximum.
(b) Profile of the z deflection across the center of the defect region shows a continuous modulation at the atomic wavelength. The defect does not appear to involve surface S atoms. (I= 2.2 nA, V= 25 mV). (From Reference 53.)

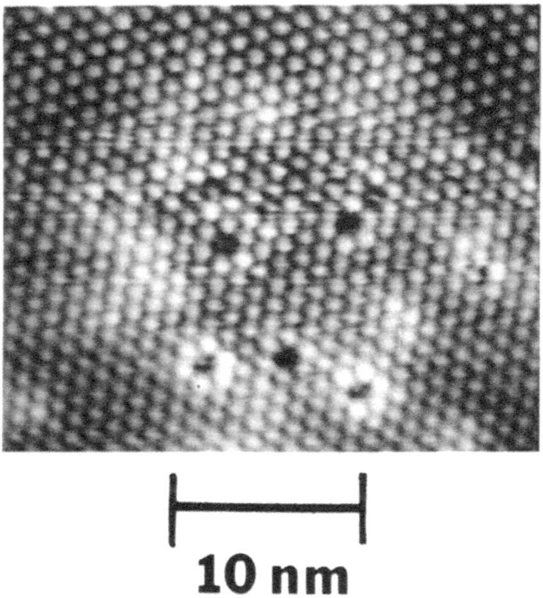

Fig. 27. Grey scale image of a STM scan on 1T-TaS_2 at 153 K. A number of defects involving missing CDW maxima are observed. Several of these involve enhancement of the surrounding CDW maxima while others do not show this enhancement. (I= 3 nA, V= 47 mV). (From Reference 53.)

CDW maximum could distort the charge transfer within the 13 atom cluster. The precise form of the defect and associated charge distribution would then depend on the number and position of the missing or substituted atoms in each 13 atom cluster. A range of different defect profiles would then be expected for random defects in the Ta layer of atoms, as is observed. 1T-VSe_2 also shows a similar type of CDW defect, but variations suggest subtle differences connected with the different CDW structure in 1T-VSe_2.

5.3 1T-VSe_2

Figure 29(a) shows a STM scan of 1T-VSe_2 at 4.2 K where three separate "black hole" defects located in the upper half of the scan. The one on the right occurs at a position close to a CDW maximum while the two on the left are slightly off the maxima. These defects are more localized than observed for 1T-$TaSe_2$ and 1T-TaS_2 and are only slightly wider than a single atom. They could result from surface Se-

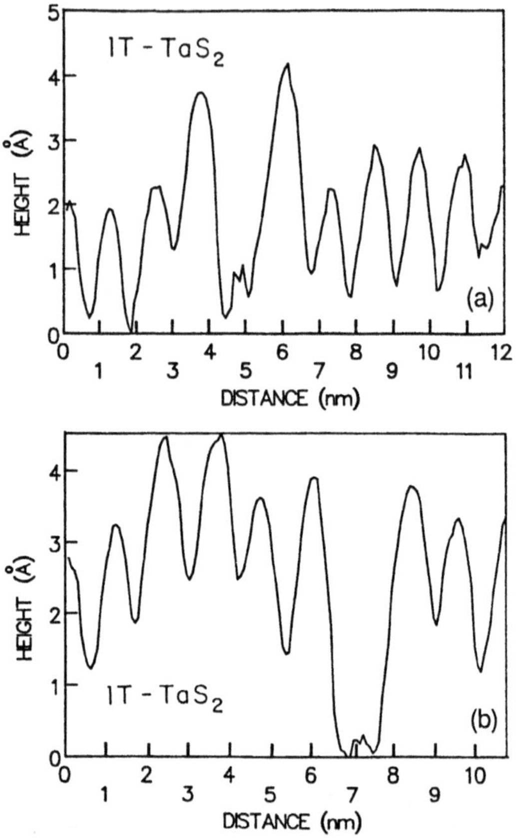

Fig. 28. Profiles of z deflection along tracks through the CDW defects shown for 1T-TaS_2 in Figure 29. All of the missing CDW maxima show a deep minimum giving a total relative z deflection of ~ 4 Å with a small structure of atomic dimensions at the minimum.
(a) Profile of defect with enhanced CDW maxima surrounding the missing one.
(b) Profile of defect without localized enhancement of the surrounding maxima.
(From Reference 53.)

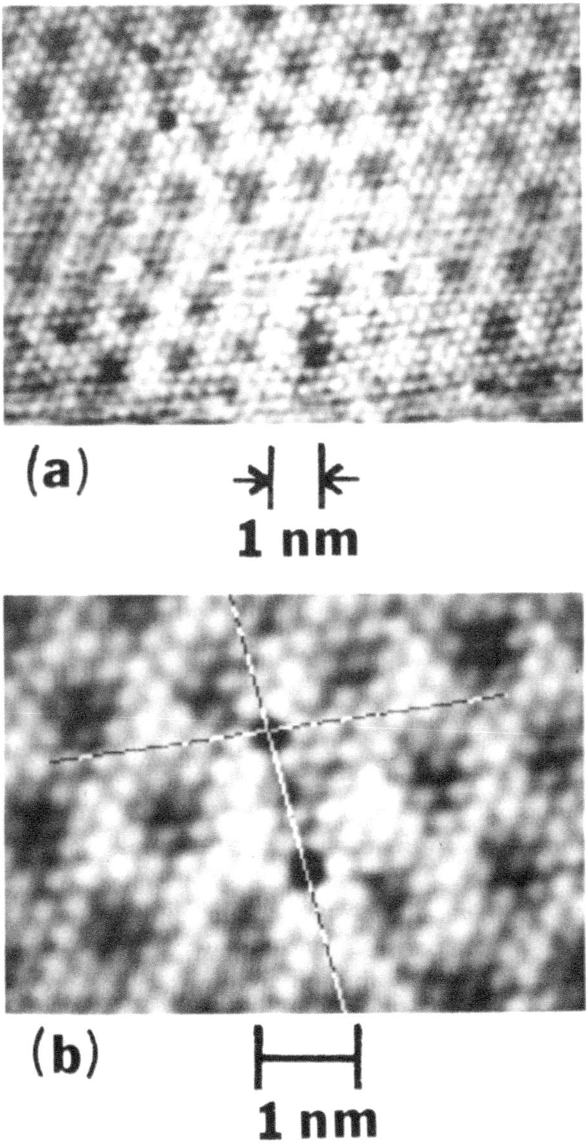

Fig. 29. Grey scale images of STM scans on 1T-VSe_2 at 4.2 K showing defects in the pattern of surface Se atoms. The defects in this case are associated with missing surface Se atoms and do not ave a specific correlations with the CDW superlattice.
(a) Large area scan showing a number of distributed defects. Three defects in the upper half of scan are marked by localized black holes.
(b) Magnified scan of two close defects near the CDW maxima. (I= 2.2 nA, V= 25 mV).
(From Reference 53.)

atom vacancies in this case since the atomic modulation is a larger fraction of the total modulation. The surface Se atoms may also participate to a greater extent in the charge transfer associated with the CDW.

A magnified scan of the two close defects is shown in Fig. 29(b) along with profiles in Fig. 30(a) which correspond to the two scan paths intersecting the defects. These scan paths emphasize the atomic modulation of ~ 1 Å while the defect represents a tip excursion of practically 4 Å. For comparison Fig. 30(b) shows a profile which crosses the CDW maxima as well as minima and also includes a defect. The maximum to minimum CDW structure is ~ 3 Å while the defect again produces a 4 Å deep hole with an absolute minimum in z deflection. The profile clearly shows that the defect has removed the LDOS associated with both the atomic and CDW modulations causing the tip to make a very close approach to the surface.

6. Mechanisms For STM Amplitude Enhancement

The profiles of STM z-deflection just discussed for the 1T phases of TaS_2 and $TaSe_2$ both show a total deflection amplitude in the range ≥ 2.0 Å as demonstrated in the STM profiles of Figs. 5(b) and 12(b). Although the atomic modulation is significantly different in the two cases, the majority of the giant amplitude in both cases is modulated at the CDW wavelength. The large corrugation cannot be understood in purely structural terms or in terms of a smooth modulation of the conduction electron density $\rho(r,E) \equiv \Sigma_\upsilon |\psi_\upsilon(r)|^2 \delta(E_\upsilon - E)$ at the position of the tip.

Tersoff[17] has suggested a straightforward mechanism peculiar to semiconductors and semimetals of low effective dimensionality. Whenever the FS collapses to a point at the corner of the surface BZ, the STM image will in effect correspond to an individual state. The nodal structure of this state gives rise to a large corrugation with the periodicity of the unit cell involved in the FS collapse.

The real CDW materials are quasi-2D with three CDWs generated in the hexagonal structure at relative orientations of 120°. The wavefunction can be expanded as six plane waves and, if the band edge falls near the BZ corners, the model will generate a hexagonal array of singular dips. In the case of the quasi-2D layer structures the FS collapse will not be complete and the nodal structure will depend on the degree of FS obliteration and the strength of mixing between $\mathbf{k}_\parallel$ and $(\mathbf{k}_\parallel + \mathbf{G})$ wavefunctions. Tersoff has calculated the nodal structure for various cases of FS collapse and mixing and four examples published in Ref. 18 are shown in Fig. 31. Fig. 31(a) shows contours of constant LDOS for total FS elimination and strong mixing by a CDW of wavelength $5a_0$. The nodal structure is extremely strong in this case. For only partial FS elimination the nodal structure is weaker, but is still quite large when the mixing is strong as shown in Fig. 31(b). With only partial elimination

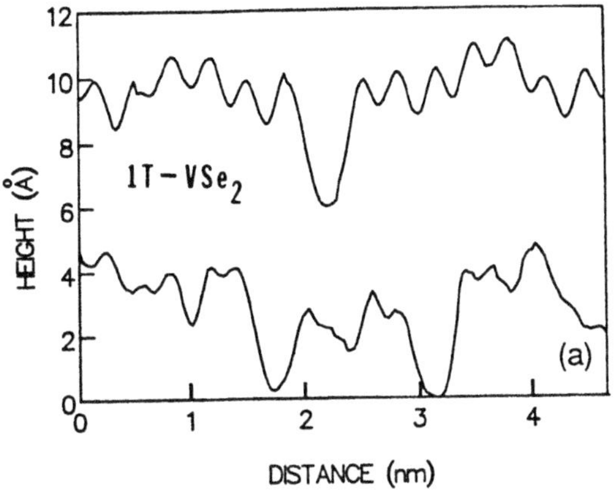

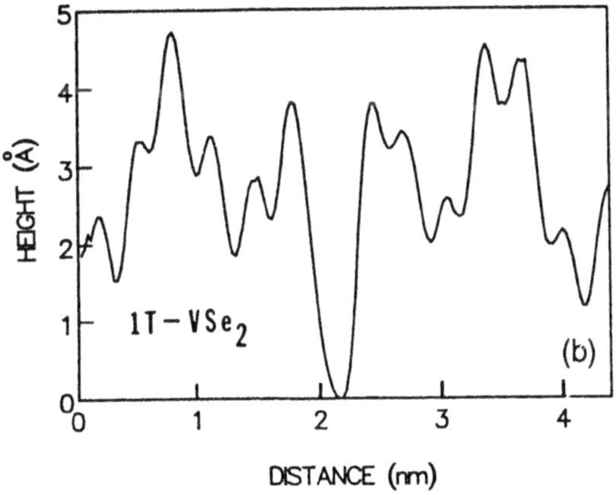

Fig. 30. (a) Profiles of the z deflections observed along tracks shown in STM scan of 29(b). Upper profile corresponds to the nearly horizontal track. Lower profile corresponds to the nearly vertical track. Both profiles show a dominant modulation at the atomic wavelength with an enhanced minimum at the position of the defect.
(b) Profile of z deflection through the defect in the upper right-hand section of 29(b) demonstrating the large localized deflection associated with the defect. (From Reference 53.)

of the FS and weaker mixing, the nodal structure is substantially reduced but still evident as shown in Fig. 31(c). The atomic modulation represents a weaker superimposed modulation on the contours at intermediate distance. The relative amplitudes of the nodal structure and the atomic modulation are essentially determined by the degree of $\mathbf{k}_\parallel$ and $(\mathbf{k}_\parallel + \mathbf{G})$ mixing and FS obliteration.

The profiles generated by the above model can clearly be adjusted to fit the whole range of experimental profiles measured for the wide variety of CDW amplitudes observed in the layer structure dichalcogenides. In the case of the 1T phase a single horizontal component of the "nesting" vector along the ΓM direction can produce good nesting for four out of the six sheets of the high temperature FS. In addition, further adjustments by considering a $\mathbf{k}_z$ component[16] may produce even better "nesting". The presence of excellent "nesting", along with the large charge

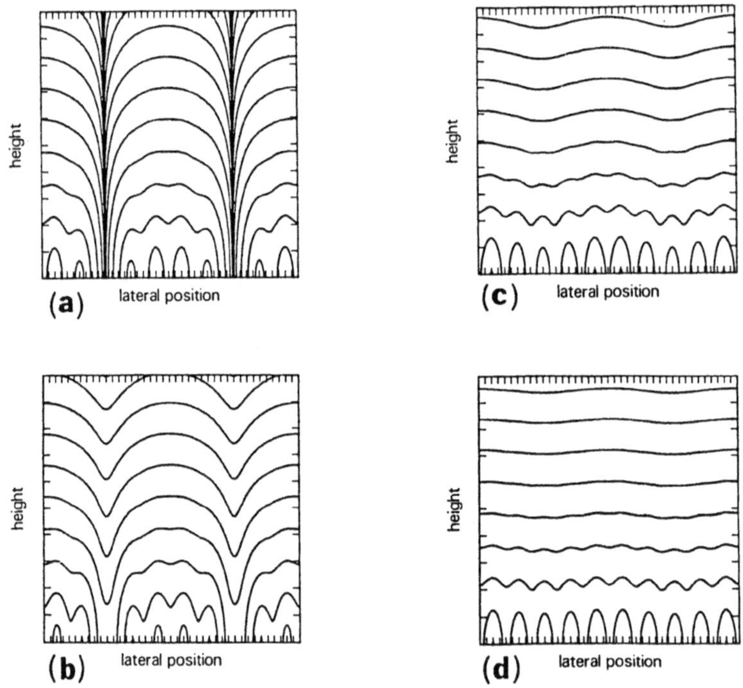

Fig. 31. Contours of constant LDOS calculated for various degrees of FS collapse and mixing of wave functions at $\mathbf{k}_\parallel$ and $\mathbf{k}_\parallel + \mathbf{G}$. They show the relative amplitudes of superimposed atomic and CDW modulation in four cases for a CDW wavelength of $5a_0$.
(a) Total FS elimination and strong mixing.
(b) Partial FS elimination and strong mixing.
(c) Partial FS elimination and weak mixing.
(d) Partial FS elimination and very weak mixing. (From Reference 18.)

transfer calculated by Smith et al.[8] in the case of 1T-TaS_2, suggest that the CDWs in the 1T phase of TaS_2 and $TaSe_2$ produce both a major reduction of the FS area and strong mixing.

7. STM and AFM of 2H Phase Layer Structure Crystals

7.1 STM OF 2H-$NbSe_2$

The CDW in 2H-$NbSe_2$ has an onset temperature of ~ 33 K and can most easily be detected by measuring the change in sign of the Hall effect. The resistivity shows only a weak anomaly at the CDW onset and 2H-$NbSe_2$ remains a good metal below the CDW onset. A superconducting transition occurs at 7.2 K and below this temperature CDWs and superconductivity coexist. The CDWs form a triple $\mathbf{q}$ structure with an approximate $3\mathbf{a}_0 \times 3\mathbf{a}_0$ superlattice. The wave vectors are incommensurate and are given by $\mathbf{q} = (1 - \sigma)\mathbf{a}_0^*/3$ with $\sigma = 0.025$ at 33 K and $\sigma = 0.011$ at 5 K as determined by neutron diffraction.[36]

The STM scans[37] of 2H-$NbSe_2$ at 4.2 K show a pattern dominated by the atomic modulation as shown in Fig. 32. The atomic modulation amplitude is ~ 0.6

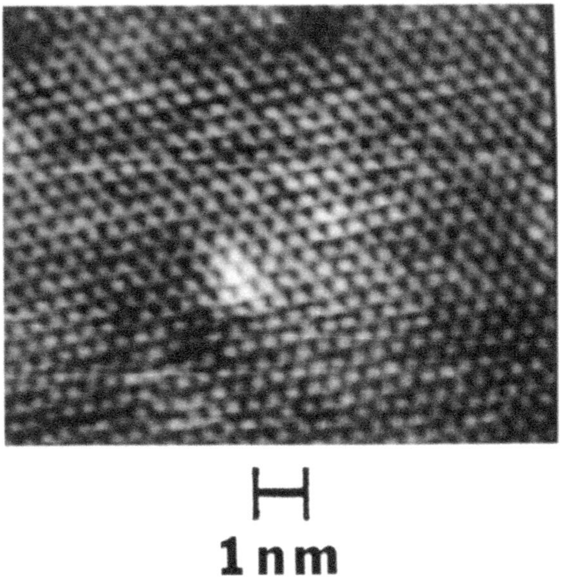

Fig. 32. Low-magnification STM image of 2H-$NbSe_2$ recorded at 4.2 K (I= 2.0 nA, V= 25 mV). Only the atomic modulation is clearly resolved at this magnification. (From Reference 37.)

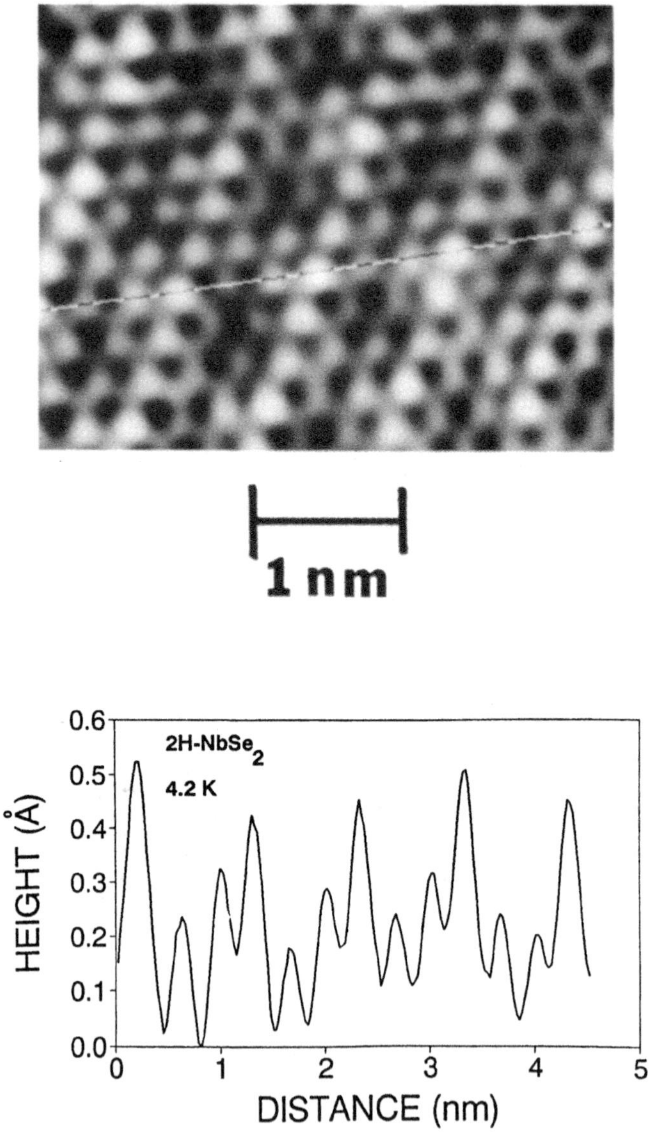

Fig. 33. (a) STM grey scale image of $2H\text{-}NbSe_2$ at 4.2 K (I= 2.0 nA, V= 25 mV). The $3a_0 \times 3a_0$ CDW modulation is clearly resolved at this magnification. The CDW modulation is centered on the Nb atoms which are displaced by $a_0/\sqrt{3}$ from the atomic modulation due to the surface Se atoms. This gives rise to the asymmetric three atom clusters near the CDW maxima. (From Reference 18.)
(b) Profile of z-deflection along the line shown in (a). The atomic and CDW modulation amplitudes are comparable.

Å and at the magnification shown the CDW modulation cannot be detected. At a higher magnification as shown in Fig. 33(a) the superimposed CDW enhancement appears on every third row of atoms and represents an additional z-deflection of ~ 0.2 Å as shown in the profile of Fig. 33(b). The CDW modulation is centered on the Nb atoms which are displaced from the atomic modulation of the surface Se atoms by $a_0/\sqrt{3}$. The superposition of this CDW modulation causes the atomic modulation pattern of Se atoms near the CDW maxima to be distorted as shown in the LDOS contour scan of Fig. 34. The appearance of enhanced three atom clusters in the gray scale plots such as Fig. 33(a) is due to this superposition of atomic and CDW modulations. The total z-deflection of the STM is approximately 0.6 Å as shown in Fig. 33(b) with equal amplitudes contributed by the atoms and the CDWs. The CDW and atomic modulations are both enhanced relative to the intrinsic spatial oscillation of the electron density expected from the periodic lattice distortion or from the atomic modulation of the electron wavefunction. The possible sources of this enhancement will be discussed in the summary and conclusions.

7.2 STM OF 2H-$TaSe_2$

2H-$TaSe_2$ is very similar to 2H-$NbSe_2$, but forms a stronger CDW with an onset at 122.3 K and goes into a superconducting phase only at temperatures below ~ 0.2 K.

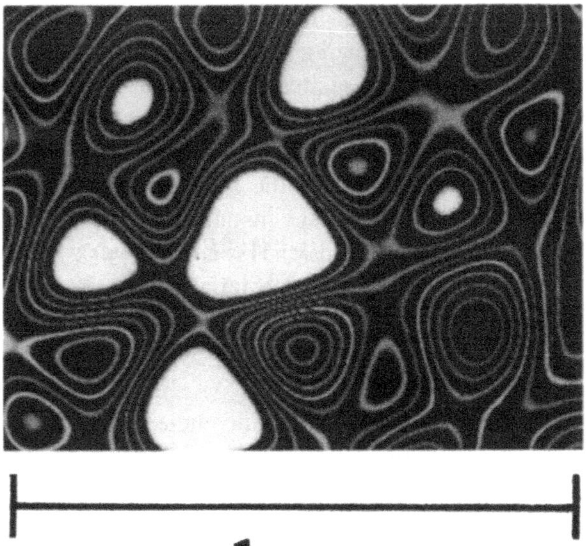

Fig. 34. High-magnification contour plot of 2H-$NbSe_2$ taken at 4.2 K showing detail of the three atom cluster near a CDW maximum (I = 2.0 nA, V = 25 mV). (From Reference 18.)

At 122.3 K a triple **q** CDW forms in an incommensurate state characterized by a second-order (or nearly second-order) transition followed by a first-order transition at ~ 90 K below which the $3\mathbf{a}_0 \times 3\mathbf{a}_0$ CDW superlattice becomes commensurate. The STM scans at temperatures of 77 and 4.2 K show the presence of a weak CDW modulation forming a $3\mathbf{a}_0 \times 3\mathbf{a}_0$ CDW superlattice. Fig. 35(a) shows a STM scan taken at 4.2 K with a weak enhancement of the z-deflection present on every third row of atoms. A profile along the track drawn in Fig. 35(a) is plotted in Fig. 35(b). The total z-deflection is only 0.2 ± 0.05 Å with a CDW enhancement of ~ 0.1 Å on every third atom. This enhanced z-deflection due to the CDW is larger than the periodic lattice distortion of the surface Se atoms which is estimated to be about 0.02 Å perpendicular to the surface. The FS and band structure undergo substantial rearrangement, but many extended sections of FS remain as detected by magnetoquantum oscillations of the magnetoresistance[38] and the susceptibility.[39] Strong $\mathbf{k}_\parallel$ and $(\mathbf{k}_\parallel + \mathbf{G})$ mixing may occur at selected points on the FS, but FS obliteration and singularities in the DOS should be much weaker than in the 1T phase of $TaSe_2$. The observed z-deflection is an order of magnitude smaller than observed for 1T-$TaSe_2$, and in this respect follows the expected trend.

Doran and Woolley[40] have carried out band structure calculations for the $3\mathbf{a}_0 \times 3\mathbf{a}_0$ CDW state of 2H-$TaSe_2$. They start with results obtained for the high temperature band structure calculated by Wexler and Woolley,[41] and show that stabilization of the CDW comes from destruction of the peak in the DOS lying just below the Fermi level. Further details can be found in the original papers.

Smith et al.[8] have also developed a model band structure calculation in order to analyze photoelectron data. The model allows them to estimate the number of Ta d electrons residing on each inequivalent Ta atom in the distorted CDW state. These are labeled "a", "b" and "c" in Fig. 36 and charge transfer occurs toward the center atom "a" located at the center of the seven atom clusters. The numbers calculated for the Ta d electrons on these inequivalent sites are $n_a = 1.014$, $n_b = 1.005$ and $n_c = 0.977$. This charge transfer is relatively weak and the STM results certainly reflect this, although an exact quantitative comparison cannot be made.

7.3 STM OF 2H-TaS_2

The 2H phase of TaS_2 forms a ~ $3\mathbf{a}_0 \times 3\mathbf{a}_0$ incommensurate CDW superlattice at low temperatures. The resistivity shows a very small anomaly at the CDW onset suggesting only a small change in the conduction electron DOS. 2H-TaS_2 has a peak in susceptibility at 80 K, a break in resistivity at 75 K, and a change in sign of the Hall resistivity between 65 and 55 K. Careful resistivity measurements by Tidman et al.[42] show an abrupt change in $\rho_\parallel$ at 75.3 K leading to the conclusion that the Fermi surface modification occurs at or below this temperature.

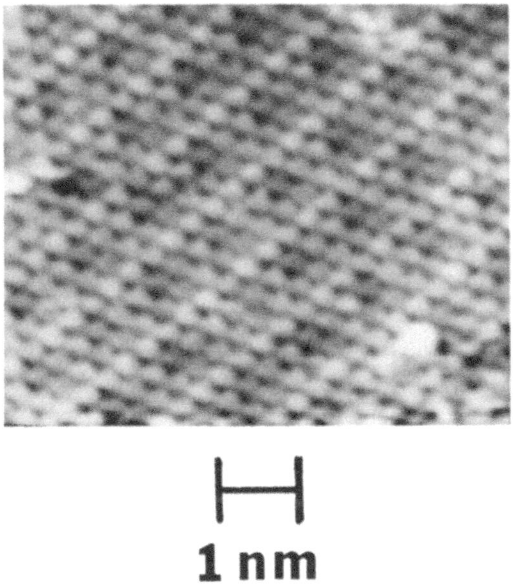

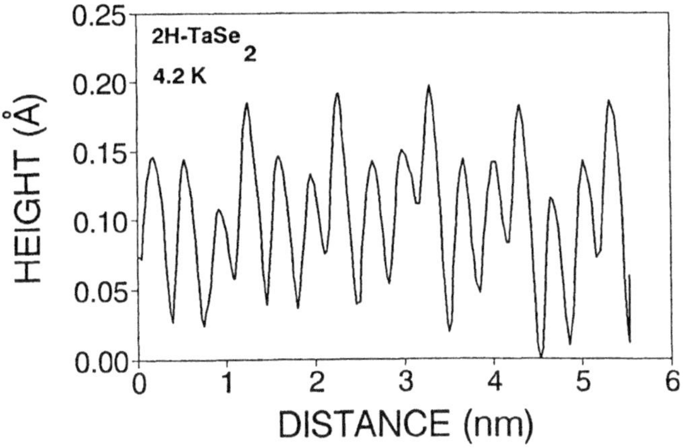

Fig. 35. (a) STM grey scale image of 2H-$TaSe_2$ at 4.2 K (I= 2.2 nA, V= 25 mV).
(b) Profile of z-deflection along a line of atoms shown in (a) revealing a weak CDW superlattice enhancement on every third atom. (From Reference 18.)

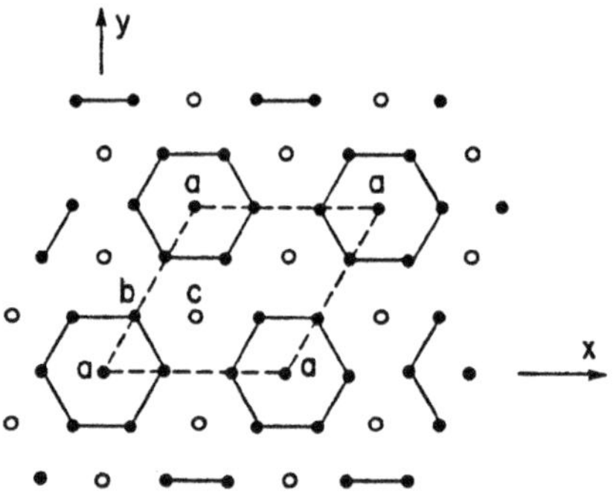

2H - $TaSe_2$

Fig. 36. Diagram of the Ta atom pattern in 2H-$TaSe_2$ showing a nine-atom cluster associated with each CDW maximum. The three inequivalent atom sites labeled a, b, and c result from a weak charge transfer from the c-type atoms toward the b-type and a-type atoms. (From Reference 8.)

Electron diffraction experiments[43] show the CDW to be incommensurate down to about 14 K. The wave vector is given by $\mathbf{q} = (1 + \delta)\mathbf{a}_0^*/3$ where $\delta = 0.017 \pm 0.009$ at 70 K. A possible slight decrease in δ as temperature decreases is suggested by the data. The departure from commensurability for 2H-TaS_2 is roughly of the same magnitude, although opposite in sign, as that of the incommensurate phases of 2H-$NbSe_2$ and 2H-$TaSe_2$.

Nuclear magnetic resonance (NMR) studies[44] of 2H-TaS_2 powder at 4.2 K indicate a locally commensurate CDW superlattice. The authors conclude that at low temperatures there is a discommensurate CDW state for 2H-TaS_2 where regions of locally commensurate CDWs are separated by domain walls or discommensurations.

At 4.2 K the STM scans show a well developed hexagonal $3\mathbf{a}_0 \times 3\mathbf{a}_0$ superlattice as shown in the grey scale image of Fig. 37(a). The charge maxima are exceptionally well defined and show a high atom at each vertex of the triple $\mathbf{q}$ CDW pattern. The seven atom cluster of S atoms near each charge maximum still shows the asymmetry connected with the phase difference due to the $a_0/\sqrt{3}$ displacement between the CDW and atomic modulations. This is again clearly demonstrated in the high-magnification contour plot of Fig. 37(b). This charge asymmetry due to the superposition is characteristic of all three 2H phases studied. A STM scan profile

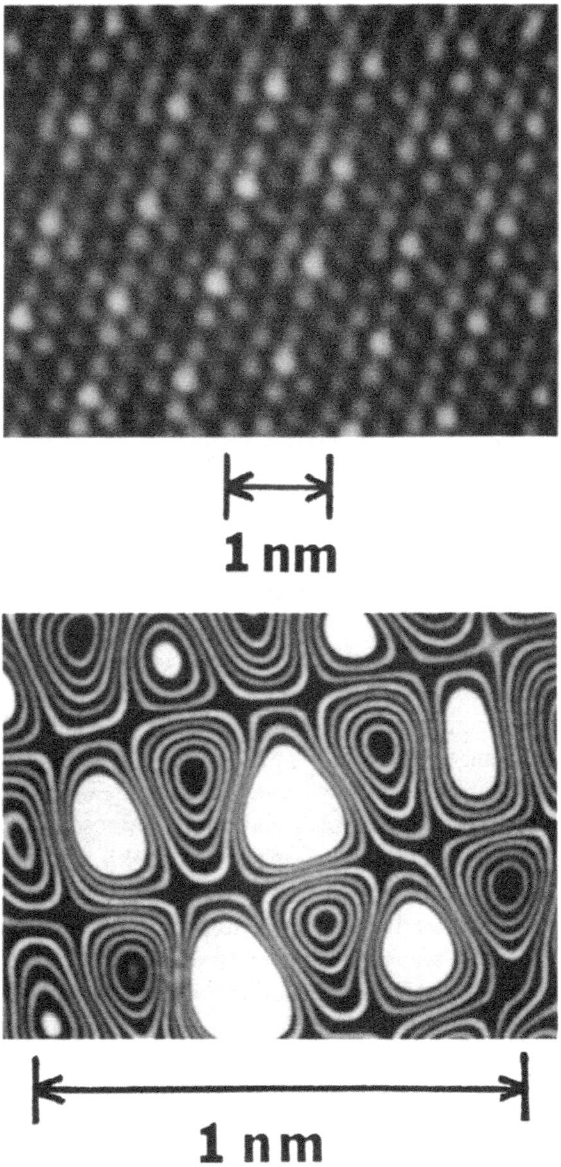

Fig. 37. (a) Grey scale STM image of $2H\text{-}TaS_2$ taken at 4.2 K (I = 2.2 nA, V = 25 mV).
(b) Contour plot showing the asymmetric seven-atom cluster of surface S atoms located above the CDW maximum occuring on the Ta atom cluster. (From Reference 18.)

of $2H\text{-}TaS_2$ is shown in Fig. 38. In this case the total z-deflection is 0.6 ± 0.1 Å compared to the value 0.2 ± 0.05 Å observed for $2H\text{-}TaSe_2$.

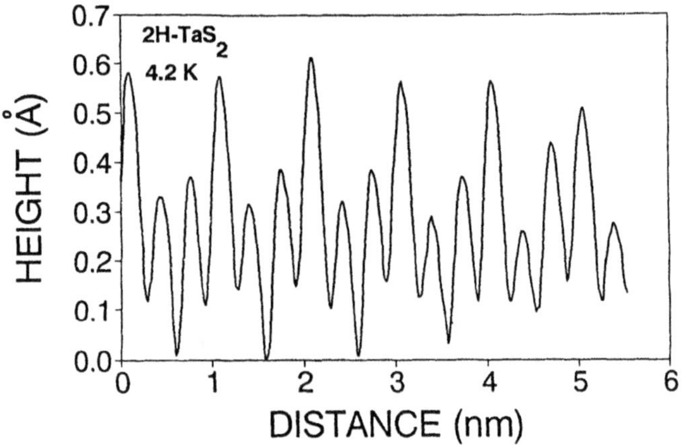

Fig. 38. A profile of the STM scan on $2H\text{-}TaS_2$ showing the CDW modulation of wavelength $3a_0$. The total z-deflection is 0.6 Å, which is larger than that observed for $2H\text{-}TaSe_2$. (From Reference 18.)

Shubnikov-de Haas measurements of the FS cross sections have been carried out by Hillenius and Coleman[45] which show a range of cross sections that are more than an order of magnitude smaller than those observed for $2H\text{-}TaSe_2$. Eleven frequencies in the range 0.04 to 6.4 MG were detected. The two highest of frequencies in $2H\text{-}TaS_2$ match the two lowest frequencies in $2H\text{-}TaSe_2$. The larger extended sections observed in $2H\text{-}TaSe_2$ are completely absent in $2H\text{-}TaS_2$. The rest are systematically smaller, thereby leading to a total FS area in $2H\text{-}TaS_2$ which is more than an order of magnitude smaller than observed in $2H\text{-}TaSe_2$.

The high temperature FSs constructed from the band structure calculations of Wexler and Woolley[41] show that the high temperature FS sections in the 2H phases of TaS_2 and $TaSe_2$ have nearly identical areas. Therefore we conclude that the CDW in $2H\text{-}TaS_2$ obliterates a much larger portion of the FS than occurs in the CDW phase of $2H\text{-}TaSe_2$. Within the Tersoff[17] model this would be expected to enhance the nodal structure at the CDW wavelength. It is also likely that a number of the extremely small FS cross sections may be located at the BZ corners and edges, leading to greater mixing of $\mathbf{k}_\parallel$ and $(\mathbf{k}_\parallel + \mathbf{G})$. This latter point cannot at present be confirmed since no detailed band calculations or models of the $3\mathbf{a}_0 \times 3\mathbf{a}_0$ CDW phase of $2H\text{-}TaS_2$ have been worked out. However, the overall evidence from the FS experiments and comparison of the STM profiles shows that these subtle electronic differences are detected in the STM scans.

8. 4Hb Phase Transition Metal Dichalcogenides

The 4Hb phase polytypes of $TaSe_2$ and TaS_2 have four three-atom-layer sandwiches per unit cell arranged with alternating sandwiches of trigonal prismatic coordination and octahedral coordination. The CDW transitions observed in these phases occur independently in the sandwiches of different coordination and maintain the characteristics associated with the pure octahedral (1T) phase and the pure trigonal prismatic (2H) phase. The CDW transitions are modified to some extent, but onset temperatures in the trigonal prismatic layers are still ~ 600 K. The octahedral sandwiches show a very strong $\sqrt{13}\,\mathbf{a}_0 \times \sqrt{13}\,\mathbf{a}_0$ CDW superlattice while the trigonal prismatic sandwiches show much weaker $3\mathbf{a}_0 \times 3\mathbf{a}_0$ CDW superlattices. The STM scans indicate that the strong CDW amplitudes observed for the 1T sandwiches have been reduced relative to the pure 1T phase, but they still provide the dominant z-deflection in the STM scans of the 1T surface. This reduction of CDW amplitude does, however, allow the STM response to provide a better resolution of the surface atomic modulation. Details of the STM results for 4Hb-$TaSe_2$ and 4Hb-TaS_2 will be discussed below.

8.1 STM OF 4Hb - $TaSe_2$

The CDWs forming in 4Hb-$TaSe_2$ are similar in structure to those forming in the pure octahedral (1T) and pure trigonal prismatic (2H) phases. They maintain the same CDW superlattice orientation as observed in the pure phases, but only the CDW in the octahedrally coordinated layer becomes commensurate with lockin to a $\sqrt{13}\,\mathbf{a}_0 \times \sqrt{13}\,\mathbf{a}_0$ superlattice at 410 K after an onset at ~ 600 K. The trigonal prismatic layer develops an approximate $3\mathbf{a}_0 \times 3\mathbf{a}_0$ superlattice at ~ 75 K, which remains incommensurate down to 10 K, as determined by neutron diffraction data.

The CDWs in the trigonal prismatic sandwiches show no order in the **c** axis direction and therefore exhibit two-dimensional correlation in the plane of the trigonal prismatic sandwich. On the other hand, the much stronger CDW in the octahedral sandwiches shows long range order along the **c** axis, indicating that it interacts across the intervening trigonal prismatic layer to form a three-dimensional CDW superlattice.

The STM scans can be carried out on a crystal with either an octahedral surface sandwich or a trigonal prismatic surface sandwich obtained by cleaving a single crystal and scanning both sides of the cleave. The STM z-deflection shows that the strength of each independent CDW has been reduced in the mixed 4Hb phase compared to the corresponding pure phases with the result that the atomic modulation is more easily resolved on the octahedral sandwich. However, the

$\sqrt{13}\,\mathbf{a}_0 \times \sqrt{13}\,\mathbf{a}_0$ CDW superlattice still makes a dominant contribution to the STM deflection.

A STM scan on the octahedral sandwich of 4Hb-$TaSe_2$ is shown in Fig. 39(a) and shows the dominant $\sqrt{13}\,\mathbf{a}_0 \times \sqrt{13}\,\mathbf{a}_0$ superlattice. The profile of z-deflection is shown in Fig. 39(b) and indicates an anomalously large CDW amplitude of ~ 3 Å similar to that observed in the pure 1T phase. The strength of the CDW in the octahedral sandwich can also be seen in STM scans on the trigonal prismatic sandwich of 4Hb-$TaSe_2$. As shown in the composite of STM scans in Fig. 40 the CDW can show a large contribution by enhancing sunflowers of surface Se atoms, even though it resides in the octahedral sandwich which lies ~ 6 Å below the surface sandwich. Adjustment of the STM scan can bring out the ~ $3\mathbf{a}_0 \times 3\mathbf{a}_0$ CDW superlattice modulation in the surface sandwich, but this remains weak relative to the atomic modulation of the surface Se atoms. (See the series of scans in Fig. 40.)

By careful adjustment of the STM it is possible to simultaneously detect both CDW modulations when scanning the trigonal prismatic sandwich. An example is shown in Fig. 41(a) where the two CDW superlattices generate coincident maxima on a super-superlattice of size $3\sqrt{13}\,\mathbf{a}_0 \times 3\sqrt{13}\,\mathbf{a}_0$. This super-superlattice is outlined by the large rhombus drawn in Fig. 41(a) and gives rise to an enhanced intensity on every third major peak in the profile of Fig. 41(b).

8.2 STM OF 4Hb - TaS_2

Crystals of 4Hb-TaS_2 show a CDW structure in which two related $\sqrt{13}\,\mathbf{a}_0 \times \sqrt{13}\,\mathbf{a}_0$ commensurate superlattices appear in the octahedrally coordinated layers after an initial CDW onset temperature of ~ 600 K. Friend et al.[46] compared **q** vectors of the CDWs found in 4Hb-TaS_2 to those observed in the pure 1T and 2H phases and concluded that the transfer of ~ 0.12 electrons per tantalum atom from the octahedral to the trigonal prismatic layers could account for the required FS changes. Below the CDW resistive anomaly at 315 K the 1T sandwiches are believed to become semiconducting while the 1H sandwiches remain metallic below the CDW transition at 22 K. Data published by Wattamaniuk et al.[47] on the temperature dependence of resistivity perpendicular and parallel to the layers support this conclusion. The two CDW transitions give rise to discontinuities in $\rho_{\parallel}$, but the behavior remains metallic down to 4.2 K. $\rho_{\perp}$ is dominated by the semiconducting 1T layers and rises to ~ 1200 $\mu\,\Omega$ cm at 4.2 K.

At 4.2 K the STM scans on the octahedral sandwiches show a strong CDW superlattice superimposed on the surface S atom pattern as shown in Fig. 42(a). As shown in the profile of Fig. 42(b) the total z-deflection is ~ 2 Å with a dominant CDW modulation amplitude of ~ 1.5 Å. The CDW forms a $\sqrt{13}\,\mathbf{a}_0 \times \sqrt{13}\,\mathbf{a}_0$ hexagonal superlattice of charge maxima and deep minima owing to the strong

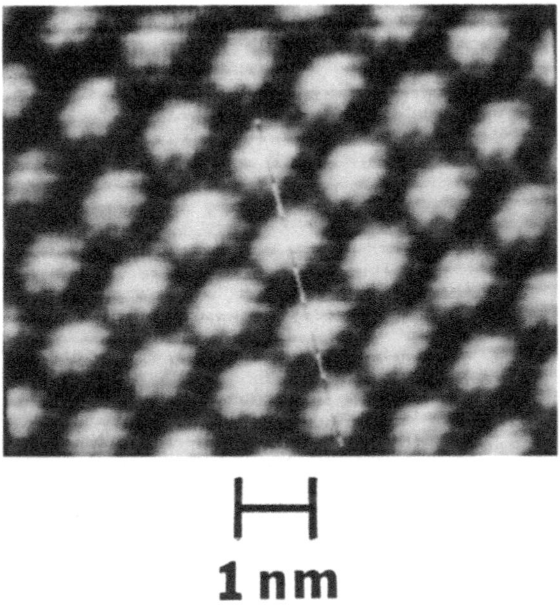

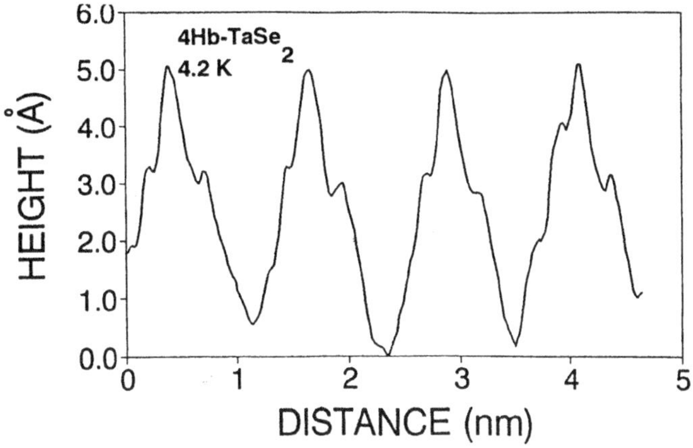

Fig. 39. (a) A STM scan on the 1T sandwich of a 4Hb-$TaSe_2$ layer at 4.2 K showing both the CDW and atomic modulations (I= 2.2 nA, V= 25 mV)
(b) Profile of z-deflection along the line shown in (a). The z-deflection of ~ 1.5 Å is dominated by the CDW in the 1T layer, but is reduced in amplitude from that observed in the pure 1T phase.

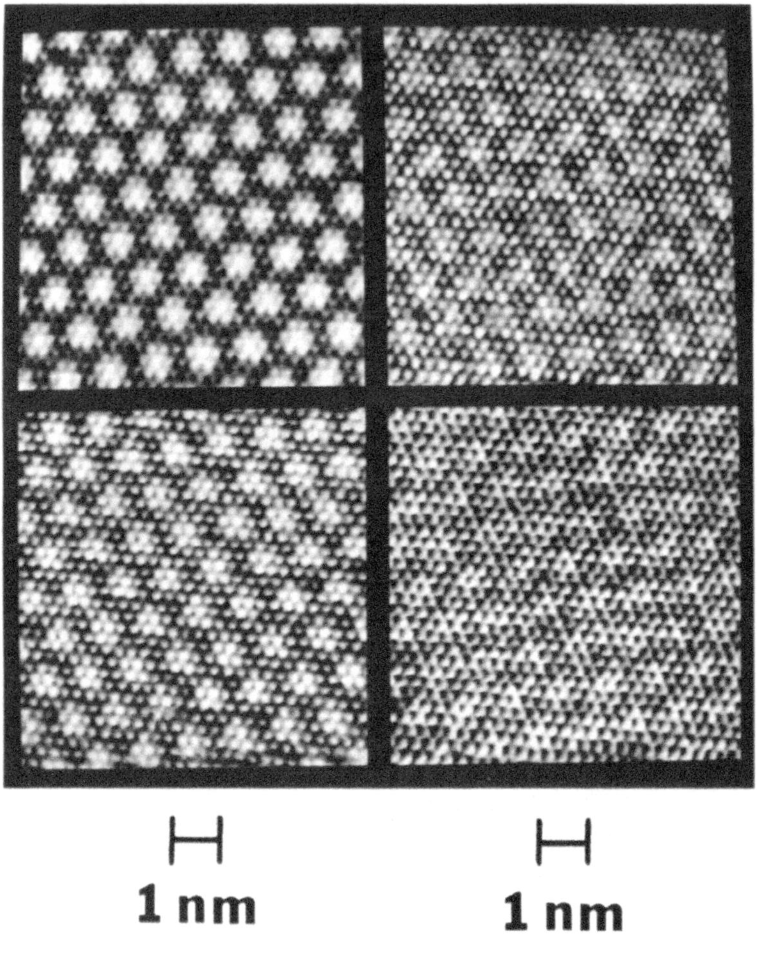

Fig. 40. STM scans on the 1H sandwich of 4Hb-$TaSe_2$ at 4.2 K. These four scans were taken during a period of ~ 2 hours and scan the crystal at identical current and bias voltage settings (I = 2.2 nA, V = 25 mV). These images demonstrate the variance in the CDW pattern seen on the 1H sandwich of 4Hb-$TaSe_2$ due to the complex superposition of the atomic modulation, the $3\mathbf{a}_0 \times 3\mathbf{a}_0$ CDW modulation from the surface 1H layer, and the $\sqrt{13}\,\mathbf{a}_0 \times \sqrt{13}\,\mathbf{a}_0$ CDW modulation from the underlying 1T layer. Each image shows good atomic resolution along with a variable contribution from the CDW modulations. The CDW contribution to each image (described in the order in which they were taken) is as follows: lower right, a $\sqrt{13}\,a_0$ modulation in one direction (slightly counterclockwise from the horizontal) and a weak $3a_0$ modulation along the atom rows closest to the vertical; lower left, an intermediate strength $\sqrt{13}\,a_0$ triple CDW; upper right, both a triple $\sqrt{13}\,a_0$ and a triple $3a_0$ CDW; upper left, a strong $\sqrt{13}\,a_0$ triple CDW. (From Reference 18.)

charge transfer and LDOS modification caused by the CDW. The superlattice pattern is oriented at the commensurate angle of 13.9°, as found in the commensurate phase of the pure 1T crystals. However, in the STM scans on the 1T sandwiches of the 4Hb phase the CDW maxima are systematically lower and the atomic modulation consistently larger than observed in the pure 1T phase where the CDW is completely dominant.

The presence of the much higher conductivity of the trigonal prismatic sandwich with a more extended FS can moderate both the effect of the strong CDW in the 1T phase layers and the possible nodal structure in the LDOS associated with CDW formation. The presence of a complex FS is confirmed by Shubnikov-de Haas oscillations observed by Fleming and Coleman[38] which detect the presence of at least 15 frequencies in the range 0.044 - 8.2 MG. Wexler et al.[48] and Doran et al.[49] have used the layer method to calculate band structure in 4Hb-TaS_2. Although the results are close to a superposition of the bands characteristic of the pure 2H and 1T phases, hybridization and electron transfer is found to occur between the two types of sandwich and could easily account for the subtle changes in STM response.

The CDW transition in the trigonal prismatic layers exhibits a reduced onset temperature of 22 K compared to the ~ 75 K observed in the pure 2H phase. STM scans on the 1H layer of 4Hb-TaS_2 are shown in Fig. 43. We have been able to detect a very weak $3\mathbf{a}_0 \times 3\mathbf{a}_0$ CDW superlattice as shown in the STM scan of Fig. 43(a) and 43(b) and in the profile of Fig. 43(c). However, in some cases the stronger $\sqrt{13}\,\mathbf{a}_0 \times \sqrt{13}\,\mathbf{a}_0$ CDW superlattice from the 1T sandwich 6 Å below is detected as enhanced "sunflowers" showing a random variation of intensity in the scan of Fig. 43(a). The weak $3\mathbf{a}_0 \times 3\mathbf{a}_0$ superlattice can also be detected in some areas of Fig. 43(a). The overall behavior is similar to the STM response observed for 4Hb-$TaSe_2$, but the CDWs appear to give a weaker modulation of the tunneling current. These results are consistent with the general trends expected from the electronic differences between 4Hb-TaS_2 and 4Hb-$TaSe_2$, but represent only qualitative observations.

9. Spectroscopy of Layer Structure Dichalcogenides

9.1 1T-TaS_2 and 1T-$TaSe_2$

The regular CDW superlattice in 1T-TaS_2 gives rather poor definition at 4.2 K; however, the spectroscopy mode which measures dI/dV versus V shows a well-defined CDW gap structure at ~ 150 mV as shown in Fig. 44. The strong change of conductance above ~ 150 mV is consistent with a large gaping of the FS. The

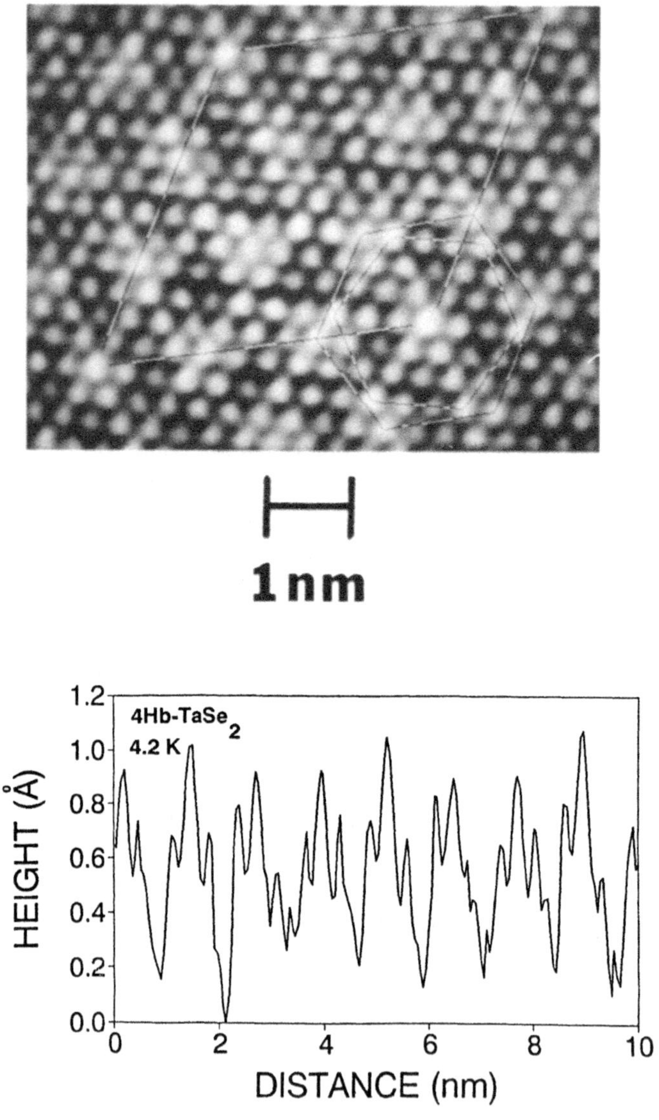

Fig. 41. (a) Grey scale image from a STM scan on the 1H sandwich of a 4Hb-$TaSe_2$ crystal at 4.2 K. Two hexagons outline the $3\mathbf{a}_0 \times 3\mathbf{a}_0$ and $\sqrt{13}\,\mathbf{a}_0 \times \sqrt{13}\,\mathbf{a}_0$ superlattices which are rotated 13.9° with respect to each other. The rhombus outlines the unit cell of the $3\sqrt{13}\,\mathbf{a}_0 \times 3\sqrt{13}\,\mathbf{a}_0$ super-superlattice generated by the superpostion of the $3a_0$ and the $\sqrt{13}\,a_0$ modulations (I= 2.2 nA, V= 25 mV).
(b) Profile of z-deflection along the edges of two adjacent $3\sqrt{13}\,a_0$ super-supercells. The three highest peaks each occur at an atom located at the corner of a $3\sqrt{13}\,a_0$ rhombus. These maxima are approximately 0.1 Å higher than the other $\sqrt{13}\,a_0$ maxima and indicate the points of phase match between the two CDW superlattices. (From Reference 18.)

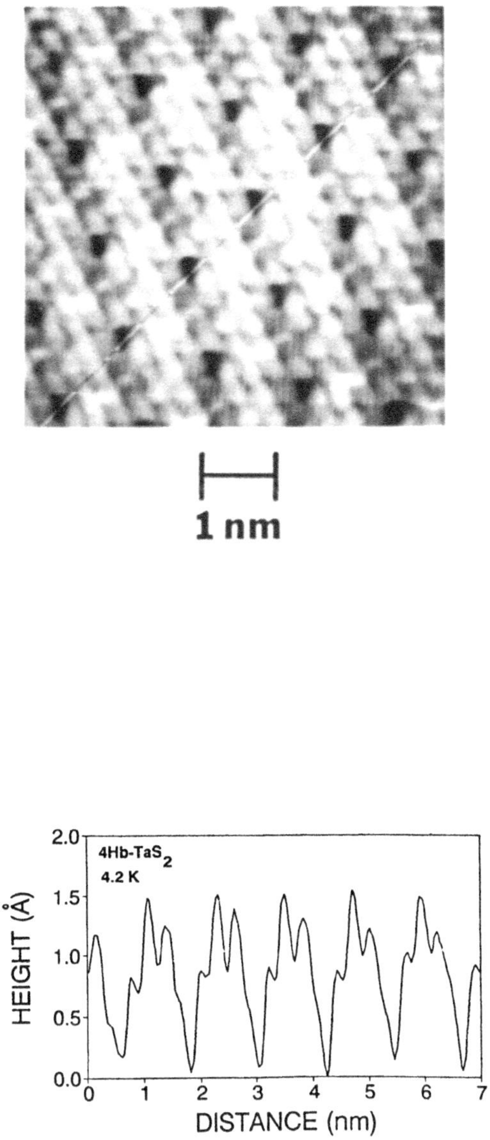

Fig. 42. (a) Grey scale image from a STM scan on the 1T sandwich of a 4Hb-TaS_2 crystal at 4.2 K. Both the CDW and atomic modulations are detected and the $\sqrt{13}\ \mathbf{a}_0 \times \sqrt{13}\ \mathbf{a}_0$ CDW superlattice generates relatively deep minima (I= 2.2 nA, V= 25 mV).
(b) Profile of z-deflection along the line shown in (a). The majority of the z-deflection is due to the deep minima. (From Reference 18.)

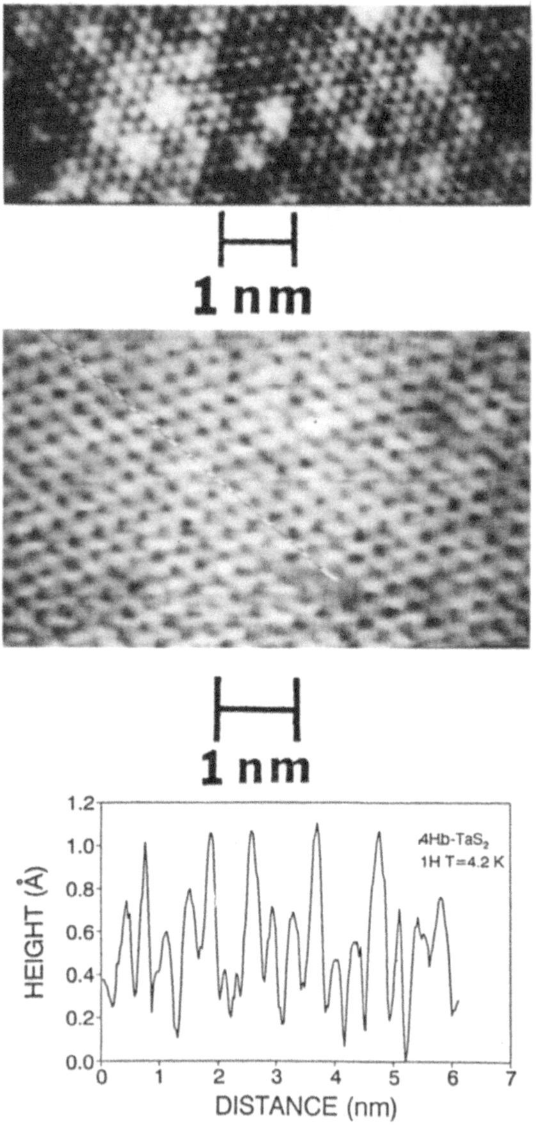

Fig. 43. (a) Grey scale image from a STM scan on the 1H sandwich of a 4Hb-TaS_2 crystal at 4.2 K. The stronger $\sqrt{13}\,\mathbf{a}_0 \times \sqrt{13}\,\mathbf{a}_0$ CDW of the 1T sandwich below the surface sandwich is detected as sunflowers of surface S atoms enhanced by the CDW in the layer below (I= 2 nA, V= 75 mV).
(b) Grey scale image from a STM scan on the 1H sandwich of a 4Hb-TaS_2 crystal at 4.2 K. A weak $3\mathbf{a}_0 \times 3\mathbf{a}_0$ CDW superlattice is visible in the image (I= 2 nA, V= 75 mV).
(c) Profile of z-deflection along the line shown in (a) showing an enhancement of every third atom due to the $3\mathbf{a}_0 \times 3\mathbf{a}_0$ CDW in the surface sandwich.

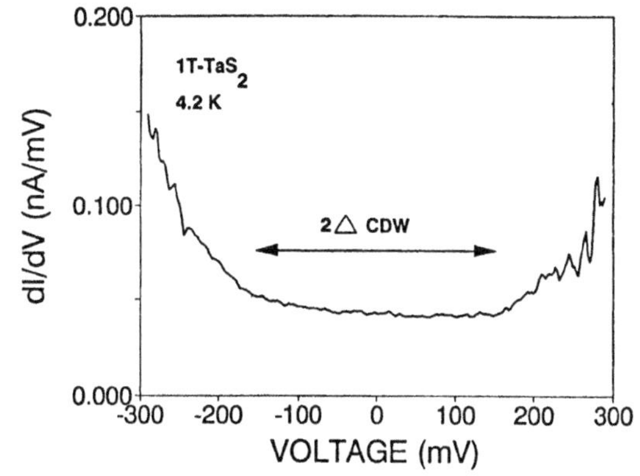

Fig. 44. Conductance (dI/dV) versus bias voltage (V) curve measured on 1T-TaS_2 at 4.2 K. The sharp decrease in conductance below ~ 150 mV is consistent with a large reduction of the FS area due to CDW gap formation. (From Reference 51.)

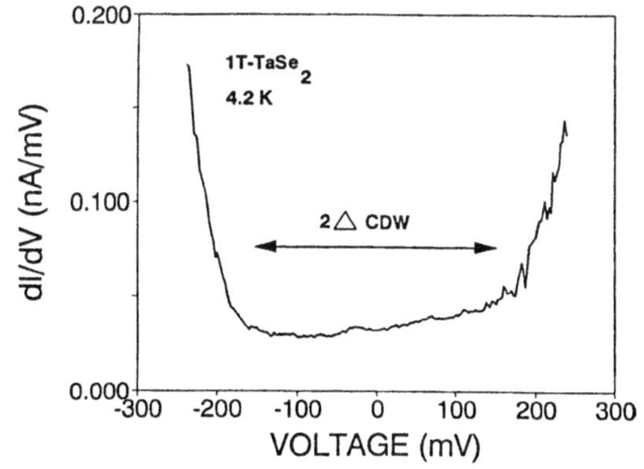

Fig. 45. Conductance versus bias voltage curve for 1T-$TaSe_2$ measured at 4.2 K. A sharp increase in conductance at ± 150 mV can be identified with the CDW gap and the resulting change in DOS above the gap edge. (From Reference 51.)

value of $2\Delta_{CDW}/k_BT_C$ = 5.8 calculated using the onset temperature of ~ 600 K is in the strong coupling regime, but is substantially smaller than that observed in the 2H phase materials to be discussed in Section 9.4.

The spectroscopy mode gives a measure of the CDW gap structure in 1T-$TaSe_2$ that is in reasonable agreement with the gap measured for 1T-TaS_2. A dI/dV versus V curve taken at 4.2 K is shown in Fig. 45 and indicates a CDW gap at ~ 150 meV. Using a CDW onset temperature of ~ 600 K this gives $2\Delta_{CDW}/k_BT_C$ = 5.8.

9.2 1T-VSe_2

Conductance data from 1T-VSe_2 at 4.2 K shows a much smaller energy gap consistent with the lower CDW onset temperature of 110 K. At 4.2 K this CDW is commensurate with a wavelength of $4a_0$. A conductance versus bias voltage curve for 1T-VSe_2 is shown in Fig. 46 with a gap structure identified at ~ 40 mV. This gives a value of $2\Delta_{CDW}/k_BT_C \simeq 8.4$, a value midway between the 1T phase Ta compounds and the 2H phase Ta compounds. The CDW in 1T-VSe_2 behaves more like the CDWs observed in the 2H phase compounds in that it has a relatively low amplitude and remains aligned along the atomic lattice. In this respect all of the CDW gaps measured by the STM show a systematic trend[50,51] consistent with the general qualitative features observed for each CDW. CDW gaps measured in the 2H phase compounds will be described in later sections and more details can be found in Reference 50 and 51.

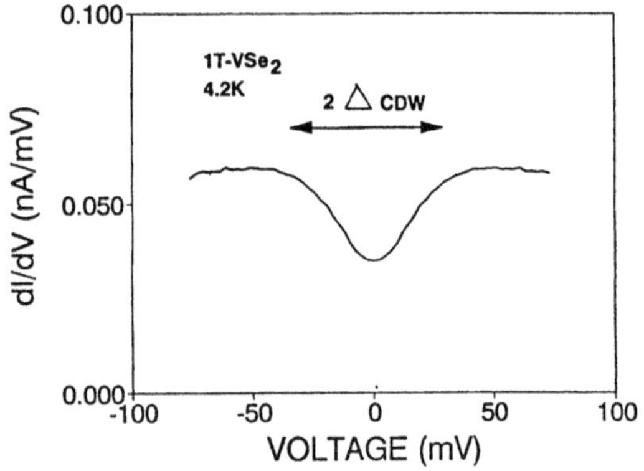

Fig. 46. Conductance versus bias voltage curve measured on 1T-VSe_2 at 4.2 K. A slope change is observed at ± 40 mV which can be identified with the CDW gap edge. (From Reference 51.)

9.3 1T-TiS_2 AND 1T-$TiSe_2$

Spectroscopy of 1T-$TiSe_2$ with the STM shows structure that could be identified with a narrow gap. A dI/dV versus V curve recorded at 4.2 K is shown in Fig. 47 and for positive bias a flat region in the DOS is observed between 0 and 0.2 V. Defects due to non-stoichiometry can produce an n-type semiconductor, which would account for the shift in the possible gap structure to the positive bias side of the curve. This structure in the conductance versus voltage curve is very similar to that observed in the STM spectroscopy of 1T-TiS_2 which is definitely a narrow band semiconductor

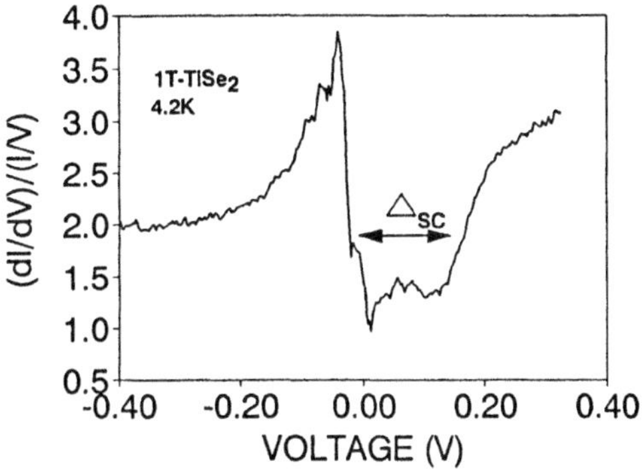

Fig. 47. Normalized conductance curve obtained on 1T-$TiSe_2$ at 4.2 K. A flat region between 0 and 0.2 V indicates the presence of a weak semiconducting gap.

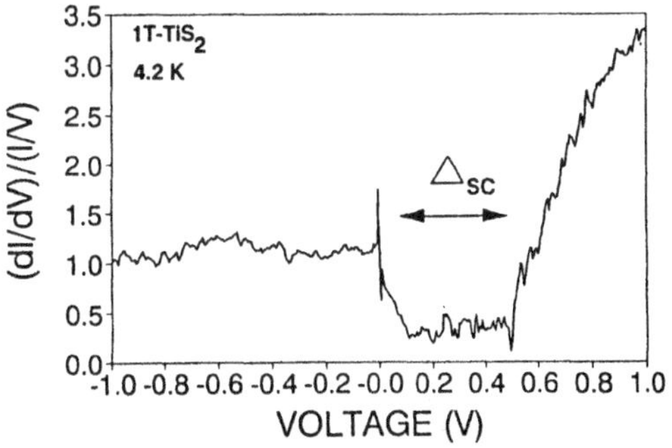

Fig. 48. Normalized conductance curve recorded on 1T-TiS_2 at 4.2 K. The flat region of low conductivity can be associated with a semiconducting gap of ~ 0.3 to 0.4 eV. (From Reference 51.)

and is discussed below. This tunneling spectroscopy data provides some confirmation of the narrow band semiconductor analysis presented by Stoffel et al.[32]

Spectroscopy measurements have been carried out on 1T-TiS_2 at 4.2 K and a typical dI/dV versus V curve is shown in Fig. 48. It shows a structure characteristic of a semiconductor with a gap of ~ 0.3 to 0.4 eV. The asymmetry suggests that the 1T-TiS_2 sample is an n-doped semiconductor with E_f near or in the Ti d conduction band due to excess Ti or defects. The flat regions in the conductance curve for positive tip bias therefore occur as the Fermi level of the tip sweeps across the gap in the TiS_2 substrate. This spectroscopy result gives reasonable confirmation of the predicted semiconductor band structure of 1T-TiS_2.

9.4 2H-$NbSe_2$, 2H-$TaSe_2$ AND 2H-TaS_2

The CDW forming below 33 K in 2H-$NbSe_2$ gaps only a small part of the Fermi surface so that the change in the density-of-states of the conduction electrons is relatively small. Nevertheless, the CDW gap can easily be detected with the STM by measuring dI/dV versus V and detecting the change in slope at the gap edge. A dI/dV versus V curve for 2H-$NbSe_2$ at 4.2 K is shown in Fig. 49 with the gap edge clearly identified at 34 mV. This gives a very large value of $2\Delta_{CDW}/k_BT_C = 23.9$ indicating that a short coherence length model as suggested by McMillan[52] is required.

The spectroscopy of 2H-$TaSe_2$ is very similar to that observed for the other 2H phase compounds. A dI/dV versus V curve is shown in Fig. 49 (upper curve). A relatively abrupt change of slope is observed at $\pm$ 80 mV and we have identified this with the onset of tunnel conductance above the CDW gap edge. Using a value of $\Delta_{CDW} = 80$ meV and $T_C = 122.3$ K gives a value of $2\Delta_{CDW}/k_BT_C = 15.2$. This is again very large and requires the short coherence length model of McMillan,[52] a conclusion reached for the CDW in all 2H phase compounds so far measured.

The spectroscopy results on 2H-TaS_2 follow the systematic behavior observed for the three 2H phase compounds measured and presented in Fig. 49. The dI/dV

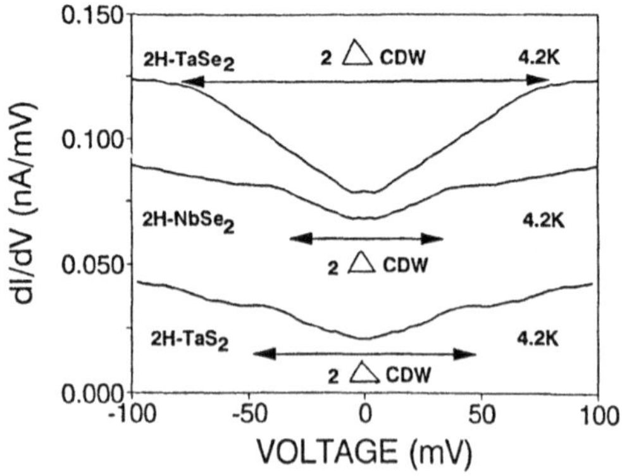

Fig. 49. dI/dV versus V plot recorded on 2H-$TaSe_2$, 2H-$NbSe_2$, and 2H-TaS_2 at 4.2 K. Slope changes can be identified with the CDW gaps and associated density-of-states (DOS) changes near the gap edges. The average gaps are estimated to be $\pm$ 80 mV, $\pm$ 34 mV, $\pm$ 50 mV respectively as indicated by arrows which indicate the region of width 2Δ centered on zero bias. (From Reference 51.)

versus V curve for $2H-TaS_2$ (lower curve) shows a weak slope change indicating that the CDW gap edge is located at a bias of $\leq$ 50 mV. With a T_C = 75.3 K this gives a value of $2\Delta_{CDW}/k_BT_C$ = 15.4, approximately the same magnitude as observed for $2H-TaSe_2$.

9.5 $4Hb-TaSe_2$ AND $4Hb-TaS_2$

Structure in the dI/dV versus V curves associated with the two separate CDWs has been observed at 4.2 K in both $4Hb-TaSe_2$ and $4Hb-TaS_2$. However, the relative structure introduced by the CDW gaps is considerably different in the 4Hb selenide versus the sulfide. A representative dI/dV versus V curve for $4Hb-TaSe_2$ at 4.2 K is shown in Fig. 50(a). Two regions, characterized by a rapid onset of conductance increase followed by a peak are clearly observed. These are characteristic of a gap edge followed by a peak in the DOS just above the gap edge. Both the STM amplitude and the associated gap structure in the dI/dV versus V curve remain well-defined indicating that the CDW strength has not been changed significantly from that observed in the pure 1T and 2H phase. Both the 1T and 1H layers remain metallic at the lowest temperatures. Using the peak positions in the curves of the type shown in Fig. 50(a) we have determined the two CDW gaps in $4Hb-TaSe_2$ to be Δ_{1T} = 140 meV and Δ_{1H} = 60 meV. Both are reduced from those observed in the pure 1T and 2H phases where the measured values were Δ_{1T} = 150 meV and Δ_{2H} = 80 meV. The calculated values of $2\Delta_{CDW}/k_BT_C$ are 5.4 and 18.5 respectively, comparable to the same values observed in the pure phases. Extremely strong coupling is observed in the 1H sandwich which requires a short coherence length model, while the 1T sandwich shows a substantially smaller value of $2\Delta_{CDW}/k_BT_C$.

These results are consistent with the overall behavior observed in the STM response and confirm the electronic modifications expected relative to the separate pure phase measurements. The sharp gap structure is consistent with the reasonably good resolution obtained for the two essentially independent CDW superlattices in $4Hb-TaSe_2$, as well as their simultaneous resolution in the CDW super-superlattice detected in Fig. 41(a).

In contrast to the 4Hb selenide the 4Hb sulfide shows a much stronger modification of the CDW structure from that observed in the separate 1T and 2H phase compounds. The $3\mathbf{a}_0 \times 3\mathbf{a}_0$ CDW superlattice onset temperature is reduced to 22 K and as discussed in the previous section the STM amplitude is extremely weak. The dI/dV versus V curves in which good CDW gap structure is observed, are completely dominated by the larger CDW gap associated with the 1T layer. An example is shown in Fig. 50(b). The high temperature CDW gap structure gives a fundamental peak at ~ 72 mV and this is followed by a harmonic series showing very strong amplitudes. The low temperature gap structure is not well-resolved and

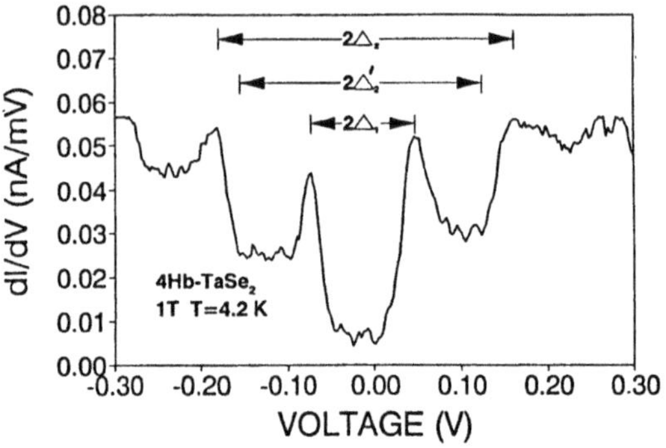

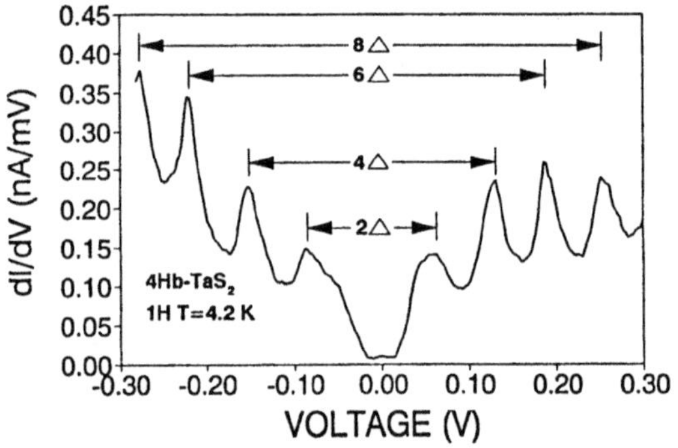

Fig. 50. (a) dI/dV curve for 4Hb-$TaSe_2$ measured at 4.2 K. Sharp peaks in the conductance associated with the CDW gaps in the crystal are apparent at ~60 mV and ~180 mV. In order to compare the high temperature CDW gap value with data taken on pure 1T-$TaSe_2$, the gap structure can be measured as existing at the point where the rapid rise in conductance occurs. This construction gives a value of ~140 mV.
(b) A dI/dV versus V curve measured at 4.2 K for 4Hb-TaS_2. The conductance peaks are identified with the high temperature CDW gap which generates peaks at the fundamental of Δ= 72 mV plus harmonics of the fundamental. The low temperature CDW gap occurs within the minimum near zero bias and is not well resolved.

lies in the region of minimum conductance near zero bias. The value is estimated from the weak structure in a number of runs to be Δ_{1H} ~ 30 meV.

The low temperature CDW gap appears to be reduced substantially from the value of 50 meV observed in the pure 2H phase, and this is consistent with the major reduction of CDW onset temperature in the 1H sandwich from 75 to 22 K. The values of $2\Delta_{CDW}/k_BT_C$ maintain the relative difference observed for the CDWs in the octahedral versus the trigonal prismatic coordination sandwiches of all of layer structure dichalcogenides, but the measured values of Δ_{1H} and Δ_{1T} are not sufficiently accurate for quantitative comparison.

The behavior of the spectroscopic response in 4Hb-TaS_2 compared to 4Hb-$TaSe_2$ may arise from the semiconducting nature of the octahedral layer in 4Hb-TaS_2. If tunneling is required between the octahedral layers as proposed by Wattamaniuk et al.[47] in their analysis of $\rho_\perp$, then the tunneling electrons may couple very strongly to the CDW gap in the octahedral layers. The existence of multiple tunnel junctions in series can easily lead to strong harmonic amplitudes.

10. Conclusions

The series of STM experiments reviewed in this chapter have conclusively established the value of the STM in detecting the real space arrangement and the intensity of charge-density wave structure in layer structure dichalcogenides. Both the atomic and CDW structure have been well-resolved, and the STM amplitude response has shown a systematic correlation with the strength of the CDW. CDW structure deduced from other types of experiments has been confirmed on an atomic scale in the real space images, and in a number of cases new and unexpected results on CDW structure have been obtained. Examples are the real space domain-like structure existing in 1T-TaS_2 at temperatures near room temperature, the simultaneous observation of the CDWs in the trigonal prismatic and octahedral sandwiches of the 4Hb phase Ta compounds, and the direct observation of possible 3q → 2q CDW structures in 1T-VSe_2.

The initial AFM experiments also demonstrate that the AFM can give good resolution of CDW structure, although the amplitude relative to the atomic amplitude is much smaller than observed with the STM. Combined STM and AFM studies should be complimentary.

A large series of spectroscopy measurements have shown that reliable information on CDW energy gaps can also be obtained with the STM. The spectroscopy mode is highly sensitive, but systematic reproducibility has been obtained. The STM can be expected to supply extremely valuable information on many aspects of electronic structure in materials.

Acknowledgements

The research described here has been supported by the U.S. Department of Energy Grant No. DE-FG05-89ER45072. The authors would like to thank P. K. Hansma, Joe Demuth, B. Giambattista, A. Johnson, and Vittorio Celli for valuable help and discussions

1. R.V. Coleman, B. Drake, P.K. Hansma, and G. Slough, Phys. Rev. Lett. **55**, 394 (1985).
2. J.A. Wilson, F.J. Di Salvo, and S. Majahan, Adv. Phys. **24**, 117 (1975).
3. G. Binnig and D.P.E. Smith, Rev. Sci. Instrum. **57**, 1688 (1986).
4. S. Alexander, L. Hellemans, O. Marti, J. Schneir, V. Elings, P.K. Hansma, M. Longmire, and J. Gurley, J. Appl. Phys. **65**, 164 (1989).
5. Microfabricated Si_3N_4 AFM tips were obtained from Park Scientific Instruments, 476 Ellis Streeet, Mountain View, CA 94043.
6. Digital Instruments, Inc., 6780 Cortona Drive, Santa Barbara, CA 93117.
7. J.A. Wilson and A.D. Yoffe, Adv. Phys. **18**, 193 (1969).
8. N.V. Smith, S.D.Kevan, and F.J. Di Salvo, J. Phys. C **18**, 3175 (1985).
9. Xian Liang Wu and Charles M. Lieber, Science **243**, 1703 (1989); Phys. Rev. Lett. **64**, 1150 (1990).
10. R.V. Coleman, W.W. McNairy, and C.G. Slough, Phys. Rev. B **45**, 1428 (1992).
11. K. Nakanishi and H. Shiba, J. Phys. Soc. Jpn. **43**, 1839 (1977).
12. B. Burk, R.E. Thomson, A. Zettl, and John Clarke, Phys. Rev. Lett. **66**, 3040 (1991).
13. J.A. Wilson, J. Phys.: Condens. Matter **2**, 1683 (1990).
14. C.G. Slough, W.W. McNairy, Chen Wang, and R.V. Coleman, J. Vac. Sci. Technol. B **9**, 1036 (1991).
15. R.E. Thomson, U. Walter, E. Ganz, J. Clarke, A. Zettl,P. Rauch, and F.J. Di Salvo, Phys. Rev. B **38**, 10734 (1988).
16. A.M. Woolley and G. Wexler, J. Phys. C **10**, 2601 (1977).
17. J. Tersoff, Phys. Rev. Lett. **57**, 440 (1986).
18. R.V. Coleman, B. Giambattista, P.K. Hansma, A. Johnson, W.W. McNairy, and C.G. Slough, Adv. Phys. **37**, 559 (1988).
19. C.G. Slough, W.W. McNairy, R.V. Coleman, J. Garnaes, C.B. Prater, and P.K. Hansma, Phys. Rev. B **42**, 9255 (1990).
20. J. Garnaes, S.A.C. Gould, P.K. Hansma, and R.V. Coleman, J. Vac. Sci. Technol. B **9**, 1032 (1991).
21. H.W. Myron, Physica B **99**, 243 (1990).
22. P.M. Williams, Crystallography and Crystal Chemistry of Materials with Layered Structure, edited by F. Levy (Reidel, Dordrecht, 1976) p. 51.
23. K. Tsutumi, T. Sambongi, T. Akira, and S. Tanaka, J. Phys. Soc. Jpn. **49**, 837 (1980).
24. J. van Landuyt, G.A. Wiegers, and S. Amelinckx, Phys. Stat. Sol. **46**, 479 (1978).
25. K.K. Fung, J.W. Steeds, and J.A. Eades, Physica B **99**, 47 (1980).

26. K. Tsutsumi, Phys. Rev. B **26**, 5756 (1982).
27. Y. Yoshida and K. Motizuki, J. Phys. Soc. Jpn. **51**, 2107 (1984).
28. D.J. Eaglesham, R.L. Withers, and D.M. Bird, J. Phys. C **19** , 359 (1986).
29. K.C. Woo, F.C. Brown, W.L. McMillian, R.J. Miller, M.J. Schaffman, and M.P. Sears, Phys. Rev. B **14**, 3242 (1976).
30. F.J. Di Salvo, D.E. Moncton, J.A. Wilson, and J.V. Waszczak, Bull. Am. Phys. Soc. **21**, 261 (1976).
31. F.C. Brown, Physica B **99**, 264 (1980).
32. N.G. Stoffel, S.D. Kevan, and N.V. Smith, Phys. Rev. B **31**, 8049 (1985).
33. K. Motizuki, Y. Yoshida, and Y. Takaoka, Physica B **105**, 357 (1981).
34. N. Suzuki, A. Yamamoto, and K. Motizuki, J. Phys. Soc. Jpn. **54**, 4668 (1985).
35. In-Whan Lyo and P. Avouris, Science **253**, 173 (1991).
36. D.E. Moncton, J.D. Axe, and F.J. Di Salvo, Phys. Rev. Lett. **34**, 734 (1975).
37. B. Giambattista, A. Johnson, R.V. Coleman, B. Drake, and P.K. Hansma, Phys. Rev. B **37**, 2741 (1988).
38. R.M. Fleming and R.V. Coleman, Phys. Rev. B **16**, 302 (1977).
39. J.E. Graebner, Solid State Commun. **21**, 353 (1977).
40. N.J. Doran and A.M. Woolley, J. Phys. C **14**, 4257 (1981).
41. G. Wexler and A.M. Woolley, J. Phys. C **9**, 1185 (1976).
42. J.P. Tidman, O. Singh, A.F. Curzon, and R.F. Frindt, Philos. Mag. **30**, 1191 (1974).
43. G.A. Scholz, O. Singh, R.F. Frindt, and A.E. Curzon, Solid State Commun. **44**, 1455 (1982).
44. H. Nishihara, G.A. Scholz, and R.F. Frindt, Solid State Commun. **44**, 507 (1982).
45. S.J. Hillenius and R.V. Coleman, Phys. Rev. B **18**, 3790 (1978).
46. R.H. Friend, D. Jerome, R.F. Frindt, A.J. Grant, and A.D. Yoffe, J. Phys. C **10**, 1013 (1977).
47. W.J. Wattamaniuk, J.P. Tidman, and R.F. Frindt, Phys. Rev. Lett. **35**, 62 (1975).
48. G. Wexler, A. Woolley, and N. Doran, Nuovo Cimento B **38**, 571 (1977).
49. N.J. Doran, G. Wexler, and A.M. Woolley, J. Phys. C **11**, 2967 (1978).
50. Chen Wang, B. Giambattista, C.G. Slough, R.V. Coleman, and M.A. Subramanian, Phys. Rev. B **42**, 8890 (1990).
51. Chen Wang, C.G. Slough, and R.V. Coleman, J. Vac. Sci. Technol. B **9**, 1048 (1991).
52. W.L. McMillan, Phys. Rev. B **16**, 643 (1977).
53. B. Giambattista, C.G. Slough, W.W. McNairy, and R.V. Coleman, Phys. Rev. B **41**, 10082 (1990).

ELECTRONIC STRUCTURE AND PROPERTIES OF LAYERED STRUCTURES

A.J. Freeman and A. Continenza
Physics and Astronomy Department and Materials Research Center
Northwestern University Evanston, IL 60208-3112, U.S.A.

and

M. Posternak
Institut de Physique Appliquée, Ecole Polytechnique Fédérale and Institut Romand de Recherche Numérique en Physique des Matériaux (IRRMA)
PH-Ecublens CH-1015 Lausanne, Switzerland

and

S. Massidda
Institut Romand de Recherche Numérique en Physique des Matériaux (IRRMA)
PH-Ecublens CH-1015 Lausanne, Switzerland

G. Benedek (ed.), Surface Properties of Layered Structures, 97–150.

1 Introduction

1.1 Background Perspective

The electronic structure of layered materials — both natural and artificial — is a subject of growing interest and investigation. This growing interest in the experimental aspects of these studies is clearly given in the other chapters of this book. Here we present some illustrations of the theoretical work on layered semiconductor structures. Because of space limitations, no attempt at completeness can be attempted; thus we confine ourselves to three examples taken from our own work.

These examples also serve to highlight the role of modern computational solid state theory in providing simulations of materials of experimental interest with which it is possible to interpret existing observations and also predictions of as yet unmeasured properties. So today, theorists are able to follow the experimentalists pace by implementing and developing new formalisms and theoretical concepts to be applied to real materials (rather than to restricted elemental models) and thereby giving precise answers and interpretations to the experimental data. Very often, they have even managed to lead the way by proposing new structures, simulating new materials for study and providing predictions and data, on the atomic scale, that are often impossible or hard to obtain experimentally.

The issues that are of particular relevant interest to this book (such as surfaces and interfaces) are many and span a wide variety of different fundamental physics aspects: structural properties, magnetic properties, electronic and transport properties (including potential barriers). On the other hand, the ultimate goal underlying the study is to achieve the capability of modifying these basic properties and eventually to "tune" them at will in order to design the "ad hoc" material for each different device application. Of course, the basic requirement for that, is the complete knowledge and understanding of each single property and how it can be affected and modified by changing several different external factors (such as: growth condition, thickness of overlayers, strain, choice of substrate, interface reactivity, etc), which remain a formidable task.

The kinds of answers that a computational approach can give are various and comprise quantities which are experimentally accessible (such as density of states, and energy level transitions which are accessible to angle-resolved photoemission spectroscopy) as well as some that are not (such as, for example, in-layer bond-length relaxations). More specifically, such studies can give precise information regarding the interface morphology (bond length, atomic distances and surface relaxation), the electronic properties (such as band gaps, energy transitions, carrier effective mass as well as magnetic moments, hyperfine fields and magnetic anisotropy, potential line up and Fermi level pinning), and the stability properties (stability of nanostructures against disproportionation in the constituents, stability against chemical reaction or diffusion at the interface). These kinds of information are of remarkable relevance since they are closely related to the final electrical and magnetic behavior of the interface. Moreover, understanding the electronic properties of a given system is at the basis of the full comprehension of its electric, thermal and mechanical properties; clearly, band structure calculations give us a powerful tool to explore in detail these properties.

The advantage of a computational approach consists in the ability to predict basic properties and to study "clean" systems in which interface intermixing or reaction is absent (and that are not so easily accessible experimentally) in order to single out each different contribution. In recent years, highly precise LDF (local density functional theory) and LSDF (local spin density functional theory) methods have been used to obtain structural, electronic and magnetic properties of many different materials, demonstrating how these calculations can give a clear understanding of the experimental results and lead to predictions on novel systems not yet made in the laboratory. In particular, the occurrence of interlayer states, the magnetization (including its anisotropic behavior), and charge and spin densities obtained from spin-polarized calculations can be compared directly with experiments (including conversion electron Mössbauer spectroscopy and other measurements of hyperfine fields). For excitations (like spin-polarized photoemission and inverse photoemissions), spin polarized photocurrents can be calculated from the LSDF results.

2 Computational Approach

To address the problem of computing the electronic structure of superlattices and interfaces we use the all-electron self-consistent full potential linearized augmented plane wave method (FLAPW) [1] within the local density approximation (LDA) and the exchange-correlation potential as parametrized by Hedin and Lundqvist [2] and Von Barth and Hedin [3]. In order to better define the framework of our study, we now discuss very briefly some of the main features and approximations which are at the basis of the LDA and of the FLAPW method.

2.1 Local Density Functional Theory

The major problem that is faced by any band structure calculation, within the Born–Oppenheimer approximation, is how to solve the complete Hamiltonian that describes the system, taking also into account the many-body electron-electron interaction. The problem is particularly complicated and hard to solve since the many-body wavefunctions are dependent on the coordinates of all the N electrons in the system. In this regard, density functional theory (DFT) (and more specifically LDA) represents a powerful tool that simplifies the many-body problem reducing it to a one-electron Hamiltonian. This is achieved by making use of the Kohn–Sham theorem [4] which states that the ground state energy of the many-body system with charge density $\rho(\mathbf{r})$ and spin density $n(\mathbf{r})$ can be expressed as a unique functional of the ground state charge and spin densities. Within this formalism, the charge and spin density are defined as follows:

$$\rho(\mathbf{r}) = \rho^{\uparrow}(\mathbf{r}) + \rho^{\downarrow}(\mathbf{r}) \tag{1}$$

$$n(\mathbf{r}) = \rho^{\uparrow}(\mathbf{r}) - \rho^{\downarrow}(\mathbf{r}), \tag{2}$$

where $\rho^{\uparrow}(\mathbf{r})$ and $\rho^{\downarrow}(\mathbf{r})$ are the spin-up and spin-down charge densities, respectively.

Moreover, the theorem establishes that the ground state energy can be expressed as a functional which is stationary with respect to variations of the charge density. As a consequence, [4] the ground state energy functional can be expressed in atomic (Hartree) units as:

$$E[\rho, n] = T[\rho] + \int V[\rho(\mathbf{r}), n(\mathbf{r})]\rho(\mathbf{r})d\mathbf{r} + \frac{1}{2}\int\int \frac{\rho(\mathbf{r})\rho(\mathbf{r}')}{|\mathbf{r} - \mathbf{r}'|} d\mathbf{r}d\mathbf{r}' + E_{xc}[\rho, n]. \tag{3}$$

where $T[\rho]$ is the kinetic energy term, $V[\rho, n]$ is the external potential that might also include the Coulomb term describing the electrostatic interaction between electrons and nuclei, the third term is the Coulomb repulsion between electrons, and finally, $E_{xc}[\rho, n]$ is the exchange correlation energy.

Using the variational principle it is possible to derive the Schrödinger equation for the spin-up and spin-down one-particle eigenstate $\Psi_i^{\uparrow(\downarrow)}$:

$$(-\frac{1}{2}\Delta + V_{eff}^{\uparrow(\downarrow)})\Psi_i^{\uparrow(\downarrow)} = \varepsilon_i^{\uparrow(\downarrow)}\Psi_i^{\uparrow(\downarrow)}, \tag{4}$$

with the constraint:

$$\rho(\mathbf{r})^{\uparrow(\downarrow)} = \sum_{i=1}^{nocc} \Psi_i^{*\uparrow(\downarrow)}\Psi_i^{\uparrow(\downarrow)}. \tag{5}$$

In equation (4) V_{eff} is given by:

$$V_{eff}^{\uparrow(\downarrow)}(\mathbf{r}) = V(\mathbf{r}) + \int \frac{\rho(\mathbf{r}')^{\uparrow(\downarrow)}}{|\mathbf{r} - \mathbf{r}'|} d\mathbf{r}' + \frac{\delta E_{xc}[\rho, n]}{\delta\rho^{\uparrow(\downarrow)}}. \tag{6}$$

These equations can now be solved self-consistently to determine the single particle energies ε_i, the wave functions Ψ_i, and therefore the charge and spin densities $\rho(\mathbf{r})$ and $n(\mathbf{r})$. The total energy E_T of the system can be calculated using Eqs. (4) and (3) as follows:

$$E_T = \sum_{i=1}^{nocc} \varepsilon_i - \frac{1}{2}\int\int \frac{\rho(\mathbf{r})\rho(\mathbf{r}')}{|\mathbf{r} - \mathbf{r}'|} d\mathbf{r}d\mathbf{r}' + E_{xc}[\rho, n] - \sum_{\uparrow,\downarrow}\int \rho^{\uparrow(\downarrow)}V_{xc}[\rho, n]d\mathbf{r} \tag{7}$$

A key feature of this theory is that Eqs. (4) and (5) must be solved iteratively in order to obtain a potential $V_{eff}[\rho(\mathbf{r})]$ which is self-consistent with both the charge and spin densities. The exact form of E_{xc} and $V_{xc} = \frac{\delta E_{xc}}{\delta\rho^{\uparrow(\downarrow)}}$ is very complicated and should include all the *local* and *non-local* effects of the many-body interaction. However, the exact expression for E_{xc} is not known and some approximations are generally made. The first approximation is made in equation (3) by considering E_{xc} as a local term. In addition to that, the exact form of E_{xc} is commonly replaced by the exchange-correlation energy per electron $\epsilon_{xc}[\rho(\mathbf{r})]$, of a homogeneous electron gas of density $\rho(\mathbf{r})$:

$$E_{xc}[\rho, n] = \int \rho(\mathbf{r})\epsilon_{xc}[\rho(\mathbf{r}), n(\mathbf{r})]d\mathbf{r}, \tag{8}$$

so that the exchange-correlation potential becomes:

$$\mu_{xc}^{\uparrow(\downarrow)} = \frac{\delta\rho\epsilon_{xc}[\rho, n]}{\delta\rho^{\uparrow(\downarrow)}} \tag{9}$$

and the total energy can be written as:

$$\begin{aligned} E_T^{\uparrow(\downarrow)} &= \sum_{i=1}^{nocc}\varepsilon_i^{\uparrow(\downarrow)} - \frac{1}{2}\iint\frac{\rho^{\uparrow(\downarrow)}(\mathbf{r})\,\rho^{\uparrow(\downarrow)}(\mathbf{r}')}{\mid \mathbf{r}-\mathbf{r}'\mid}d\mathbf{r}\,d\mathbf{r}' + \\ &+ \int\rho^{\uparrow(\downarrow)}(\mathbf{r})\,(\epsilon_{xc}^{\uparrow(\downarrow)} - \mu_{xc}^{\uparrow(\downarrow)})\,d\mathbf{r}. \end{aligned} \tag{10}$$

The assumption leading to Eq. (10), is referred to as the local density approximation (LDA). It drastically simplifies the problem of solving the Kohn–Sham equations (4), (5) and has been proved to be particularly successful in understanding the ground state properties of many systems. In particular, it has been remarkably successful in obtaining good agreement with experimental quantities such as magnetization, neutron form factors, hyperfine fields, lattice parameters, bulk modulus, cohesive energy, and Fermi surface properties. The major failure of such an approximation is that it does not predict correctly the excited states of the system such as the band gap in semiconductors and insulators and all the elementary excitations of many-electron systems as measured, for example, by photoemission experiments. Most of the discrepancies with these experiments are believed to be caused by the self-interaction and other non-local contributions to the correlation potential which can be included as a correction to the calculation.

2.2 The FLAPW Method

In order to solve the one-electron Schrödinger-like equations (Eq. (4), (5)), obtained within the LDA approximation, we use the FLAPW method [1]. This is a variational method in which the wave function of each energy state i of wave vector $\mathbf{k}$ (belonging to the first Brillouin zone) is assumed to be a Bloch wave function:

$$\Psi_i(\mathbf{k}, \mathbf{r}) = e^{i\mathbf{k}\cdot\mathbf{r}}\mathrm{u}_i(\mathbf{k}, \mathbf{r}). \tag{11}$$

where u_i $(\mathbf{k},\mathbf{r})$ has the lattice periodicity. $\Psi_i(\mathbf{k}, \mathbf{r})$ is then expanded variationally as:

$$\Psi_i(\mathbf{k}, \mathbf{r}) = \sum_j C_j\Phi_j(\mathbf{k} + \mathbf{G}_j, \mathbf{r}). \tag{12}$$

The functions $\Phi_j(\mathbf{k}+\mathbf{G}_j,\mathbf{r})$ are assumed to have different expressions depending on the real space region they refer to. The real space is in fact divided into two regions for bulk materials (spheres of radius R_s surroundings the atoms, and interstitial regions) and into three regions (spheres, interstitial and vacuum) for surfaces and thin films; the wave function $\Phi_j(\mathbf{K}_j,\mathbf{r})$, (taking $\mathbf{K}_j=\mathbf{k}+\mathbf{G}_j$) is then expressed as:

$$\Phi_j(\mathbf{K}_j,\mathbf{r}) = \begin{cases} \frac{1}{\sqrt{\Omega}} e^{i\mathbf{K}_j\mathbf{r}} \\ \qquad if \quad \mathbf{r} \in \text{interstitial;} \\ \sum_{lm}[A_{lm}(\mathbf{K}_j)u_l(E_l,r) + B_{lm}(\mathbf{K}_j)\dot{u}_l(E_l,r)]\, Y_{lm}(\hat{r}) \\ \qquad if \quad \mathbf{r} \in R_s. \\ \sum_q[A_q(\mathbf{K}_j)u_q(E_v,z) + B_q(\mathbf{K}_j)\dot{u}_q(E_v,z)]\, e^{i(\mathbf{k}+\mathbf{K}_q)\mathbf{r}} \\ \qquad if \quad \mathbf{r} \in \text{vacuum.} \end{cases} \tag{13}$$

In Eq. (13) Ω is the unit cell volume, $\mathbf{G}_j$ is a 3-dimensional reciprocal vector and the coefficients A_{lm}, B_{lm} are determined requiring the continuity of the wave function Φ_j and of its spatial derivative across the boundaries between the different regions. The third relation in the Eq. (13) is valid in the case of film geometry, where $\mathbf{K}_q$ denotes a two-dimensional (i.e. parallel to the surface) reciprocal lattice vector. In this case the wave vector $\mathbf{k}$ belongs to the two-dimensional first Brillouin zone. The function u_l is solution of the radial Schrödinger equation:

$$[-\frac{\partial^2}{\partial r^2} + V_{eff}(r) - E_l - \frac{l(l+1)}{r^2}]\, r\, u_l(E_l,r) = 0 \tag{14}$$

and $\dot{u}_l(E_l,r)$ is its energy derivative. $V_{eff}(r)$ is the spherical average of the potential inside muffin-tin spheres (see later).

Similarly, in the vacuum, A_q and B_q are determined from continuity requirements on the wave function and its spatial derivative at the boundaries and $u_q(E_v,z)$ ($\dot{u}_q(E_v,z)$ is its energy derivative) is solution of the equation:

$$[-\frac{\partial^2}{\partial z^2} + V_{eff}(z) - E_v + (\mathbf{k}+\mathbf{K}_q)^2]u_q(E_v,z) = 0. \tag{15}$$

$V_{eff}(z)$ is the component of the effective potential in the vacuum which depends only on the distance perpendicular to the surface and, finally, E_v and E_l are energy parameters for the vacuum and the spheres, respectively.

The coefficients C_j in Eq. (12) are determined by using the variational principle and solving the secular equation obtained from the one-electron Kohn–Sham equation (4):

$$(H_{ij} - \varepsilon_i S_{ij})C_j = 0 \tag{16}$$

where $H_{ij} = <\Phi_i \mid H \mid \Phi_j>$ are the Hamiltonian matrix elements and $S_{ij} = <\Phi_i \mid \Phi_j>$ are the overlap integrals.

Equation (16) also determines the energy eigenvalues, ε_i and can be solved by simply diagonalizing the matrix $H_{ij} - \varepsilon_i S_{ij}$. This is a remarkable advantage over the conventional APW [5] method and it is due to the linear expression chosen for the wave function in terms of the radial function and its energy derivative. This choice, in fact, makes the wave function, and therefore the matrix elements H_{ij} and S_{ij}, independent of the energy eigenvalues that can be calculated allowing for a solution of the secular equation (16) by standard diagonalization techniques. The output charge density is then calculated from the summation over all the occupied energy states ε_i over the entire Brillouin Zone (B.Z.) as:

$$\rho(\mathbf{r}) = \sum_{i=1}^{n_{occ}} \int_{B.Z.} d\mathbf{k} \Psi_i^*(\mathbf{k}, \mathbf{r}) \Psi_i(\mathbf{k}, \mathbf{r}). \tag{17}$$

In order to consider a general form for the potential (i.e. without any shape approximation), we use an expansion for the effective potential (V_{eff}) which is analogous to the one chosen for the wave function and calculate it by solving Poisson's equation. The total potential is:

$$V_{eff}(\mathbf{r}) = V_{mt}(\mathbf{r}) + V_{warp}(\mathbf{r}) + V_{ns}(\mathbf{r}), \tag{18}$$

where V_{mt} is the spherical part of the potential inside the muffin-tin spheres, V_{warp} is the potential in the vacuum and interstitial regions (including non-constant terms) and V_{ns} are the non-spherical terms of the potential inside the spheres. The representation chosen is, therefore:

$$V(\mathbf{r}) = \begin{cases} \sum_i e^{i\mathbf{G}_i \cdot \mathbf{r}} V_{G_i} & \text{if } \mathbf{r} \in \text{interstitial} \\ \sum_{lm} V_{lm}(r) Y_{lm}(\hat{r}) & \text{if } \mathbf{r} \in R_s \\ \sum_q V_q(z) e^{i(\mathbf{K}_q)\cdot \mathbf{r}} & \text{if } \mathbf{r} \in \text{vacuum.} \end{cases} \tag{19}$$

The numerical solution of the Poisson's equation is particularly difficult due to the muffin tin regions where the core charge is rapidly oscillating; however, since the potential in the interstitial region depends only on the multipole expansion of the charge inside the muffin-tin, it is possible to construct a well converged Fourier expansion of the muffin-tin charge density that has the same multipole expansion as the original one. The Poisson equation can now be easily integrated and gives the exact potential in the interstitial and on the sphere boundaries. By using the value of the potential on the spheres as a boundary condition, the problem is now reduced to solving a differential equation with known boundary conditions. The Poisson equation can be exactly solved in each sphere to obtain the potential due to the real charge. This procedure, developed by Weinert [6] and implemented into the FLAPW method [1], allows the solution of the Schrödinger equation for a general potential without any shape approximation to the potential itself nor to the charge density.

Within local density functional theory, the one-electron equation (4) has to be solved self-consistently in the charge density and the potential; this implies an iterative procedure that from a guessed input charge density, generates the potential by solving the Poisson's equation. The output charge density is obtained from the solution of the Kohn–Sham equation (4); the input and output charge densities are then mixed through an attenuated feedback method to obtain a new input charge, and the procedure is repeated until good convergence is obtained.

It is important to point out that our calculations include the contribution to the potential coming from all the electrons in the system, including the core electrons. In fact, a fully relativistic Dirac equation is solved for the core states at each iteration, taking into account the variations of the total potential inside each atom caused by the readjustment of the valence charge. This is of crucial importance to correctly reproduce the ground state properties of metals and materials that have shallow closed core shells. The valence charge, on the other hand, is treated semi-relativistically; spin-orbit coupling is usually either neglected or evaluated in a perturbative approach once self-consistency is reached.

3 Free-electron Interlayer States and Related Surface States in Layered Materials

3.1 Electronic Interlayer States in Graphite and Graphite Intercalation Compounds

Until recently, experimental and theoretical studies of pure graphite and alkali-metal graphite intercalation compounds (AGIC) had provided the following commonly accepted description of their valence energy band structures. [7] (i) In graphite the sequence of bands with increasing energy is σ and π bonding, π and σ antibonding, with the Fermi energy, E_F, in the middle of the π bands. (ii) In AGIC's, an additional strongly dispersed band is present close to E_F. It is called the "alkali band" since it is thought to arise from alkali s electrons. In LiC_6 its bottom is ~ 1.7 eV above E_F while in KC_8 its exact location relative to E_F is still not well established. [7,8,9] We show here that this traditionally accepted energy band structure of graphite is incomplete above E_F and that the interpretation of the "alkali band" in AGIC's is incorrect. [10] We find that the AGIC *additional* band is actually an interlayer free-electron band which pre-exists above E_F in pure graphite.

We have studied three model two-dimensional systems. (i) A Li layer sandwiched between two graphite layers (C_6–Li–C_6 system) with the same crystallographic arrangement as in bulk LiC_6, (ii) the same system but without the alkali atoms (C_6–C_6 system) for several interlayer distances, and (iii) a graphite monolayer (C_6). Electronic energies and wavefunctions were obtained with the all-electron local-density full-potential self-consistent linearized augmented plane-wave (FLAPW) method for thin films. [11]

The C_6–Li–C_6 system has D_{6h} symmetry and its unit cell contains twelve C atoms and one Li atom. Its band structure at the Γ point (Fig. 1a) is consistent with existing energy bands for LiC_6. [12] The levels below E_F correspond to the σ and π bonding states of graphite. Due to the presence of the C_6 layers, these levels appear as pairs of states, one symmetric and one antisymmetric with respect to the Li plane. The energy separations are larger for π states than for σ states. Let us consider the $\Gamma_1^+ - \Gamma_2^-$ pair of empty states found just below vacuum level [13]. Their charge densities are given in Fig. 2a and Fig. 2b. The Γ_2^- state (Fig. 2a), which is located 3.2 eV above E_F, has 97% of its charge in the vacuum region. It corresponds to a surface state of a LiC_6 crystal with finite thickness and has not been found in bulk calculations. The Γ_1^+ state is located 2.2 eV above E_F , in agreement with the absolute energy of the bottom of the "Li-band" of bulk LiC_6. [14] 38% of its charge is in between the two graphite layers (Fig. 2b).

The amount of interlayer charge of the Γ_1^+ state is rather low for a state supposed to originate from Li 2s orbitals. Furthermore, this charge is strongly non-spherical around the Li atoms (the s-like charge is less than 1% within the Li APW sphere with radius 1.36 a.u.). We have therefore inspected the logarithmic derivative of the s radial solution within the Li APW sphere and found that the Li 2s orbital actually gives rise to a resonance at higher energy, above vacuum zero. This high energy location of Li $2s$ states is not surprising if one considers that the average radius of the Li $2s$ orbital (3.9 a.u.) is about half the carbon interlayer distance (3.5 a.u.). Hence, the orthogonality repulsion with lower lying σ and π graphitic states will raise considerably the energy of the Li $2s$ states and will also reduce their localization. FLAPW studies of a C_8–K-C_8 sandwich yielded similar results. The K orbitals give rise to a resonance just above vacuum zero, [15] while the Γ_1^+ state is at lower energy, just below E_F. These results show that the nature of the Γ_1^+ state close to E_F in C_6–Li–C_6 and C_8–K–C_8 is more complex than the generally accepted picture of a predominantly alkali valence state.

As a decisive test, we removed the Li atoms from C_6–Li–C_6, keeping the carbon sites unchanged and studied the resulting system consisting of two graphite layers with *AA* stacking. Its unit cell contains four C atoms, but in order to make the comparison with C_6–Li–C_6 easier, we considered the three times larger unit cell corresponding to a C_6–C_6 double layer. The backfolded energies at Γ (Fig. 1b) are very similar to those of the C_6–Li–C_6 sandwich, apart from the fact that some splittings have disappeared because of the higher translational symmetry. The two bands above E_F are still there. Hence their existence is independent of the Li atoms. The Γ_1^+ state now lies somewhat higher in energy than in C_6–Li–C_6. Its charge density (Fig. 2c) shows that it is an interlayer state which is not localized to the carbon atoms but which exhibits free-electron character in the planes parallel to the layers.

A question arises concerning the bonding character of the state in the direction perpendicular to the layers. Is the electron confined between the two layers, or is it already bound to a single graphite layer? We have therefore computed the band structure of a graphite monolayer, keeping the in-plane lattice parameter unchanged. Again, for an easier comparison with C_6–Li–C_6, we consider a C_6 monolayer whose backfolded energies at Γ are displayed in Fig. 1c. The bound state is seen to exist for a single graphite layer. Its charge density (Fig. 2d) shows again that it is not localized to the C atoms but is rather bound to the neutral graphite monolayer by its attractive short-range potential. Since the potential extends symmetrically from the layer, two states are actually bound to C_6, one symmetric and one antisymmetric. The Γ_1^+ interlayer state found for C_6–C_6 is the bonding combination between the two Γ_1^+ single-layer states. The Γ_2^- state of C_6–C_6 (which becomes a surface state when the slab thickness is increased) is simply the antibonding partner of the same Γ_1^+ single-layer state. The antisymmetric state bound to the C_6 monolayer gives rise to resonances of the C_6–C_6 system.

We have also carried out one-dimensional model calculations which show that the bonding and antibonding interlayer states found in C_6–C_6 and C_6–Li–C_6 depend on the strength of the planar averaged attractive electrostatic potential U_e of the graphite layer, and on the distance d between carbon planes, but not on the crystallographic structure of the system. In particular, we have shown [10] that interlayer states also exist at the d value of pure graphite regardless of the particular layer stacking. We have therefore predicted the existence in graphite of a previously unrecognized band in the range of energies of the π

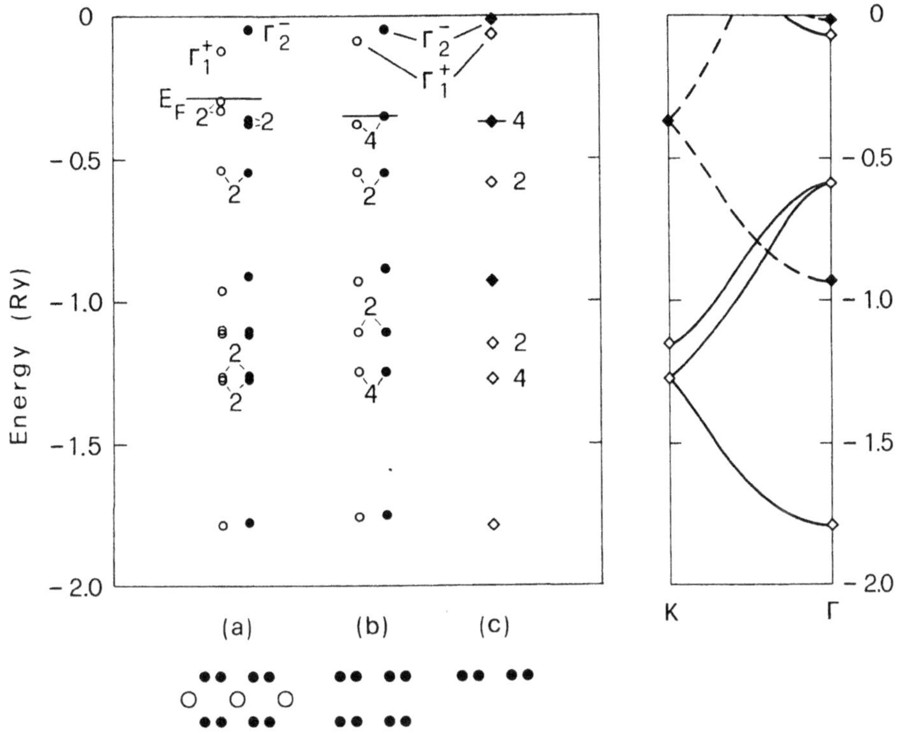

Figure 1. Band eigenvalues at Γ for (a) C_6–Li–C_6, (b) C_6–C_6 and (c) C_6 monolayer. Open (solid) symbols represent symmetric (antisymmetric) states with respect to z-reflection symmetry. Numbers give the degeneracy of the levels. The Fermi energies are indicated by horizontal lines. Crystallographic structures are schematically given at the bottom. The energy bands for a graphite monolayer along the $\Gamma - K$ direction are given at the right side.

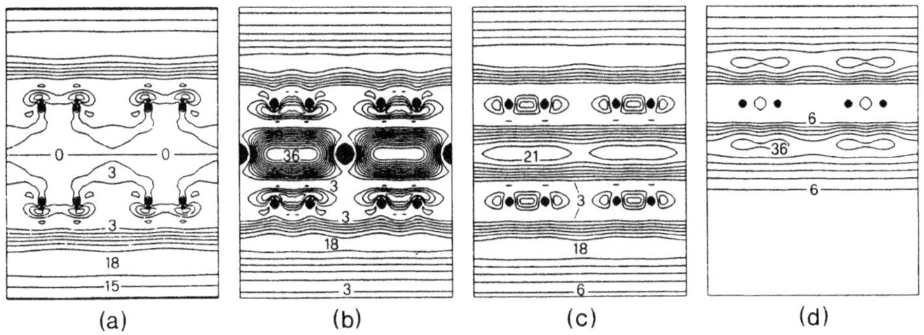

Figure 2. Charge-density contour plots for the Γ_2^- state of (a) C_6–Li–C_6 and the Γ_1^+ state of (b) C_6–Li–C_6, (c) C_6–C_6, and (d) C_6. Contours are given in multiples of 3×10^{-3} electron/Å^3 in (a), (b) and (c), and of 6×10^{-3} electron/Å^3 in (d).

antibonding states. Apparently such an interlayer band could not be detected in previous LCAO band structure calculations since they were based only on $2s$ and $2p$ carbon orbitals. [16] We find this band to be clearly present (but not interpreted) in the band structures obtained by pseudopotential [17] and cellular [18] methods. This band has apparently also escaped definite experimental detection, until recent inverse photoemission [19,20] and secondary-electron emission spectroscopy measurements. [20,21] The main reason is that the relevant states are confined to the interlayer region and therefore have small overlap with various kinds of filled levels which are initial states for excitation processes (*i.e.,* $1s$ C states for X-ray absorption and σ and π bonding states for UV reflectivity measurements). In addition, dipole selection rules forbid transitions to this interlayer band at several points of the Brillouin zone.

The results presented here demonstrate that: (i) A new band is present in pure graphite just above E_F. It originates from interlayer bonding states which exhibit free-electron character in the planes parallel to the layers. (ii) The so-called *Li-band* in LiC_6 actually originates from interlayer bonding electron states orthogonalized to the Li $1s$ core (see Fig. 2b). The occurrence of this band is not correlated to the presence of the Li atoms. (iii) The Li $2s$ valence orbitals give rise in LiC_6 to electron states located at higher energy than those resulting from the interlayer bonding states in agreement with earlier LCAO results. [17] (iv) Conclusions similar to (ii) and (iii) also apply to K intercalated graphite as well as to other AGIC's.

Finally, we lay stress on the fact that the existence of interlayer states is not restricted *a priori* to AGIC's. They may occur in other types of GIC's, as well as in other layered compounds provided that the layers are sufficiently electronegative. Some examples will be discussed in Section III.3.3 and in Section IV.

3.2 Electronic Surface States in Layered Materials: Graphite

Inverse photoemission measurements by Fauster *et al.* [19] have provided the first detailed information on low-energy unoccupied states of graphite. Two kinds of features have been observed: (i) Structure arising from states with a strong dispersion in the direction perpendicular to the basal plane (c-axis). Such states, described by several authors [10,17,18,22] were discussed in Section III.3.1 and interpreted by us as three-dimensional interlayer states [10]. (ii) Structure at lower energy (3.6 eV above the Fermi energy E_F), with negligible c-dispersion. The question was raised [19] whether feature (ii) is due to indirect transitions into the bottom of a bulk band or to the presence of a surface state.

In this section, we present a consistent theoretical picture of surface states in graphite. We find unoccupied surface states 3.8 eV above the Fermi level, which are split off from the bulk free-electron interlayer band. [23] They provide a convincing interpretation of the inverse photoemission results of Fauster *et al.* [19]. This proved to be the first report of surface states in layered materials, where bands with strong c-dispersion, from which surface states could derive, are not a common feature.

We have studied the formation of a surface state for a sequence of isolated graphite thin films with an odd number N of layers ranging from 1 to 25. Considering that the interlayer coupling in graphite is weak, we assume bulk graphite lattice parameters ($a = 4.64875$ a.u. and $c = 6.33061$ a.u.) and AA stacking for all systems. The existence of surface states should not be affected by these assumptions. Electronic energies and wave functions of the systems with $N < 5$ have been computed using the FLAPW method for thin films [11]. Thicker slabs and the reference infinite crystal have been studied with a model potential derived from the FLAPW results.

As expected, the characteristic occupied σ- and π-bonding states are obtained for the one-, three-, and five-layer systems and will be of no further concern here. The eigenvalues at Γ of all unoccupied states below the vacuum zero are displayed in the (lower) right side of Fig. 3. Going from 1 to 5 layers, the two lowest eigenvalues, which have opposite z-reflection symmetry, converge to the energy of a surface state. Their nature as surface states becomes evident from their single-state charge densities given in Fig. 4 for the monolayer, and in Figs. 5(a) and 5(b) for $N = 5$. The two states shown in Figs. 5 (a) and (b) have a high charge density outside the surface layer. However, for this film thickness they still interact, as is indicated by the difference in their energies (-70 mRy versus -62 mRy) and wave functions. Fig. 5(c) shows the charge density of the next state (-5 mRy) which is even under z-reflection. This state has interlayer bulk character and should correspond to the bulk band-edge. Thicker slabs are necessary to obtain (i) negligible interaction between surface states and (ii) information on bulk states in the same energy range. Fortunately, the states of interest have strong free-electron-like character parallel to the atomic planes (cf. Figs. 4 and 5), and thus lend themselves to a simplified treatment by a model potential; a FLAPW treatment of slabs consisting of more than five layers would be prohibitively time-consuming.

Our one-dimensional model is defined from the three-dimensional FLAPW potential and wave functions calculated for the monolayer. These are projected through x–y planar averaging onto the 1-dimensional electrostatic potential $U_e(z)$ and wavefunctions $\Phi^{\pm}(z)$ (the $\pm$ refers to parity under z-reflection). Then non-local norm-conserving monolayer potentials $U^{\pm}$ are obtained by constructing pseudo wave functions from the functions $\Phi^{\pm}$ near the atomic planes and inverting the Schrödinger equation [24]. We write the effective potentials $U^{\pm}$ as the sum of the local potential U_e and an orthogonality repulsion part: $U^{\pm} = U_e + U^{\pm}_{orth}$. The potentials U^{+}, U^{-} and U_e are displayed in the insert on the right side of Fig. 3. Potentials of multilayer films cannot be constructed as superpositions of these monolayer model potentials, due to mutual layer-layer polarization which modifies U_e in-between the layers. Planar averaged FLAPW potentials for three- and five-layer slabs show, however, that the interior electrostatic potential U_i (cf. Fig. 3), though different from U_e, does not depend on the number of layers N, and that the exterior potential is the same as for the monolayer. Therefore, the effective potentials for multilayer slabs are taken as the sum of an electrostatic part equal to U_i inside the film and U_e outside, and of a non-local orthogonality potential equal to the superposition of $U^{\pm}_{orth}$ centered on each

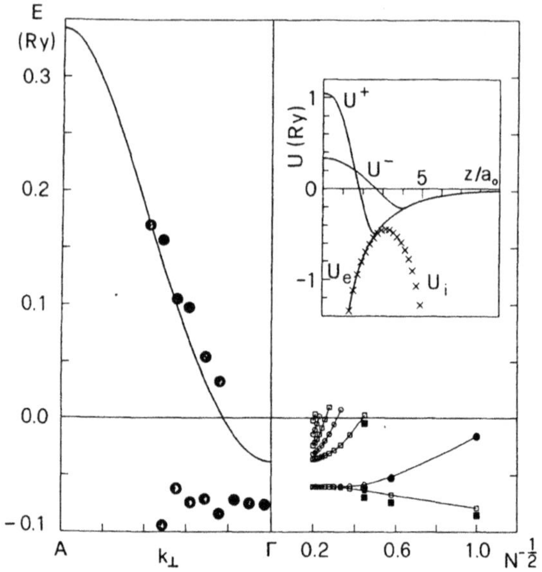

Figure 3. Right panel: band energies at Γ of the empty states below vacuum zero for N-layer graphite slabs. Even (odd) states are indicated with squares (circles). Solid (open) symbols indicate FLAPW (model) results. Left panel: $\Gamma - A$ dispersion of the interlayer band for the infinite model graphite crystal. Solid circles represent data by Fauster *et al.* [19]. Insert: model potentials U_e, U_i, U^+ and U^- derived from planar averaging of FLAPW results.

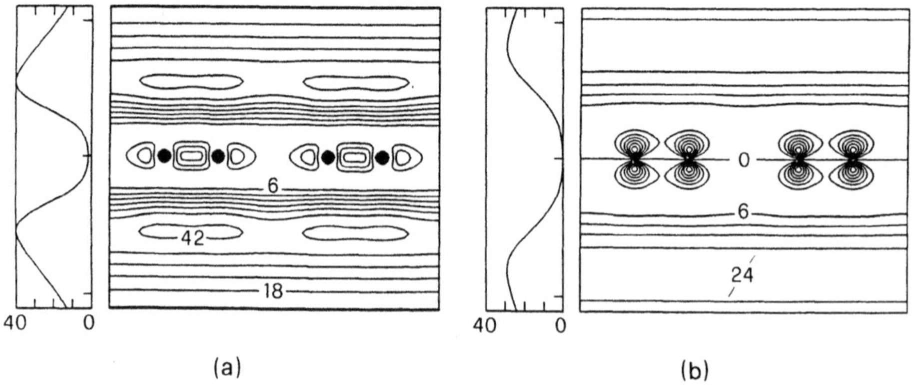

Figure 4. Contour maps of the FLAPW charge densities of the (a) even and (b) odd empty state below vacuum zero for a graphite monolayer. Charge densities are in units of 10^{-3} e/Å^3 and are represented in a plane perpendicular to the layer. The linear plots give the corresponding model densities. The tick marks on the vertical axis indicate multiples of the c parameter.

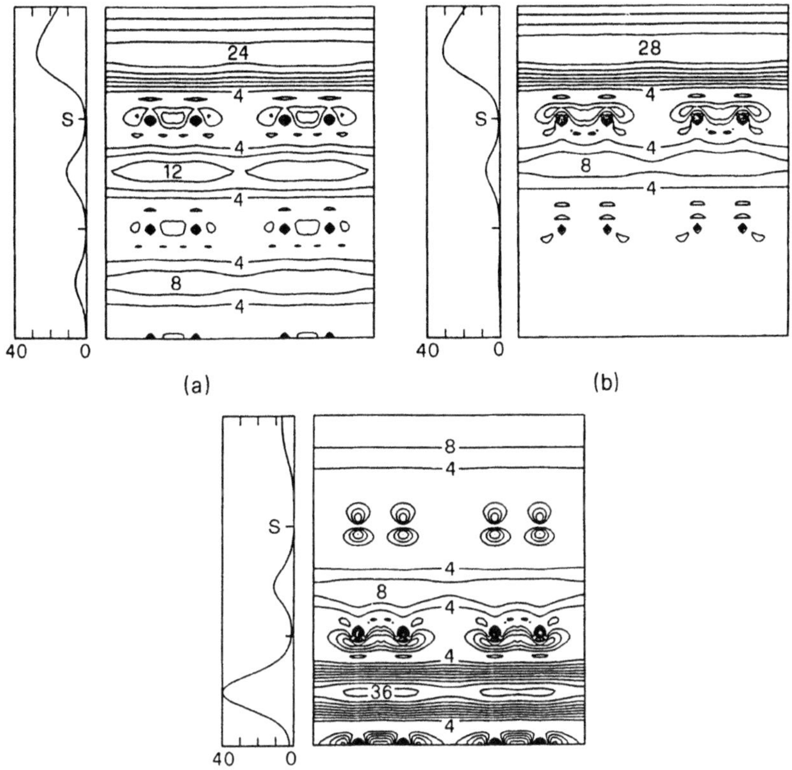

Figure 5. FLAPW charge densities of the empty states below vacuum zero for a five-layer graphite slab. (Same notation and units as in Fig. 4.) (a) Lower even state, (b) odd state and (c) upper even state.

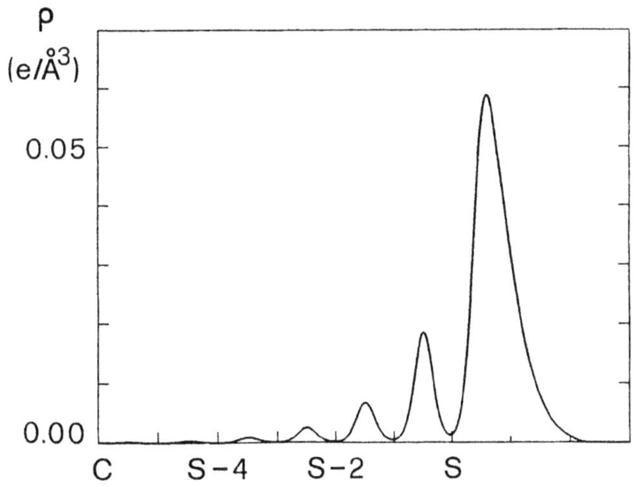

Figure 6. Charge density of the right-hand-side surface state for a thirteen-layer graphite slab. The surface layer is indicated by S and the central layer by C.

layer site.

This non-local model potential has been used to study films with an odd number of layers ($N < 25$) and the infinite periodic system. We use plane waves as basis functions for the periodic system and uniformly distributed Gaussians of given width for the films. The model reproduces within 7 mRy the FLAPW energies (cf. Fig. 3). Comparison of their charge densities for $N = 1$ and 5 is displayed in Figs. 4 and 5. The resulting energies for $1 < N < 25$ and for the infinite system are collected in Fig. 3. The slab results (right-hand side of Fig. 3) clearly show (i) the formation of a pair of surface states for thick systems, and (ii) the condensation of levels at higher energies corresponding to the formation of the bulk interlayer band projected onto the two-dimensional Brillouin-zone. This result is confirmed by the infinite system calculation (solid line, left-hand side of Fig. 3), which shows the existence of a bulk band with large c−dispersion. The bottom of this band corresponds to the discretely spaced levels obtained for the finite films. For $N > 13$ the interaction between the surface states on the two surfaces of the slab is negligible and the eigenvalues of the even and odd partners become quasi-degenerate (their splitting is 0.1 mRy for $N = 13$). The eigenvalues are located in an absolute energy gap, so that they are true surface states at Γ. Fig. 6 shows the charge density of the surface state for $N = 13$; it extends about three layers into the bulk.

The experimental data of Fauster *et al.* [19] are reproduced in the left-hand side of Fig. 3, using E_F as reference energy. Comparison with our results shows that: (i) the experimental structure which is independent of photon energy (i.e. non-dispersed final states) corresponds to transitions into the surface state. (ii) in agreement with the findings of Fauster *et al.* [19], Maeda *et al.* [20], and Takahashi *et al.* [21], the experimental structure which is strongly dependent on photon energy corresponds to transitions into the highly dispersed interlayer band lying above the surface state.

To understand the existence of the surface state, one has to consider (i) the large extension in the c direction of the states bound to a single layer (Fig. 4 shows that their charge density is still large at distances from the layer of the order of the c parameter), and (ii) the strong orthogonalization potentials felt by an electron at each atomic layer. Orthogonalization effects on such extended wave functions are stronger in the interior than in the vacuum due to the additional orthogonalization to the layers which are adjacent to the surface layer. The states with lowest energy will then extend into the vacuum, where the effective potential is more attractive. The same argument explains why all bulk interlayer states are found at higher energies than the surface state. According to our results, the energy gap between surface and bulk states is ~ 0.3 eV.

The bulk interlayer band has a width of 0.38 Ry, *i.e.* much larger than that of a free-electron (0.25 Ry). Such strong dispersion originates from the non-locality of the effective model potential, which in turn is due to the different orthogonality conditions with respect to σ- and π-valence states. As a result, symmetric states (*e.g.*, Γ states) sample on the average a more attractive potential than antisymmetric states (e.g., A states). This non-locality considerably affects the band width. In fact, assuming a local orthogonality potential equal to U^{+}_{orth}, the band width reduces to 0.04 Ry.

Finally, it should be mentioned that the electronic states of the highly dispersed band have been interpreted [22] as Bloch combinations of empty $3s$ carbon orbitals. This interpretation, however, is in contradiction with the symmetry of our wave function at the A point. We find that the bulk state at A is odd under the reflection σ_h with respect to the atomic planes. It is impossible to construct such a state from a linear combination of atomic $3s$ orbitals, since the latter are even under σ_h.

Summarizing, we have predicted the existence and have investigated the origin of surface states in a layered material, graphite. Unoccupied surface states are associated with the presence of a strongly dispersed interlayer band, whose width originates from the non-locality of orthogonalization effects. The absence of orthogonality outside the surface plane leads to the formation of a surface state at lower energy. These findings provide a consistent interpretation of the data by Fauster *et al.* [19]. Surface states derived from interlayer bands could exist in other layered materials, their energies being controlled by the interlayer spacing and the strength of orthogonalization effects.

3.3 Electronic Interlayer States in Hexagonal Boron Nitride

Hexagonal boron nitride (h–BN) is a layered compound which is isoelectronic to graphite. Both crystals consist of a stacking of two-dimensional arrays with honeycomb structure, and are characterized by strong intralayer bonds and weak interlayer interactions. The electronic states are also similar and, in a simple LCAO picture, the sequence of their energy bands with increasing energy is σ and π bonding, π and σ antibonding, the Fermi energy E_F being in the middle of the π bands. The major difference between the two materials is that the π-bonding and π-antibonding bands overlap weakly at the Brillouin zone boundary in graphite, which is thus a semimetal, whereas these bands are separated by an energy gap of several eV in BN, which is therefore an insulator.

It has been mentioned in Section III.3.2 that the LCAO method with a minimal basis set does not provide the correct energy band structure of graphite above E_F. In fact, when a set of basis functions with sufficient variational freedom is used, additional unoccupied bands appear, corresponding to interlayer states which exhibit free-electron character parallel to the atomic planes, and result from bonding and antibonding combinations of surface states bound to single graphite layers. [10,23] The existence of interlayer states has been demonstrated for pure graphite using inverse photoemission [19,20] and secondary-photoelectron [20,21] spectroscopy. Because of the similarity between graphite and h–BN, the question arises whether interlayer states exist also in the latter material. Indeed, we have found that unoccupied interlayer states do exist in the lower portion of the conduction bands. [25] This result confirms our earlier finding for graphite and LiC_6, and indicates that, as expected, interlayer states appear to be a general feature of layered materials. [10]

Hexagonal BN has D_{6h} symmetry; its unit cell contains two formula units and extends over two consecutive layers. The experimental values of the lattice parameters at equilibrium are $a_0 = 2.504$ and $c_0 = 6.66$ Å, [26,27] which are very close to those of graphite. Electronic energies and wave functions of three-dimensional h–BN have been calculated with a bulk version of the FLAPW method. [1] The energy-band structure calculated at the equilibrium lattice parameters, shown in Fig. 7, yields a valence band dispersion in close agreement with all previously reported results (see *e.g.*, Robertson [28]). However, we find two additional conduction bands not yet reported in the literature for this material.

The first additional empty band is the lowest unoccupied state at Γ and presents the characteristics of an interlayer band. Indeed, as is clear from Fig. 7, this interlayer band has a strong parabolic behavior around its minimum at Γ, which is located at 4.7 eV above the top of the valence band at H, and is even under reflection with respect to the atomic planes. According to the notation of Robertson, [28] this state belongs to the Γ_1^+ representation. Since the unit cell contains two formula units, this band folds back at the upper hexagonal face of the Brillouin zone and gives rise to a second interlayer band at higher energy. The higher interlayer state at Γ (10.0 eV above the valence band top) is odd with respect to the reflection operation and has Γ_3^+ symmetry. The interlayer character of these Γ states is clearly evident in the single-state charge-density plots of Fig. 8. The structures close to the N nuclei are the results of the effects of orthogonalization to the occupied states. These plots indicate that most of the charge is situated in the interstitial region (the calculated values are 86% and 67% for the Γ_1^+ and Γ_3^+ states, respectively). The charge distributions presented in Fig. 8 are similar to those of the interlayer states in graphite, [10,22] apart from a weaker interlayer character (the calculated values in graphite are 92% and 74% for the Γ_1^+ and Γ_3^+ states, respectively). Apparently, the reason why interlayer states have not been reported earlier for h–BN is the use of insufficient basis sets of atomic orbitals in all previous calculations.

We have extended the study of the electronic properties of h–BN by investigating its surface properties. We have seen in Section III.3.2 that surface states originating from bulk interlayer states have been found experimentally and theoretically in graphite. [19,23] Since our bulk band structure shows that interlayer bands also exist in h–BN, we investigated the possible existence of surface states in this compound by performing one-dimensional model calculations on slabs with different thicknesses.

The results of the self-consistent FLAPW bulk calculations were used to construct a one-dimensional periodic model potential, in order to reproduce the k_z dispersion of the interlayer bands. The procedure followed in the derivation of the potential is analogous to the one described above for the graphite slabs, and details are given by Catellani *et al.* [29]. The soundness of our procedure is shown in Fig. 9, where we compare the results of the model for infinite one-dimensional periodic systems with FLAPW interlayer-band dispersions of the crystals with $c = c_0$ and $c = 1.75c_0$. In order to study surface properties, it is of course also necessary to extend the model potential into the vacuum region.

This method has been used to study BN-slabs with an odd number N of layers ($1 \leq N \leq 35$). The eigenvalues of bound states for these slabs are reported in Fig. 9.

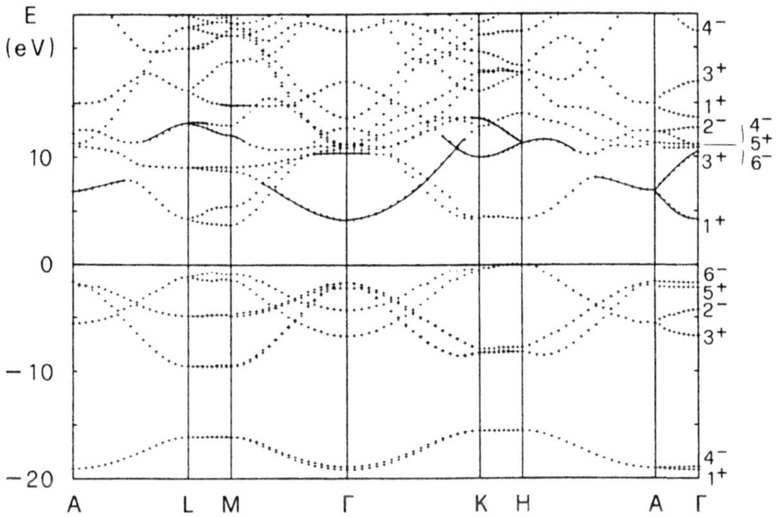

:e 7. Energy-band structure of hexagonal BN. Electron states at the zone center lassified following the notation used by Robertson [28]. Solid lines serve to identify iterlayer bands, which show several (anti)crossings with σ and π bands. The zero :rgy is taken as the top of the valence band (point H of the Brillouin zone). The ıum of the conduction band occurs at M.

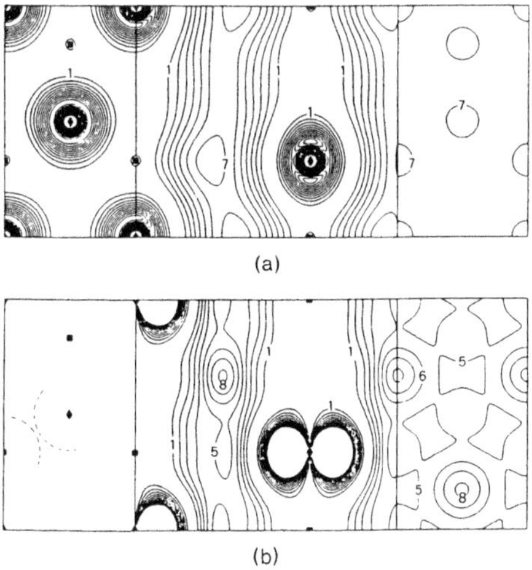

3. Charge-density contour plot for (a) the symmetric Γ_1^+ and (b) the antisymmetric rlayer states. The charge densities are displayed in the basal atomic plane (left-hand in the vertical plane through a B–N bond (central panel), and in the horizontal isecting the interlayer space (right-hand panel). Contour values are given in units :/Å^3, with subsequent contours separated by 0.01 e/Å^3. Diamonds (◇) and squares icate nitrogen and boron sites, respectively. The dotted lines in the left panel of cate the augmented plane-wave (APW) spheres used in this work.

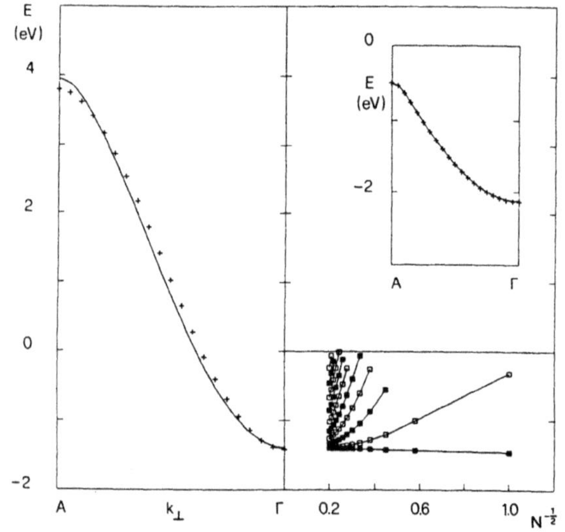

Figure 9. Right panel: energies of bound states in N-layer model BN slabs with interlayer distance $c = c_0$. Even (odd) states are indicated by solid (open) squares. Left panel: lowest energy band for the infinite one-dimensional model BN crystal, compared with three-dimensional FLAPW band energies of interlayer states (crosses) calculated along the $\Gamma - A$ direction for the crystal with $c = c_0$. The same comparison for the crystal with $c = 1.75\, c_0$ is given in the insert.

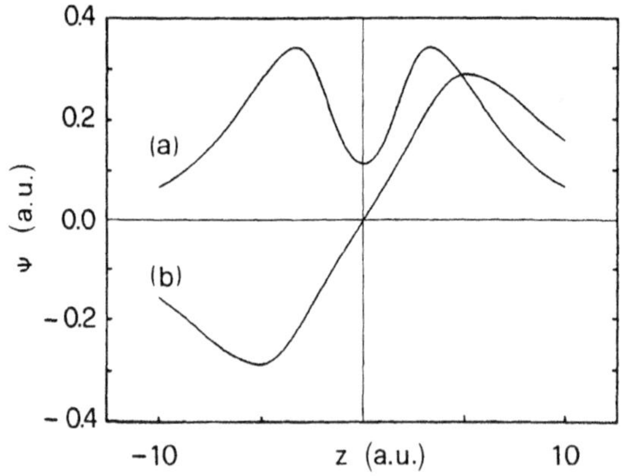

Figure 10. Wave functions Ψ of the symmetric (a) and antisymmetric (b) states bound to the isolated model BN monolayer. All values are in atomic units.

With increasing thickness, the energies converge towards the projected band structure of the infinite periodic system. Contrary to the case of graphite, [23] no surface state splits off at lower energy.

Further evidence that no surface state exists in h–BN is obtained by studying the evolution of the wavefunction of the ground state in slabs with increasing thickness. Two states, one symmetric (ground state), and one antisymmetric, are bound to the isolated monolayer. Their wave functions are displayed in Fig. 10. In Fig. 11, we give the charge density of the lowest bound state in the 9-layer and the 15-layer slabs. We see in the sequence $N = 1, 9, 15$ that the ground state becomes closer and closer to a Bloch state, and its amplitude does not decay appreciably inside the slab.

The absence of surface states in $h--$BN, despite its close similarity with graphite, can be related to the more pronounced atomic character that bulk interlayer states have in the former case. In fact, for graphite, the probabilities to be in the atomic spheres are 8% and 26% for the symmetric Γ_1^+ and the antisymmetric Γ_3^+ interlayer states respectively, while the corresponding values for BN are 14% and 33% (1% and 0% in the boron sphere, 13% and 33% in the nitrogen sphere, respectively; the larger amount of charge in the nitrogen spheres is due to the greater electronegativity of this element compared with boron). Since bulk interlayer states in h–BN are more atomic in nature, they are less likely to give rise to surface states, which extend considerably into the vacuum region.

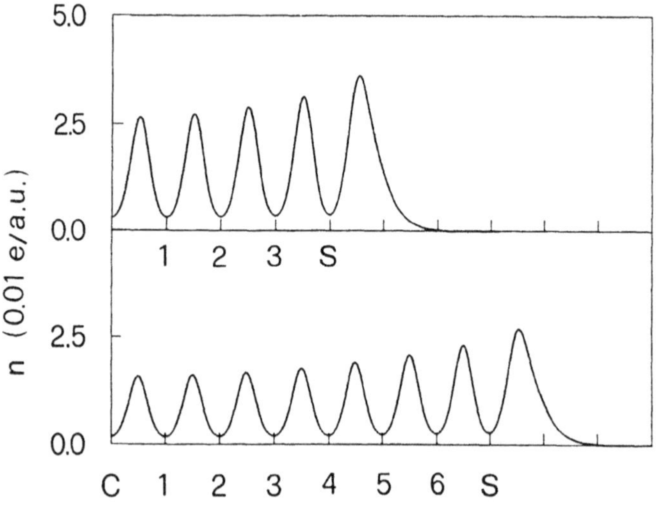

Figure 11. Charge density of the ground state in a 9-layer (upper) and a 15-layer (lower) model BN slab, in units of 0.01 e/a.u. The surface layer is indicated by S and the central layer by C.

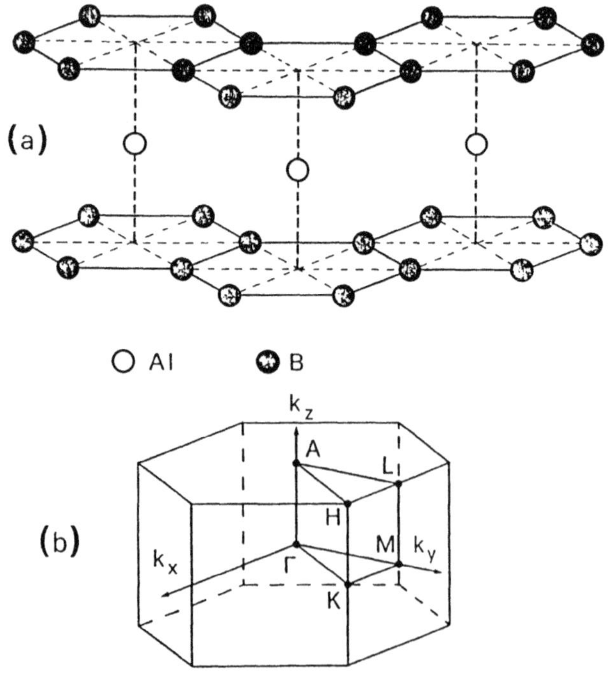

Figure 12. (a) Crystallographic structure of AlB_2 and (b) corresponding Brillouin zone.

4 Compounds with the AlB_2 Structure

The family of compounds with the hexagonal C32 structure (also known as the AlB_2 structure) contains many members [26,30] with formula MX_2, where X = B or Ga and M can be a simple metal (di- or tri-valent), a transition metal or a rare earth element. Among these systems, the transition metal diborides have properties such as high melting temperature, hardness and chemical stability which make them important for technological applications at high temperatures [31]. All the members of this family are metallic [32,33].

The MX_2 structure is closely related to that of graphite and of intercalated graphite systems [34]. The X atoms form graphite-like two-dimensional networks with AA stacking and which alternate with hexagonal layers of M atoms. The latter occupy the centers of all the hexagonal prisms defined by the X atoms. According to a model proposed for the bonding in these compounds [35], the cohesion of the planes of the X atoms is due to the σ bonds between the X sp^2 hybrids, and the MX bonding comes from the overlap between the X p_z and the metal orbitals. This model would justify a charge transfer to the X atoms.

The electronic properties of the compounds of the AlB_2 family have received only limited attention untill now, which was mostly concentrated on transition metal diborides. From the experimental side, the x-ray photoemission spectrum [36], the Fermi surface [37] and the optical properties [38] were studied for ZrB_2. The x-ray emission and absorption spectra were investigated for TiB_2 [39,40] and other diborides [41,42]. Measurements of the specific heat at low temperatures were performed for several transition metal diborides by Tyan et al. [43] and by Castaing et al. [44]. Juretschke and Steinitz [33] measured the Hall coefficients and resistivity for a series of transition metal diborides; resistivity measurements were also performed by Dempsey [45]. Silver and Bray [46] studied the nuclear magnetic resonance on TiB_2. Great attention was devoted to the problem of charge transfer in these compounds; in particular there are two conflicting models for the transition metal diborides; some experiments are interpreted in terms of a charge transfer from boron to the metal [33,43,45], while other experiments seem to indicate the opposite conclusion [46]. According to the latter, the B^- negative ions form networks isoelectronic to graphite.

The band structure has been calculated for only a few systems. Among the simple metal diborides, BeB_2, MgB_2, and AlB_2 have been studied by the OPW method [42,47], and AlB_2 by the LCAO method [48]. These studies show the presence, in these materials, of σ and π bands similar to those of graphite. Among the transition metal diborides, ZrB_2 and CrB_2 have been studied by the APW and KKR methods [36,49,50] and the first row transition metal diborides ScB_2, TiB_2, CrB_2 and MnB_2, by the LCAO method [51]. Graphite-like σ and π bands were also found in these compounds. Regarding the problem of charge transfer, the self-consistent calculation on ZrB_2 [50] indicates that in this material a large fraction of the electronic charge is dispersed in the interstitial regions, excluding a severe charge inbalance. The electronic properties of $CaGa_2$ were investigated with the empirical pseudopotential method [52]. Interlayer states, similar to those found in graphite, LiC_6 [10] and BN [25] were also found in this compound.

In the following, we will provide a consistent description of the electronic properties of the compounds with AlB_2 structure, as obtained using the FLAPW method [1]. In particular, we will examine BeB_2, AlB_2, $CaGa_2$, and TiB_2, concentrating our attention on the chemical bonding, charge transfer and on the existence in these compounds of interlayer states. For a more detailed discussion see Ref. [53].

4.1 Crystallographic Structure

The crystallographic structure of AlB_2 [26] (space group D^1_{6h}) is illustrated in Fig. 12, which shows its similarity with that of the graphite intercalated compounds. The unit cell contains one Al atom at the origin of the coordinates and two B atoms in the positions $\mathbf{d}_1 = (\frac{a}{2}, \frac{a}{2\sqrt{3}}, \frac{c}{2})$, $\mathbf{d}_2 = -\mathbf{d}_1$. The values of the lattice parameters a and c for the MX_2 compounds studied are given in Table 1, together with the sphere radii used in the FLAPW calculations.

4.2 Simple Metal Diborides

Figures 13 and 14 show the energy bands of AlB_2 and BeB_2, respectively, along the main symmetry lines of the Brillouin zone. The similarity between these band structures and that of two-dimensional graphite (see Fig. 1 and Ref. [22]) is evident, with three σ-bonding bands and, at higher energy, the π-bonding and antibonding bands. This analogy with graphite, together with the symmetries of Bloch states formed by the atomic orbitals, allows one to establish a correspondence between energy bands and atomic orbitals. The group of bands with symmetries Γ_1^+ and Γ_6^+ at the Γ point are formed by the bonding combinations of the $2s$ (Γ_1^+) and $2p_x$, $2p_y$ (Γ_6^+) orbitals of the two B atoms per unit cell. The corresponding antibonding combinations form the Γ_3^- and Γ_5^- bands, with s and p character respectively at Γ. The bonding and antibonding combinations of the B $2p_z$ orbitals form the π bands, with symmetries Γ_2^- (bonding) and Γ_4^+ (antibonding) at the Γ point. The Al s states give rise to Γ_1^+ bands at the Γ point; they will therefore participate in the formation of the three Γ_1^+ bands. Since the interlayer states also transform according to this symmetry, these bands will be examined in detail by analyzing the corresponding wave functions.

The dispersion of the energy bands along k_z ($\Gamma - A$ direction in Fig. 13) is related to the inter-plane (Al–B and B–B) interactions. The σ-bonding bands, corresponding to states localized in the B planes, are weakly dispersed along this direction. Relatively larger dispersion is shown by the s-bonding Γ_1^+ band, due to the interaction between the s orbitals of Al and B. The π-bonding band is strongly dispersed along the k_z direction. This dispersion is much larger than the corresponding one in graphite (6.1 eV in AlB_2 and about 2 eV in graphite). This is probably due to the larger inter-plane overlap between p_z orbitals

Table 1. Values of the lattice parameters a and c of the compounds studied (after Ref. [26]). The radii of the atomic spheres used in the FLAPW calculations are also given. All values are in atomic units.

MX_2	AlB_2	BeB_2	$CaGa_2$	TiB_2
a	5.688	5.558	8.155	5.728
c	6.166	5.425	8.155	6.106
R_M	2.57	2.27	3.27	2.42
R_X	1.64	1.60	2.35	1.65

Table 2. Total self-consistent valence charge contained inside the APW spheres of the M and X atoms and in the interstitial region (ρ_I). The values of ρ_X refer to a single X sphere. The values in parenthesis give the decomposition for the overlapping-atom charge densities.

MX_2	AlB_2	BeB_2	$CaGa_2$	TiB_2
ρ_M	2.32	1.65	1.19	2.83
	(2.82)	(1.93)	(1.38)	(3.24)
ρ_X	1.75	1.66	1.74	1.76
	(1.53)	(1.47)	(1.69)	(1.55)
ρ_I	3.18	3.03	3.33	3.64
	(3.12)	(3.13)	(3.24)	(3.64)

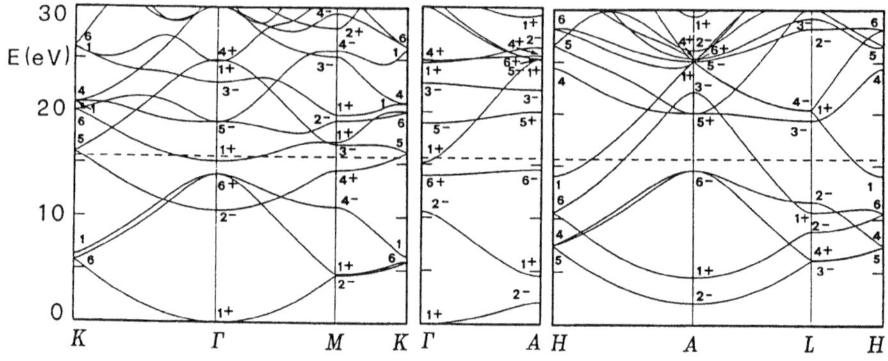

Figure 13. Self-consistent energy bands of AlB_2 along the main symmetry axes. The Fermi level is indicated by a dashed line. Symmetry notations follow G.F. Koster, *Properties of the Thirty-two Point Groups* (M.I.T. Press, Cambridge, Massachusetts).

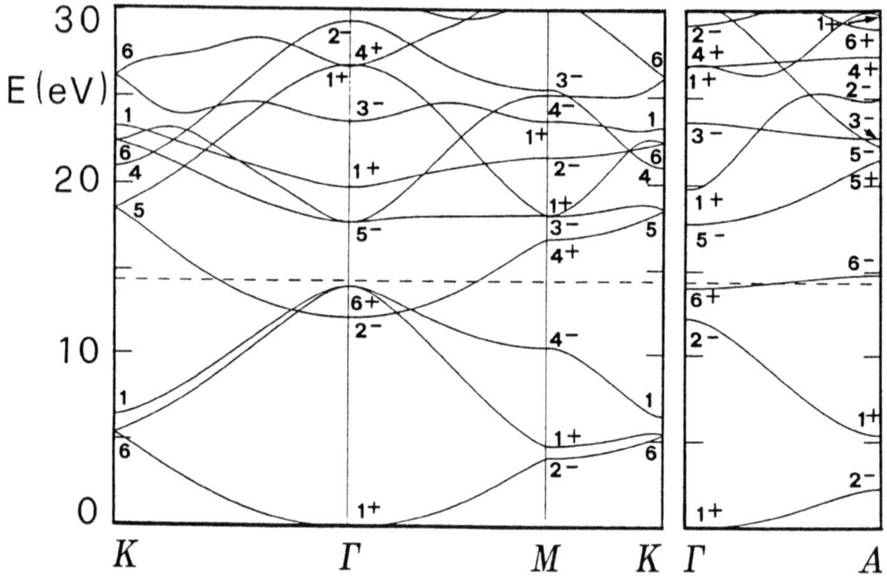

Figure 14. Self-consistent energy bands of BeB_2 along the main symmetry axes. The Fermi level is indicated by a dashed line.

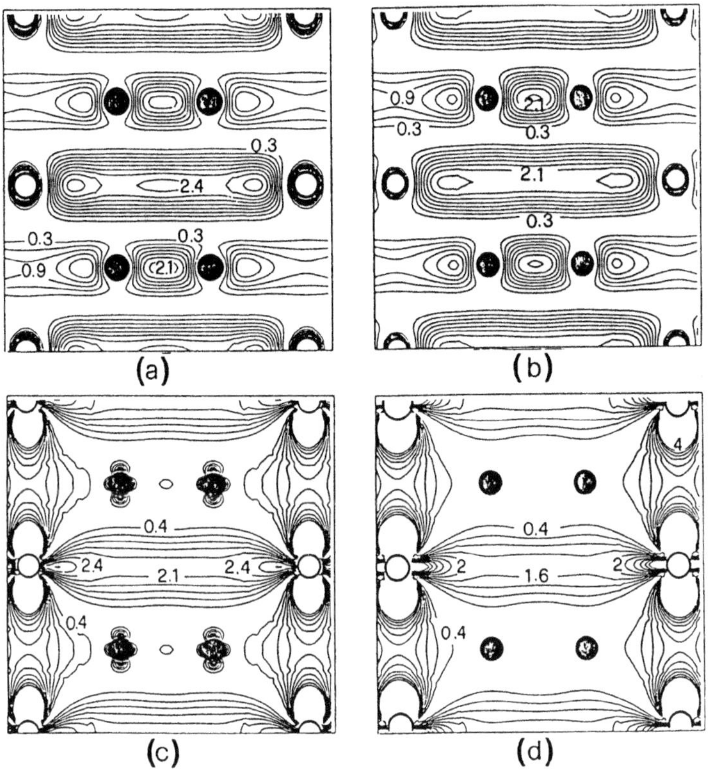

Figure 15. Contours plots of the charge density for the interlayer Γ_1^+ state in (a) AlB_2, (b) BeB_2, (c) $CaGa_2$, and (d) TiB_2. The density is represented on a plane parallel to the c-axis and passing through the X–X bonds. M and X atomic positions are indicated by open and full circles, respectively. Contours are given in units of electrons/unit cell.

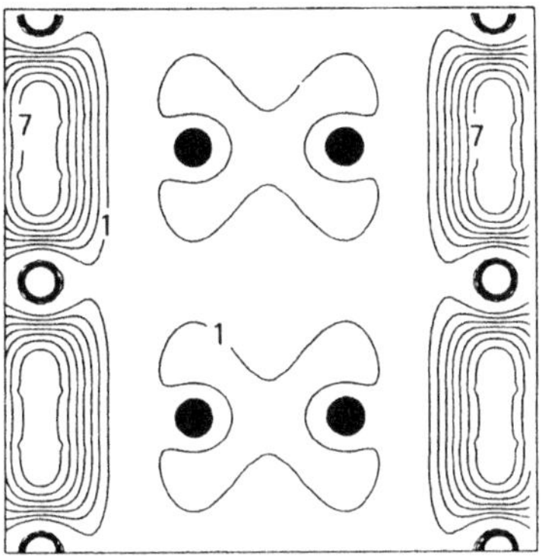

Figure 16. Contours plots of the charge density for the higher energy Γ_1^+ interstitial state of BeB_2. Same geometry, units and atomic-site symbols as in Fig. 15.

in AlB_2. The distance between the B planes in this compound is in fact $d = c = 6.18$ a.u., to be compared with $d = c/2 = 6.34$ a.u. in graphite; correspondingly, the average radii of the $2p$ atomic orbitals of B and carbon are 2.205 and 1.715 a.u., respectively [54]. This effect is, however, strongly enhanced by the presence of the Al atoms. The B $2p_z$ – Al $3s$ hybridization is in fact symmetry forbidden at the Γ point but it is allowed along the Γ–A direction. The analysis of wave functions actually shows a large mixing as one moves along the Γ_2^-–A_1^+ band. In order to investigate more closely the effect of the Al atoms, we performed a self-consistent calculation with the same crystallographic structure but without the Al atoms. Correspondingly, the dispersion of the Γ_2^-–A_1^+ band decreases from 6.1 to 4.1 eV.

The interaction between the $2s$ orbitals of Al and B is justified by their large overlap. The sum of the corresponding average radii of these orbitals (5.23 a.u. [54]) is in fact larger than the distance between the Al and B atomic sites in the AlB_2 crystal (4.50 a.u.). The interaction will consequently raise the energy of the Al s states. Let us now examine the Γ_1^+ level just above the Fermi energy. Its charge distribution is highly non-spherical around the Al atoms; the s and d-like charges inside the Al muffin-tin spheres are 0.26 and 0.19 electrons, respectively. The contour plots of the charge density for this band, at the Γ point, are shown in Figs. 15(a) and 15(b) for AlB_2 and BeB_2, respectively, in a vertical plane passing through the B–B bonds and six Al (Be) atoms; this band corresponds mainly to interlayer states, orthogonalized to the Al s core states, similar to the interlayer states recently found in graphite and graphite intercalated compounds [10] as well as in hexagonal BN [25], and in $CaGa_2$ [52]. The third Γ_1^+ band also contains a contribution from the Al s states; the Al atomic spheres contain 0.25 s-like and 0.27 d-like electrons.

A third Γ_1^+ band appears at still higher energies in both AlB_2 and BeB_2. The contour plot of the charge density of this Γ_1^+ state of BeB_2 is shown in Fig. 16. A similar plot describes the corresponding Γ_1^+ state in AlB_2. In both materials the upper Γ_1^+ band corresponds to a different kind of interstitial state since it exhibits free-electron-like behaviour in the vertical channels defined by the hexagons of B atoms. This kind of interstitial state has been identified previously in $CaGa_2$ [52].

The energy band structure of BeB_2, represented in Fig. 14, shows strong similarities with that of AlB_2. There are, however, differences, mainly due to the presence of an extra electron per unit cell in AlB_2. The width of the valence band is about 1.5 eV larger in AlB_2 than in BeB_2. This fact introduces important modifications to the nature of the states at the Fermi level: (i) the σ-bonding bands, which are completely filled in AlB_2 now cross the Fermi level near the A point in BeB_2; (ii) the Γ_1^+ interlayer band, which is partially filled in AlB_2, is empty in BeB_2; (iii) the π-bonding states are only partially occupied in BeB_2, while they are almost completely occupied in AlB_2. The π-antibonding states are partially filled in both materials. The Γ_1^+ interlayer band is found at higher energy in BeB_2, where it lies above the Γ_5^- p_σ antibonding states. We also notice that the dispersion of σ-bonding bands along the k_z direction is slightly larger in BeB_2 than in AlB_2, probably due to the smaller interplane spacing in BeB_2.

In Figs. 17(a) and (b) we show the density of states of AlB_2 and BeB_2. In AlB_2, the Fermi level, which lies just above a minimum of the density of states, cuts the Γ_1^+ interlayer band close to the Γ point and the π-bonding and antibonding bands. The occupied states are dominated by the σ-bonding bands, formed by the sp^2 B hybrids. The π states and the Al orbitals are partially filled. In BeB_2 the interlayer band is completely empty while the p_σ bonding band crosses the Fermi level around A.

The electronic structure of AlB_2 has been calculated by Armstrong et al. [48] with the LCAO–MO method and by Tupitsyn [47] with the OPW method. The band structure calculated by Armstrong et al. shows important differences with respect to our results; in particular, some bands are missing in their study, which can be identified as the interlayer bands. It is not surprising that bands with interlayer character are not well described in an LCAO calculation, since a limited basis with s and p atomic orbitals cannot reproduce inter-atomic, free-electron states. The ordering and the shape of the energy bands calculated by Tupitsyn [47] are similar to ours. The upper Γ_1^+ band is however missing in Tupitsyn's work, and the lower band with the same symmetry is found at slightly higher energy in his study. The width of his valence band is about 2 eV larger than that obtained in our work. Tupitsyn [47] also reports the result of x-ray spectra on AlB_2 from which one can derive an estimate for the width of the valence band. The measured band-width is 15 eV, in good agreement with our results.

The energy bands of BeB_2 have been previously investigated with the OPW method by Tupitsyn et al. [42] and by Tupitsyn [47]. The shape of their valence band structure is in overall agreement with our study; some conduction bands are however missing in their study, namely the B σ-antibonding band (Γ_3^-) and the highest Γ_1^+ band. The full width of their valence bands is 15 eV, which is about 1 eV larger than our result, 14.4 eV.

The measured width of the B and Be K-emission bands [42] are 12.8 and 12.1 eV, respectively. Since the lowest lying s-like valence states are excluded by symmetry in such an emission process, only the width of the π-like valence bands must be considered; the theoretical valence band-width can therefore be compared with the experiment only through the calculation of the transition probabilities, which is outside the scope of the present work. In any case, our theoretical prediction for the width of the B K-emission band would be in between the full width of the valence band (14.4 eV) and the difference between the Fermi level and the lowest lying π-like state at the zone boundary, which is about 10 eV. Our results thus compare well with the experimental results.

4.3 $CaGa_2$

Figure 18 shows the dispersion of the energy bands of $CaGa_2$ along the main symmetry lines of the Brillouin zone and confirms also for this compound the presence of σ-bonding bands. In the conduction region, the energy bands of $CaGa_2$ are rather complex, as compared to those of simple metal diborides. We can however use symmetry arguments at the high symmetry points in order to associate the bands with the originating atomic orbitals.

The two pairs of degenerate states Γ_5^+ and Γ_6^+ (not present in simple metal diborides) correspond to the Ca $3d_{xz,yz}$ and $3d_{xy,x^2-y^2}$ orbitals, respectively. The interaction of the latter with the Ga σ-bonding states will be analyzed later. The Ca $4s$ and $3d_{z^2}$ orbitals form Γ_1^+ bands at the Γ point; therefore, their symmetry is not sufficient, even at Γ, to identify them precisely.

The Ca d bands are quite flat in the $k_z = 0$ plane, an indication of weak in-plane interactions. This is consistent with the small size of the $3d$ orbitals compared with the $CaGa_2$ in-plane lattice parameter, $a = 8.155$ a.u. (as an indication, the average radius of the $3d$ orbitals of Sc, next to Ca in the periodic table, is 1.68 a.u. [54]). Going to the zone boundary, these states interact with the Ga states, due to the Ca $d_{xz,yz}$ – B π interaction; the shape of π bands ($\Gamma_2^- - K_5 - \Gamma_4^+$), which was simple in BeB_2 and AlB_2, shows anticrossings in this compound. The d_{xy,x^2-y^2} orbitals interact at Γ with the σ-bonding states and at the zone boundary (for instance at K_6), with the σ-antibonding states.

Along the k_z direction ($\Gamma - A$ in Fig. 18), we notice that the Ga σ-bonding bands have small dispersion. Since these dispersions are related to the Ga σ–Ca s (Γ_1^+) and Ga p_σ–Ca d_{xy,x^2-y^2} couplings (Γ_6^+), they indicate that these states hybridize weakly with each other. The Ga π bands on the other hand have large dispersion along the Γ–A direction, due to the large Ga p_z–Ga p_z and Ga p_z–Ca inter-plane interactions.

More detailed information on the bonding properties of this compound can be obtained from the analysis of the electronic wavefunctions at the Γ and A points. The lowest lying band is mainly formed by the Ga s-bonding states, with a small contribution of the Ca s states. The two Γ_6^+ levels are due to the Ga p_σ and Ca d_{xy,x^2-y^2} orbitals with the Ca d states contributing mainly to the higher Γ_6^+ bands. The π-bonding band is weakly hybridized with the Ca p_z at Γ_2^- and, very strongly, with the Ca s orbitals at the lowest A_1^+ level. This interaction is responsible, together with the Ga–Ga inter-plane interaction, for the large dispersion of this band. It should also be noticed that the coupling of the π-bonding states with the Ca d_{z^2} orbitals is small. Throughout the Brillouin zone, the π-bonding states hybridize also with the Ca $d_{xz,yz}$ orbitals, for instance along the Γ_2^-–K_5 band.

The Γ_1^+ band just above the Fermi level contains the contribution of the Ca $3d_{z^2}$ orbitals, with a very small contribution of the Ca s and Ga σ-bonding states. The Ca s and d contributions are inversed in the higher Γ_1^+ state. An inversion of the energy ordering of the Ca $4s$ and $3d$ orbitals with respect to what happens in isolated atoms can be related to the different size of these orbitals; when in fact the Ca atoms are compressed by the neighboring Ga atoms, the $4s$ orbitals move up in energy more rapidly than the more localized $3d$ orbitals. If we now consider the contour plots of the charge density corresponding to the first Γ_1^+ level above the Fermi level E_F, shown in Fig. 15, we can see that, besides the contribution of the Ca d_{z^2} orbitals, it shows free-electron character, with charge density spread in the regions defined by the Ga layers. Its character would have been therefore very similar to that of the corresponding states in simple metal diborides, if it were not

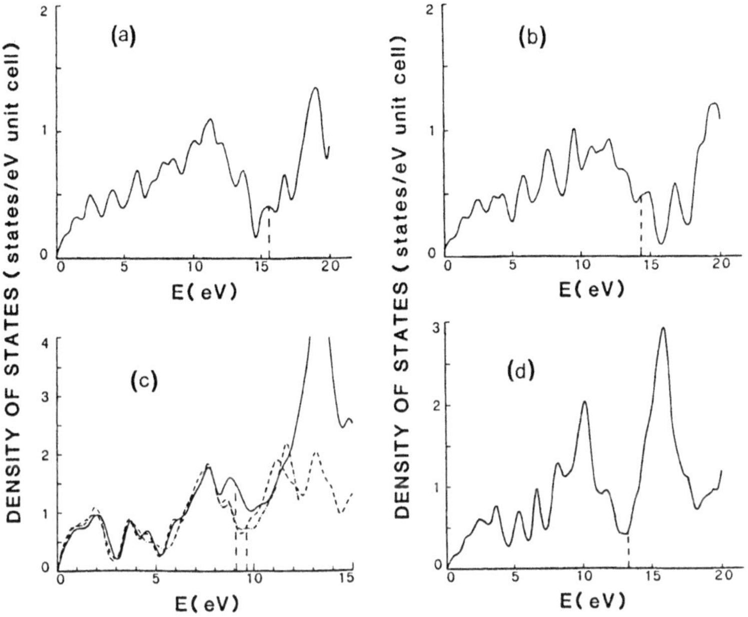

Figure 17. Density of states of (a) AlB_2, (b) BeB_2, (c) $CaGa_2$, and TiB_2 in units of states/(eV–unit cell). The Fermi level is indicated by a vertical dashed line. The DOS of $CaGa_2$ in (c) are calculated as follows: FLAPW method and energy eigenvalues calculated on an uniform grid corresponding to 75 inequivalent k-points (full curve); empirical pseudopotential method (EPM), same uniform mesh based on 75 k-points (dashed line); EPM, but with a random 3000 k-point mesh (dotted-dashed line).

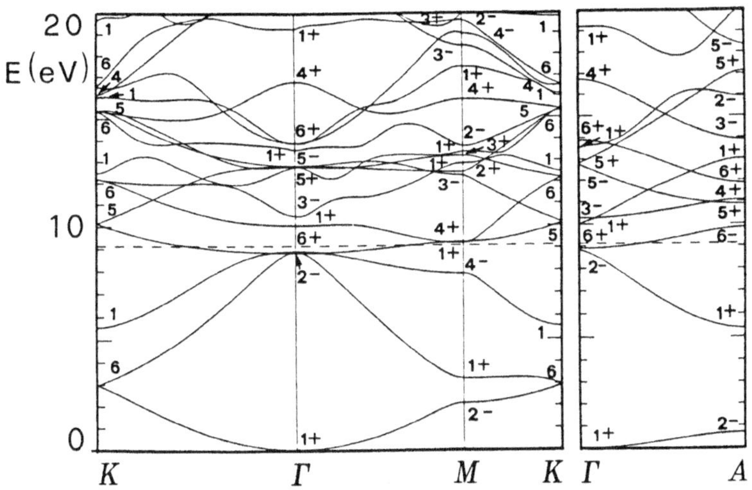

Figure 18. Self-consistent energy bands of $CaGa_2$ along the main symmetry axes. The Fermi level is indicated by a dashed line.

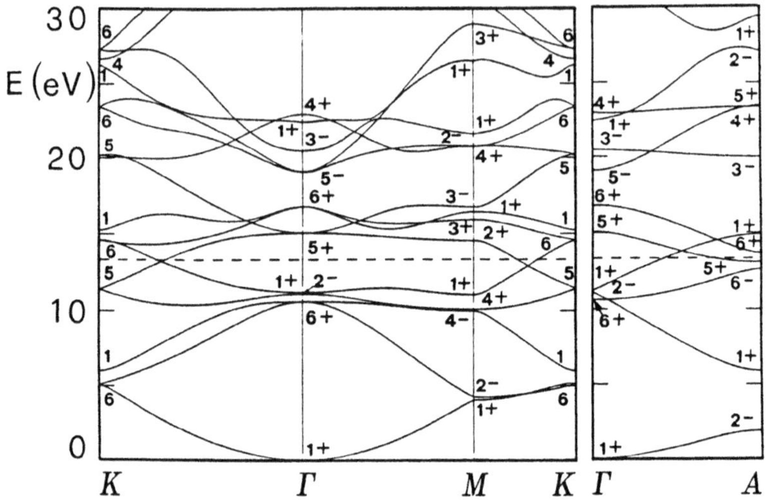

Figure 19. Self-consistent energy bands of TiB_2 along the main symmetry axes. The Fermi level is indicated by a dashed line.

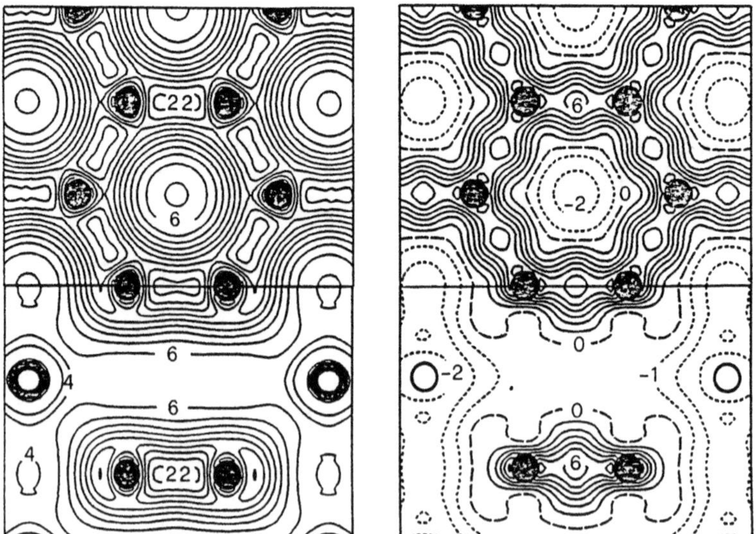

Figure 20. (a) Contour map of the self-consistent valence charge density of AlB_2 on the same plane as in Fig. 15 and on a horizontal plane containing the B atoms; (b) difference between the self-consistent and overlapping-atom charge densities represented on the same planes. The full lines correspond to positive contour values. Units and symbols for the atomic sites are the same as in Fig. 15.

for the presence of the Ca d states. In order to verify the interlayer nature of this band, we performed a self-consistent energy-band calculation on the ideal system obtained from $CaGa_2$ by taking out the Ca atoms and leaving the lattice spacings unchanged. This analysis leads to the following main conclusions: (i) the σ-bonding bands are slightly modified when the Ca atoms are taken out of the crystal, indicating a small contribution to these bands from the Ca states; (ii) the two Γ_1^+ interlayer bands exist independently of the presence of the Ca atoms, in agreement with the conclusions of an empirical pseudopotential method (EPM) study, and with similar studies in Li intercalated compounds [10] and in AlB_2.

The Fermi level in $CaGa_2$ crosses π bands and, near the A point, the σ-bonding bands. The σ bonding bands, almost completely filled in the ground state, dominate the occupied states. The π-bonding and antibonding bands are partially filled. In Fig. 17(c) we represent the density of states of $CaGa_2$. For comparison, we also show the density of states obtained by calculating the energies over the same mesh of 75 **k**-points by the empirical pseudopotential method. The results of the two methods are in good agreement in the low energy region of the valence band. The higher valence states and, mainly, the conduction states are quite different, due to the presence in the FLAPW calculation of the Ca d bands, which give rise to the narrow peak above E_F. In order to test the influence of the restricted mesh of **k**-points used on the density of states, we also report in Fig. 17(c) the density of states obtained by calculating the energies over a mesh of 3000 random **k**-points by the pseudopotential method. A comparison of the two pseudopotential curves shows that the 75 **k**-point mesh is sufficient to reproduce the main features of the density of states.

4.4 TiB_2

The energy bands of TiB_2 are shown in Fig. 19. In the $k = 0$ plane the three lowest bands (Γ_1^+, Γ_6^+) are similar to the σ-bonding bands of graphite, and can therefore be attributed to the bonding combinations of the sp^2 hybrids of B. However, these states can interact throughout the basal plane with the Ti s states, while at Γ the p_σ bonding states hybridize with the Ti d_{xy,x^2-y^2} orbitals. As a result, we notice that the width of the σ-bonding bands is smaller in TiB_2 than in AlB_2, which has almost the same lattice spacings. This latter interaction is confirmed by the relatively large energy dispersion of the p_σ-bonding band along the k_z direction, compared for instance to the same dispersion in AlB_2. The Ti d_{xy,x^2-y^2}–B p_σ interaction is in fact symmetry-forbidden at A (where the B and Ti states form the A_6^- and A_6^+ bands, respectively) and allowed at Γ, where they transform according to the same Γ_6^+ symmetry.

The Ti s and d_{z^2} orbitals participate to the formation of bands with Γ_1^+ symmetry at the Γ point. In particular, the Γ_1^+ band just below the Fermi level has a dominant contribution from the Ti d_{z^2} orbitals (at Γ, it can be decomposed as 3% Ti s and 62% Ti d inside muffin-tin spheres) and has also interlayer character, as can be seen in Fig. 15(d).

The upper band with the same symmetry is a hybrid of Ti s and d_{z^2} orbitals, with a contribution from the B s states. The interaction between the Ti and B s states raises up the energy of the latter, so that Ti s states mostly contribute to the upper Γ_1^+ state and, also, to the lowest state with the same symmetry.

The B π states cannot interact with the Ti states at the Γ point but, going toward the zone boundary, they interact with the Ti d_{xz} and d_{yz} orbitals (which form, at Γ, the Γ_5^+ states). As a consequence, the graphite-like π structure, well defined in the simple metal diborides, is rather complex in TiB_2, as in $CaGa_2$. The B π-bonding states interact, along the Γ–A direction with the Ti s orbitals and, away from Γ in the basal plane, with the Ti $d_{xz,yz}$ states. At the A point, the lowest lying A_1^+ level can be decomposed (inside muffin-tins) as Ti $s^{0.23}$ $d^{0.04}$–B $p^{0.12}$, while the lowest K_5 states can be decomposed as Ti $d^{0.51}$–B $p^{0.10}$. This interaction accounts for the different shape of the Γ_2^-–K_5 π bonding band in TiB_2 with respect to what is observed in the simple metal diborides (see Figs. 13–14).

In the crystal ground state, the B σ-bonding and the B π-bonding bands are completely filled. The π-antibonding band and the Γ_1^+ band, mainly formed by the Ti d states are partially occupied. The states at the Fermi level are mainly formed by the Ti d orbitals. In Fig. 17(d) we represent the density of states of TiB_2. The Fermi level lies in a region with low density of states, which separates a bonding from an antibonding part. The two main peaks, just below and above the Fermi energy have a strong contribution from the Ti d bands.

The electronic properties of TiB_2 have been calculated by the LCAO method by Armstrong [51]; his energy bands show some important differences with respect to our results; in particular, the full width of the valence band is about 5 eV larger, and the Ti d bands are narrower than ours. A comparison of our results with those obtained by Johnson, Harmon, and Liu [50] for ZrB_2, with the KKR method (the only self-consistent calculation reported, up to now, for a transition metal diboride), shows an overall similarity between the energy bands of the two materials, and in particular that in both compounds: (i) the B-derived bands hybridize strongly with the metal orbitals; (ii) the states at the Fermi level are dominated by the metal states; (iii) the Fermi level occurs in a region with low density of states.

The coefficient of the linear term in the specific heat expansion at low temperature γ is proportional to the value of the density of states at the Fermi level. The value of γ for the 3d transition metal diborides was measured by Castaing et al. [44], who found that γ is a minimun for TiB_2. Qualitatively the same results was found by Tyan et al. [43], who found that, among a number of transition metal diborides, TiB_2, ZrB_2, and HfB_2 show the lowest value of γ. The experiment therefore indicates for TiB_2 a low value of the density of states at the Fermi level, in agreement with our results.

The Ti $L_{II,III}$ emission spectrum of TiB_2 [39] indicates the presence of a doublet at 1.7 and 3.1 eV below the Fermi level. Our calculated density of states show the presence of a peak about 3 eV below E_F with appreciable Ti d character, in qualitative agreement

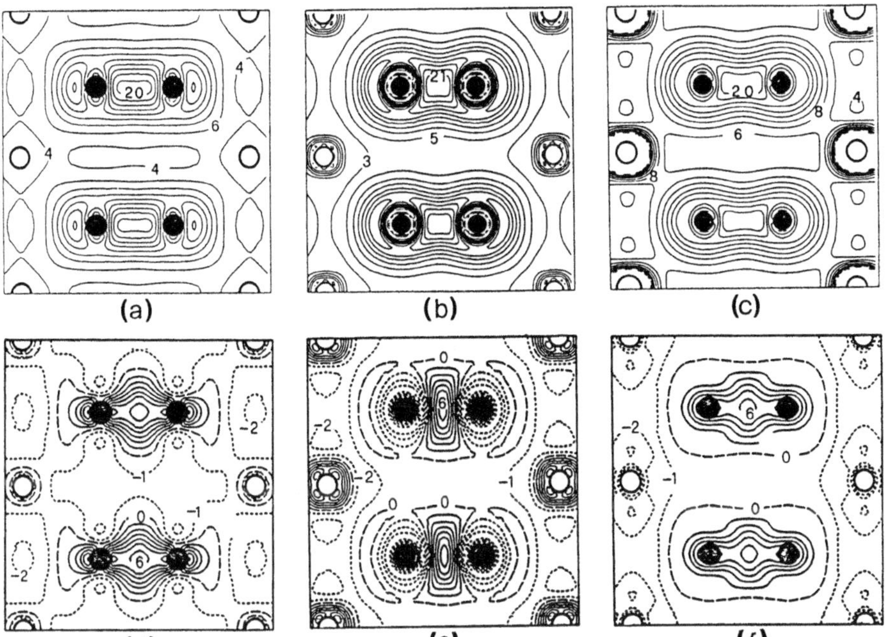

(a) (b) (c)

(d) (e) (f)

Figure 21. Contour plots of the valence charge densities of (a) BeB_2, (b) $CaGa_2$, and (c) TiB_2. Plots (d), (e), and (f) give the difference between the self-consistent crystal density and the overlapping-atom charge density for BeB_2, $CaGa_2$, and TiB_2, respectively. Units, geometry and symbols are the same as in Fig. 15.

(a) (b) (c)

Figure 22. Contour map of the self-consistent valence charge density for the systems obtained from AlB_2 by removing: (a) the Al atoms, or (b) the B atoms and keeping the lattice parameters unchanged. (c) Differences between the self-consistent valence charge density of AlB_2 and the sum of the two densities given in (a) and (b). The densities are represented in a vertical plane passing through the B–B bond. Units, geometry and symbols are the same as in Fig. 15.

with the experiment. The x-ray B K spectra of TiB_2 [41] (and of other transition metal diborides), show the presence of an energy gap between the beginning of absorption and the short wave length of the emission spectra; the same gap was not observed in the Ti $L_{II,III}$ spectra. This can be explained assuming that the low lying empty states are mainly formed by the Ti d states, which are excluded by symmetry from the B K spectra. Our band-structure calculation supports this conclusion, the lowest lying empty state with appreciable B p character being M_3^- (see Fig. 19).

4.5 Charge Transfer

The problem of charge transfer in the compounds of the AlB_2 family has received attention in the literature, particularly for transition metal diborides, where two conflicting models have been proposed in order to explain the different experiments. [33,43,45,46] From the theoretical side, only one self-consistent calculation has been reported [50], up to now, in which the problem of charge transfer has been examined in detail.

In this subsection, we compare the self-consistent crystal charge density ($\rho(\mathbf{r})$) with the charge obtained by overlapping neutral atoms ($\rho_{sup}(\mathbf{r})$). Figure 20 shows the contour plots of the valence charge density of AlB_2 in the plane of B atoms and in the vertical plane cutting the B–B bonds and including the Al atoms. This figure clearly shows that the σ bonds between the X atoms constitute the main feature of the crystal charge density, confirming our previous considerations on the chemical bonding in this compound. We also show in Fig. 20 the difference between $\rho(\mathbf{r})$ and $\rho_{sup}(\mathbf{r})$ in the same planes, while in Fig. 21 we show the same quantities for the other MX_2 compounds studied, in the vertical plane considered above. These figures clearly show that going from the overlapping-atoms to the self-consistent charge density, the charge accumulates in the region of the bonds between the X atoms, coming from the metal atoms. This demonstrates a charge transfer from the metal to the X atoms. In order to give a measure of this charge transfer, we show in Table 2 the values of the total charge contained inside the M and X atomic spheres for $\rho(\mathbf{r})$ and $\rho_{sup}(\mathbf{r})$. Table 2 shows that the transfer is not big; furthermore, an important fraction of the valence charge is spread out in the interstitial regions, and cannot therefore be assigned to any particular atom. This is in agreement with what was found in Ref. [50] for ZrB_2; however, in that case a small charge transfer was found from the B to the Zr atoms, i.e., opposite to our results.

Finally we provide more evidence of the small charge transfer from M to X atoms in the case of AlB_2 using a different decomposition of the self-consistent crystal charge density $\rho(\mathbf{r})$. In Fig. 22 we represent the charge densities $\rho_{Al}(\mathbf{r})$ and $\rho_B(\mathbf{r})$ of the systems obtained from AlB_2 by removing all B atoms or all Al atoms, respectively, and keeping the lattice parameters unchanged. The differential charge density $\rho(\mathbf{r})-\rho_{Al}(\mathbf{r})-\rho_B(\mathbf{r})$ (see Fig. 22c) shows that also in this case the electrons leave the Al spheres and the interstitial region to accumulate close to the B atoms. Notice however that the charge pile-up in the middle of the B–B bond is stronger for the system without Al (Fig. 22a) than for AlB_2.

4.6 Concluding Remarks

We have presented in this section the electronic structures of BeB_2, AlB_2, $CaGa_2$, and TiB_2. The following conclusions can be drawn: (i) The chemical bonding in these materials is characterized by the strong sp^2 bonds between the X atoms and by the interaction between the M and X orbitals. (ii) In all these compounds there is a charge transfer from the metal to the X atoms; however, the transfer is not large. Furthermore, an important fraction of the charge is spread in the interstitial region. (iii) The simple-metal diborides have band structures very similar to each other, and similar to those of two-dimensional graphite. In TiB_2 the graphite-like σ and π bands are strongly modified by the hybridization between the B and Ti orbitals. (iv) Interlayer states, similar to those recently found in graphite, Li-intercalated graphite and hexagonal BN also exist in the MX_2 compounds. (v) The states near the Fermi level are mainly formed by the B π states in the simple metal diborides, whereas in TiB_2 the Ti d states are dominant. The Fermi level lies at minimun of the density of states in TiB_2 while in the other compounds studied the density of states at E_F is larger.

5 Semiconductor Interfaces and Heterojunctions

5.1 Background Perspective

Among the various properties of layered semiconductors that have been the object of careful and thorough investigation, the properties of junctions (metal-semiconductor and semiconductor-semiconductor heterojunctions) are still not fully understood because of the complexity and the variety of the problems related to them. These systems strictly fall into the layered materials category since they are artificially obtained via a layer by layer growth, and much of their properties and characteristics depends on the delicate and remarkable inter-play and balance of both bulk and interface effects. The interest in these particular issues is obviously related to the rectifying properties that metal-semiconductor junctions and heterojunctions in general show and to their importance for device applications; the aim of the great majority of the studies is therefore focussed on the effort to single out the basic factors that can affect the potential barrier height at the interface.

The two interfaces just mentioned are basically different, from an electronic point of view, but show many similarities and, for this reason, have been often studied together using very similar models.

The main feature of a metal-semiconductor junction is that, usually the Fermi level of the metal is "pinned" (i.e. fixed) in a narrow energy range and its energy position is then independent of the doping of the semiconductor. When the two materials are put in contact, some charge will flow from the metal into the semiconductor; as a result, a small interfacial dipole is generated together with a bending of the bands which extends for several atomic layers inside the semiconductor. As a result of this charge rearrangement, metal electrons flowing toward the conduction band in the semiconductor have to overcome a potential barrier (the so-called Schottky barrier) determined by the difference in energy between the bottom of the semiconductor conduction band (for an n-type semiconductor) and the common Fermi level.

A very similar phenomenon occurs at the interface between two semiconductors; the main difference is due to the presence of a gap on both sides of the interface. The relevant quantity in this case is the relative position of the band gaps which determines the potential steps (valence band offset and conduction band offset) that the different carriers (electrons and holes) have to overcome in order to permit conductivity across the interface.

In both cases, Schottky barriers and band offsets, the relative alignment of the potential in the two different materials in contact is determined by some intrinsic bulk properties (i.e. the position of the Fermi level or of the conduction band with respect to the vacuum level) and by the charge readjustment at the interface (interface dipole). The way these two different contributions affect the final line-up and determine the barrier height is not a trivial matter and is the central object of most of the studies.

The model we use to determine the line up takes into account bulk properties as well as interface properties, and is conceptually very close to the main idea underlying the XPS experimental procedure to determine the offset in solids. First, a complete self-consistent calculation is performed for the particular interface under study; then, from this calculation, the relative energy position of two core levels of different atoms (which are far enough from the interface so that the bulk properties are completely recovered) is evaluated. This value is used as reference energy level with respect to which the valence band maximum (or the Fermi level) of the bulk compounds is aligned. This procedure has been used successfully to predict and explain the valence band offset in different structures [55,56] and seems to take into account both the bulk effect and the interface dipole contribution to the band line up. However, as we will see, there are still many factors that need to be accommodated for in the model and that can greatly affect the final band line up. These problems are mainly related to how close to the real interface is the model junction that we calculate.

As a last remark, let us point out the geometrical assumptions we make in order to approach the interface problem. Due to the chemical nature of the compounds we are interested in, the study of a free surface is neither feasible nor realistic: as is well known, cleaved semiconductor surfaces are not stable and they always undergo a "reconstruction" process in which all the unsaturated bonds are somehow "readjusted" by oxidation, bond dimerization or other physical and/or chemical reactions. Furthermore, wide technological applications are based on semiconductor superlattices and quantum wells — artificial structures in which two or more different semiconductors are grown layer by layer, one on top of the other in a regular sequence. We will therefore concentrate on this peculiar geometry, in which interface characteristics can still be extracted from an essentially bulk calculation.

5.2 Metal-Semiconductor Interfaces

We now briefly discuss a few selected examples of the systems we have recently studied; we will focus, in particular, on the Fe/ZnSe [57,58] and Fe/GasAs (001) systems that have been recently grown by G. Prinz and collaborators [59,60,61,62,63,64] via MBE.

Due to its magnetic (easy switching) [59,60,61,62,63,64] and optical properties (ZnSe has a wide band gap and therefore it is a good high frequency emitter), the Fe/ZnSe system appears to be very useful for lasers and other magneto-optical applications. In particular, the breakdown of the 3-dimensional crystal symmetry has been shown to remarkably enhance the magnetic properties of several ferromagnetic materials; therefore it is expected that magnetic film incorporated into semiconductor structures can be used to realize planar microelectronic configurations with novel magneto-optical properties. Furthermore, the Fe/ZnSe interface presents particularly interesting features (such as small mismatch of their lattice constants and high stability with respect to chemical reaction [63]) that makes it highly desirable for a theoretical calculation and for a consequent meaningful comparison with experiment.

We focussed on the electronic and magnetic properties of several superlattices of

Fe on ZnSe with different layer thickness and Fe/GaAs. In particular, superlattices of $Fe_n/(ZnSe)_m$ (001) for $n = 1, 3$ and $m = 1, 2$ and the monolayer Fe/GaAs were investigated using structures that are basically obtained from the zincblende structure, replacing one cation plane with one or more Fe planes in the [001] direction. The atomic arrangement of the Fe atoms is such that four Fe unit cells are needed to fit on a (001) zincblende plane; the mismatch in this configuration is only 1.1% and 1.4% for ZnSe and GaAs, respectively.

The interface characteristics of both the Fe–Se and Fe–As interfaces were determined from a total energy calculation. The difference between the two is remarkable and confirms the experimental results. While the interface between Fe and Se is well established, clean and non-reactive, the Fe–As interface is highly reactive. We found that the equilibrium distance between Fe and Se is approximately the average of the Fe–Fe and Zn–Se bond length and agrees with the experimentally measured tetragonal distortion induced by Fe alloying in the $Zn_{1-x}SeFe_x$ dilute magnetic semiconductor [63].

In the case of Fe/GaAs we found that the Fe–As interface bond length deviates quite significantly from that in bulk GaAs and Fe, leading to a much shorter bond length. We found $d_{Fe-As} = 2.30$ Å to be compared with $d_{Fe-Se} = 2.46$ Å. Such different behavior correlates well with experiment: while Fe forms a well defined surface on ZnSe, it interdiffuses with the GaAs substrate forming intermixed compounds.

The magnetism of the Fe in the 1×1 Fe/ZnSe superlattice (SL) showed a remarkable enhancement of its magnetic moment (to $\mu = 2.6\ \mu_B$) about 20% of its bulk value [57]. Due to the different site coordination of Fe in the different atomic layer this enhancement decreased as the number of Fe layers is increased: for three Fe layers, the Fe magnetic moment decreases to its bulk value $\mu = 2.2\ \mu_B$ [58]. Furthermore, due to the high directionality of the Se bonds, we observed a different magnetic behavior in the in-plane [110] direction with respect to the $[1\bar{1}0]$. This anisotropy is shown to decrease as the Fe thickness is increased, in agreement with the experiment. In order to better illustrate the atomic site differences, we show in Fig. 23 a schematic diagram of the different structures considered. The calculated magnetic moments for the Fe sites in the thicker superlattices, see Table 3, are strongly reduced with respect to the results calculated for the 1×1 superlattice not only for the central layer, but also for the interface layers with the exception of the Feb2 site; however, they are still larger when compared with the magnetic moment of bulk Fe. The trend of a reduced total magnetization as a result of an increased number of Fe layers has been also experimentally observed [64].

The enhanced magnetization observed on the Feb2 site is due to the different site coordination between Feb1 and Feb2 [58] since the latter site has a smaller number of next nearest-neighbors (its closest neighbor along the (001) direction is Seb, which is 4 atomic layers away). The consequence of this peculiar site coordination is that the electrons with negative polarization are mainly located in the interstitial region and far away from the muffin-tin sphere, thus enhancing the total magnetic polarization of that site.

A comparison of the results obtained for the 1×1, 3×1 and the 3×2 superlattices, clearly shows that the magnetic moments of the larger superlattices are reduced even for the

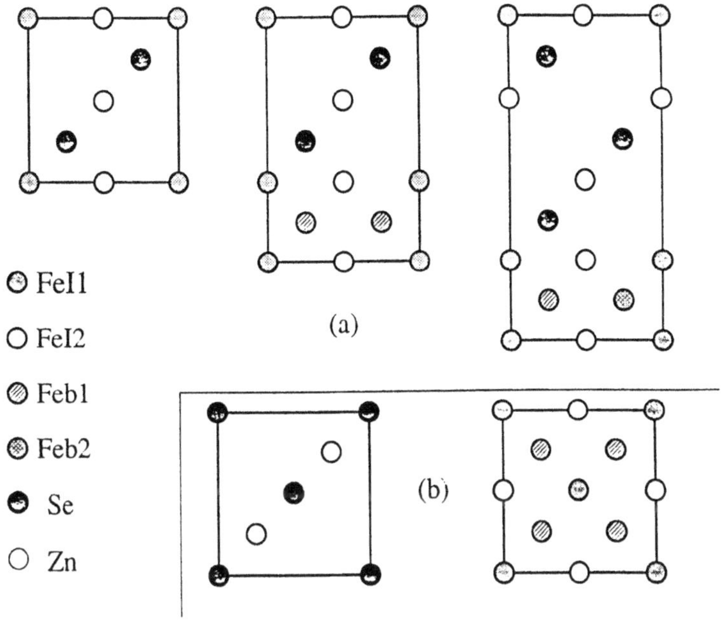

Figure 23. Projections on (a) the yz plane and (b) the xy plane of the atomic positions for the 1x1 (Fe/ZnSe), 3x1 (Fe_3/ZnSe) and 3x2 ($Fe_3/(ZnSe)_2$) superlattices.

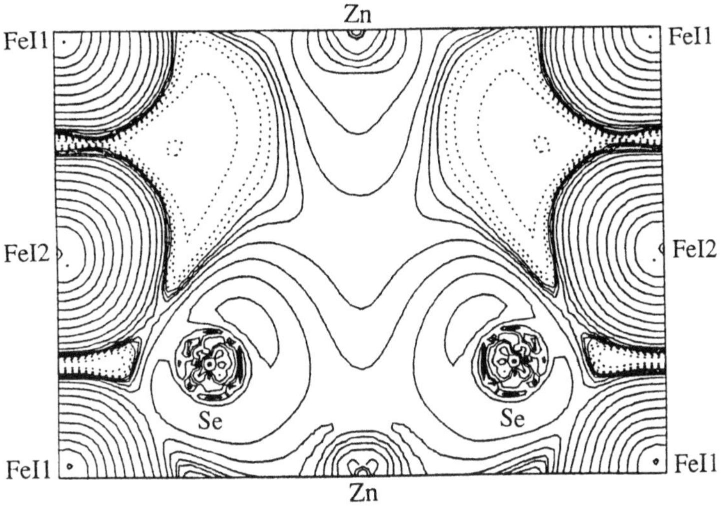

Figure 24. Spin density distribution in the (110) plane of the Fe/ZnSe superlattice. Solid (dashed) lines indicate contours of positive (negative) spin density in units of $10^{-4}e/(a.u.)^3$. Successive contours shown differ by a factor of 2.

Table 3. Hyperfine field in (kGauss) and magnetic moment μ in (μ_B) for the Fe-sites in the superlattices considered.

	μ (μ_B)	Hyperfine field (kG) core	$C.E.$	H_c(tot)	H_c/μ(kG/μ_B)
1 × 1					
FeI1	2.76	−386	+218	−169	61.23
FeI2	2.67	−371	+119	−252	94.38
3 × 1					
Feb	2.25	−318	−70	−389	172.89
FeI1	2.17	−307	+14	−293	135.02
FeI2	2.22	−312	−17	−329	148.20
3 × 2					
Feb1	2.25	−311	−98	−409	181.78
Feb2	2.66	−370	+28	−342	128.57
FeI1	2.33	−321	+9	−312	133.91
FeI2	2.40	−334	+1	−333	138.75
Fe − slab[a]					
surface	2.98	−398	+143	−252	84.56
center	2.25	−291	−75	−366	162.67
Fe-bct					
Fe	2.13	−299	−48	−347	162.91

[a] Ref.[65]

Table 4. Schottky barrier height (Φ_B), Pauling electronegativity (χ^a) and work function (W^b) for different metal-ZnSe junctions.

metal	χ	W(eV)	Φ_B (eV)
Au	2.4	5.15	1.29[c] − 1.36[d]
Pt	2.2	5.7	1.4[d]
Cu	1.9	4.7	1.1[d]
Co	1.8	4.63	1.72[c]
Fe	1.8	4.5 − 4.3	1.09[e]

[a] Ref. [67]
[b] Ref. [68]
[c] Ref. [71]
[d] Ref. [69]
[e] present work

Fe interface layers as the Fe thickness is increased (i.e., as the interface effect diminishes). If the direct interaction between adjacent layers of Fe were the major contribution to the suppression of the surface-like magnetic enhancement of Fe at the interface, then still considerably increased magnetic moments for these layers, compared to the bulk layer, should result. This does not seem to happen since the magnetic moments in the two layers (bulk and interface) are very close (only a 2% difference). This is due to the fact that as the number of Fe layers increased (from 1 to 3), the Fe interface sites have an increased coordination number; as a consequence, their behavior is more bulk-like than surface-like.

The interaction among non-adjacent Fe layers through the semiconductor interlayer might also contribute in suppressing the Fe magnetic properties. This might be confirmed by (i) the noticeable delocalization of the charge with negative spin polarization into the semiconductor region for the 3×1 SL, and (ii) by the fact that the interface Fe-sites have higher magnetic moment in the 3×2 than in the 3×1 superlattice. From the spin density distribution plots (see Figs. 24, 25 and 26), it is clear that in the 3×2 case the magnetism is much more strongly confined to the Fe region than in the 3×1 case. This results in a complete suppression of any magnetic interaction between non-adjacent Fe layers and in an enhancement of about 7% (0.17 μ_B) in the magnetic moment at the interface. Actually, in addition to the highly anisotropic spin density distribution, the 3×1 superlattice has a negative spin density which spreads from the interface layer over the entire crystal and is not localized only in the zincblende hollow sites as it is observed in the 1×1 case (Fig. 24).

The results for the Fermi-contact term of the hyperfine field separated into core and conduction electron (CE) contributions are shown in Table 3 and compared with the calculated results obtain for a 7 layer Fe slab [65] and bulk Fe in the same crystallographic arrangement as in the superlattices studied (i.e. body centered tetragonal, Fe-bct). As expected [66], the core contribution scales with the magnetic moment and the CE contribution is strongly reduced in going from the 1×1 to the 3×2 indicating that the outermost Fe s-like charge is affected by the indirect polarization with the polarized d-states. The slightly positive value of the C.E. contribution for FeI1 is a result of its more favorable bonding position compared with FeI2, whereas the positive CE contribution found for Feb2 is the result of a strong positive polarization caused by the enhanced magnetic moment. The hyperfine field values (as well as the magnetic moment) calculated for Feb1 and Feb are very close to those calculated for the central layer of the Fe slab and for bct-Fe, showing that the bulk behavior is well recovered in both cases.

We also studied the electric properties of the interface and in particular the Schottky barrier height. The planar average charge density in the thicker (3×2) superlattice as a function of the (001) coordinate revealed that about 0.4 electrons are transferred at the interface from the Fe layers into the ZnSe side, and that the bulk charge densities are well recovered in both sides of the interface in the inner layers.

By considering the core levels as reference levels we were able to determine the Schottky barrier height (SBH), in analogy with the well known procedure used by XPS experiments. In this scheme, the discontinuity (Φ_B) between the Fermi level in the Fe side and

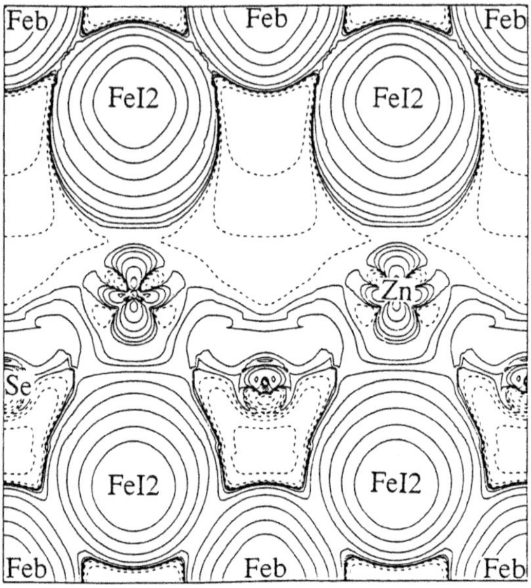

Figure 25. Spin density distribution in the (110) plane of $Fe_3/(ZnSe)$. Solid (dashed) lines indicate contours of positive (negative) spin density in units of $10^{-4}e/(a.u.)^3$. Successive contours shown differ by a factor of 3.

Figure 26. Spin density distribution in the (110) plane of $Fe_3/(ZnSe)_2$. Solid (dashed) lines indicate contours of positive (negative) spin density in units of $10^{-4}e/(a.u.)^3$. Successive contours shown differ by a factor of 3.

the conduction band in the semiconductor side of the interface is evaluated by calculating the difference between an Fe and a Se core level, as obtained from the superlattice calculation and using the corresponding binding energy of these same levels in the reference bulk systems (Fe-bct and ZnSe). We found that the potential barrier at the interface is $\Phi_B = 1.09 \pm 0.09$ eV. Since, to the best of our knowledge the Schottky barrier height for such a junction has not yet been measured, we compare our result with some measurements and data [67,68,69,70,71] obtained for metal-ZnSe interface, such as Au [70], Co [71], Cu [69], and Pt [69] junctions on ZnSe, shown in Table 4.

The Schottky model predictions are also shown in Table 4 for comparison; since the substrate is the same, one would expect a higher SBH for the metal with higher work function. From Table 4, we see that this general trend is actually quite well reproduced by the ZnSe surface, if we exclude the Co and Pt cases where chemical interactions between metal and substrate cannot be neglected. However, even for the Au and Fe cases, the bare Schottky model does not hold exactly (we recall that the ZnSe electron affinity $\chi =$ 4.06 eV).

Experimental evidences [70] for the low reactive Au/ZnSe interface showed that the work function changes accordingly to the band bending, supporting the idea that the barrier is essentially determined by the metal work function and the semiconductor affinity. Moreover, they also showed that the deviation from the Schottky model predictions is related to a dipole formation at the ZnSe surface and to the metal electronegativity.

In our case, due to the lower electronegativity of Fe with respect to Au, we expect Fe to behave differently from Au. In particular, since we found that charge flows from Fe into the more ionic ZnSe substrate, we expect a dipole contribution that tends to increase, rather than decrease, the metal work function. Within this sort of qualitative consideration, our calculation is in reasonable agreement with the experiment and is consistent with the picture that describes the SBH as dependent on two different contributions combined: (i) the intrinsic properties of the two compounds (i.e., metal work function, electron affinity, and/or metal electronegativity and semiconductor ionicity) and (ii) the characteristics of the interface (dipole formation, interface states). Unfortunately, the determination of such contributions is still not trivial overall if effects due to the non-perfect abruptness of the interface have to be taken into account.

Let us make a final remark: although LDA has been shown to successfully reproduce the band-offset at semiconductor heterojunctions, we cannot rule out a possible failure in the present case where the different dielectric and screening properties of the two materials at the interface (metal and semiconductor) can lead to a non-exact cancellation of non-local effects which are usually neglected within LDA. This would imply that, in order to obtain more realistic results, non-local effects have to be considered in the superlattice calculation. This task has been only partially addressed [72] so far, and should be more carefully explored.

5.3 Semiconductor-Semiconductor Interfaces

Technological interest in layered semiconductors is focussed on device applications while scientific interest is mainly devoted to understanding some characteristic properties, such as potential barriers, carrier properties and band gaps, and how these can be modified by changing different external factors, such as epitaxial growth, strain effects, junctions and doping. A complete knowledge and understanding of these complex issues is, in fact, at the basis of achieving the ability to "tune" basic properties "at will" and designing "ad hoc" materials for device applications.

For examples, it has been shown that the valence band offset is independent on the interface orientation for homopolar and unstrained interfaces (GaAs/AlAs being the "classical" and most studied example). Unfortunately this does not hold for heteropolar and strained interfaces, where the interface morphology and internal strain plays a crucial role in determining how the charge redistributes at the interface and the consequent shape of the potential across the junction. This is among the most important issues still under theoretical investigation and for which the experiments have not been able to give precise answers. It must be noticed, in fact, that due to the particular nature of these phenomena (which are so much dependent on the detailed interface morphology on the atomic monolayer scale), the experiments are often not capable of resolving all the details and to single out each individual problem.

On the other hand, these kinds of information are partially obtained through accurate computational studies. We have performed extensive calculations of several different semiconductor heterojunctions in order to explore some of the crucial issues involved in the potential line-up problem; our studies on GaAs/AlAs [55], InAs/InP [73,74] and recently α–Sn/CdTe together with several other theoretical studies performed by other groups (see Refs. [56,75,76,77] just to name a few of them) have shown that such calculations can give remarkable insights into interface formation, stability and potential line-up.

In order to better illustrate how a theoretical study fits into the more general framework, we discuss, here, an example of how external factors (in this case the strain) can affect the transport and electronic properties of a particular heterojunction and how our calculations relate and compare with experiments. We will briefly present some recent results on heterostructures based on In compounds, since these have recently attracted increasing interest because of their optical and transport properties [78,79,80]. The InAs–InP system in particular, has been the focus of many experimental works; the mismatch between the two semiconductors is not particularly high (only 3.2%) and, moreover, it can be adjusted "at will" when using InP substrates, by adding small amounts of Ga or P to the InAs matrix. The growth of $InAs_{1-x}Ga_x$/InP by metal-organic-chemical-vapor-deposition (MOCVD) and by molecular beam epitaxy (MBE) has been reported by many groups [78,79,80,81] — as has the growth of InAs/InP quantum wells [79].

In our study, we focussed on the InAs/InP (001) systems taking into account the effect of strain. To this end, we first studied the properties of the pure binary compounds and then performed total energy calculations in order to determine all the unknown lattice parameters of the 1 x 1 structure,[73] i.e., the in-plane lattice constants, and the In-P and In-As distances (for the so called free standing mode (FSM) case). In order to compare with the available experiments and to study the effect of strain on the electronic properties of this structure, we also studied InAs as if grown on an InP substrate (InP-sub); in this latter case, the in-plane lattice constant and also the In-P bond length are considered fixed and equal to their InP equilibrium values, whereas the In-As bond length is determined by a total energy calculation.

We found that in both cases the ultra-thin 1 x 1 superlattice is unstable (the heat of formation is ~ 20 meV/4-atoms and 30 meV/4-atoms, respectively) and that the major contribution to this instability is due to the excess of elastic energy associated with the deformation of the two bulk semiconductor compounds. in the case in which no constraints are assumed on the in-plane lattice constant, the ordered phase of In_2AsP is seen to follow Vegard's law (within 0.02%), whereas the equilibrium anion-cation distances in the superlattice follow the bond length conservation rule already observed in several other semiconductor alloy systems. We calculated that the charge transfer at the interface is rather small (less than 0.04 electrons are transferred from the InAs side into the more stable InP bonds) and the corresponding energy term is negative (i.e., it Constitutes a stabilizing energy term in the formation of the superlattice).

The study of the effect if strain on the electronic properties of the 1 x 1 structures revealed that the hole effective masses and the transport properties are very sensitive to the uniaxial distortion,[73] making this system a good candidate for tailoring the band discontinuity, and the carrier properties. In particular, in the case of no-substrate, the top of the valence band shows a heavy mass for the in-plane hole and a light longitudinal hole mass. In the case of an InP substrate, the situation is reversed leading to a configuration in which the hole mobility is enhanced for two-dimensional transport. In this latter case, in-plane transport is favored over longitudinal transport by a factor of 3 in the effective mass. This effect is also reflected in the charge density distribution at the top of the valence band. From a study of the strain effects in the thicker superlattices we were able to show that the hole becomes more and more confined within the InAs region as the in-plane strain is increased and that the band gap of the structure increases as the strain is relieved.

Total energy calculations performed on thicker superlattices aimed at exploring the effect of bond length relaxation on the layers away from the interface, showed that the interface contribution to the potential line-up is not significantly affected by the atomic positions in the layers underneath the interface layer,[74] provided that the interface atomic positions are left unchanged, and that the bulk characteristics are very well recovered in the layers just below the interface.

By looking at the charge profile of the planar charge density and comparing it with the same profile in the pure bulk compounds, we were able to establish that the charge readjustment at the interface is rather small, that it does not significantly contribute to

Table 5. Core energy differences and corresponding valence band offset, ΔE_v, (in eV) for the $n = 2,3$ superlattices considered. ΔE_v^{so} denotes the spin-orbit corrected ΔE_v values. The results corresponding to no substrate constraint are referred to as FSM (free standing mode).

	core state	2 × 2				3 × 3			
		Δb	ΔE_b	ΔE_v	ΔE_v^{so}	Δb	ΔE_b	ΔE_v	ΔE_v^{so}
FSM	1s	0.02	0.21	0.23	0.31	0.01	0.21	0.22	0.30
	2s	0.01	0.21	0.22	0.30	04.02	0.21	0.23	0.31
	3s	0.02	0.21	0.23	0.31	0.02	0.21	0.23	0.31
InP-sub.	1s	0.03	0.39	0.42	0.50	0.03	0.39	0.42	0.50
	2s	0.02	0.40	0.42	0.50	0.03	0.40	0.43	0.50
	3s	0.02	0.40	0.42	0.50	0.03	0.40	0.43	0.50

the potential line-up nor it is sensibly affected by the strain conditions.

The study of the potential discontinuity at the interface was performed by looking at the relative alignment of the core levels in both sides of the interface and comparing them with their respective positions in the bulk compounds. In the structures considered, there is one In atom at the interface (In(i)) and two independent "bulk" In sites on the two sides of the interface: In(bAs) and In(bP) respectively. The evaluation of the band offset is performed by considering the relative alignment of the In(bAs) and In(bP) core levels (Δb) and the binding energy differences (ΔE_b) of the same core levels (relative to the valence band top) as calculated in the pure binary compounds. The total valence band offset is therefore:

$$\Delta E_v = \Delta b + \Delta E_b \tag{20}$$

We found that the relative alignment of different core levels in the two sides of the interface was almost the same, resulting in a very small contribution to the overall potential line-up (Δb); this term is one order of magnitude smaller than the potential step due to the so-called "band" term (i.e. the relative alignment of the two valence band tops in the pure constituents ΔE_b). This is shown in Table 5 where we report our results regarding the valence band offset in these structures and the values corrected for the spin-orbit interaction (which lifts the top of the valence band in the InAs site by 0.08 eV, i.e. $\frac{1}{3}$ of the difference of the spin-orbit corrections in InAs and InP [82]). The strain is seen to mainly affect the energy position of the valence band maximum in the two constituents (i.e. the term ΔE_b) rather than the interface dipole potential (i.e. the term Δb).

These results are very interesting since they show that the effect of the charge and potential readjustment at the interface is very small and that they do not change significantly as the strain conditions and the atomic positions are varied. On the other hand, the most important contribution to the band offset comes from the binding energy difference of these levels in the bulk materials which are much more sensitive to the strain conditions.

Finally, the present results show that InAs/InP junctions are particularly suitable for the design of junctions with tailored valence band discontinuities, since such a discontinuity can be easily estimated from a knowledge of quantities mainly related to the pure bulk compounds, the interface contribution to the band line-up being negligible.

The calculated valence band offset, obtained by considering the core levels as reference energy levels, is in very good agreement with the experimental results available to date for the strained InAs/InP interface [74]. It perfectly matches the XPS result by Waldrop et al. [81] if we take as binary reference structures for the evaluation of the ΔE_b term the same equilibrium structures considered in the experiments (in this last case we obtain $\Delta E_v = 0.31 \pm 0.03$ eV).

Acknowledgments

Work at Northwestern University was supported by the National Science Foundation (through the Northwestern University Materials Research Center, Grant No. DMR88-21571, and a computing grant from its Division of Advanced Scientific Computing at the National Center for Supercomputer Applications, University of Illinois, Urbana/Champaign). Work in Lausanne was supported by the Swiss National Science Foundation, Grants no 2.926-0.85 and 20-5446.87. The work on diborides by S.M. was performed at ISAS (Trieste, Italy) during completion of his PhD. We are grateful to some of our colleagues for close collaboration, especially A. Baldereschi, A. Catellani, M. Weinert and E. Wimmer, on part of the work described here.

References

[1] H. J. F. Jansen and A. J. Freeman, Phys. Rev. B **30**, 561 (1984).

[2] L. Hedin and B. I. Lundqvist, J. Phys. C **4**, 2064 (1971).

[3] U. Von Barth and L. Hedin, J. Phys. C **5**, 1629 (1972).

[4] W. Kohn and L. J Sham, Phys. Rev. **140**, A1133 (1965); L. J. Sham and W. Kohn, Phys. Rev. **145**, 561 (1966).

[5] J. C. Slater, Phys. Rev **51**, 846 (1937).

[6] M. Weinert, J. Math. Phys. **22**, 2433 (1981).

[7] P. Pfluger and H.J. Güntherodt, *Festkörperprobleme: Advances in Solid State Physics*, edited by J. Treusch (Vieweg, Braunschweig, 1981), Vol. **21**, p. 271 and references therein.

[8] H. Kamimura, Ann. Phys. (France) **11**, 39 (1986).

[9] T. Takahashi, N. Gunasekara, T. Sagawa and H. Suematsu, J. Phys. Soc. Jpn **55**, 3498 (1986).

[10] M. Posternak, A. Baldereschi, A.J. Freeman, E. Wimmer and M. Weinert, Phys. Rev. Lett. **50**, 761 (1983); Mat. Res. Soc. Symp. Proc. **20**, 117 (1983).

[11] E. Wimmer, H. Krakauer, M. Weinert and A.J. Freeman, Phys. Rev. B**24**, 864 (1981).

[12] N.A.W. Holzwarth, S. Rabii and L.A. Girifalco, Phys. Rev. B**18**, 5190 (1978).

[13] The three systems studied in this section have D_{6h} symmetry. The irreducible representations of this group are labeled according to the notations of G.F. Koster, J.O. Dimmock, R.G. Wheeler and H. Statz, *Properties of the thirty-two point groups*, M.I.T. Press, Cambridge, (1963).

[14] According to Ref. 12, the bottom of the Li band in LiC_6 is $\sim$ 1.7 eV above E_F. Notice however that, assuming rigid bands, we calculate that E_F (LiC_6) E_F (C_6–Li–C_6) + 0.4 eV since the concentration of Li atoms in LiC_6 is twice that in C_6–Li–C_6.

15] Observation of K 4*s* states in KC_8 at about this energy has been reported *e.g.* by J.J. Ritsko and C.F. Brucker, Solid State Commun. **44**, 889 (1982).

16] L. Samuelson and I.P. Batra, J. Phys. C**13**, 5105 (1980), and reference therein.

17] W. van Haeringen and H.G. Junginger, Solid State Commun. **7**, 1723 (1969).

[18] C.P. Mallett, J. Phys. C**14**, L213 (1981).

[19] Th. Fauster, F.J. Himpsel, J.E. Fischer and E.W. Plummer, Phys. Rev. Lett. **51**, 430 (1983).

[20] F. Maeda, T. Takahashi, H. Ohsawa and S. Suzuki, Phys. Rev. B**37**, 4482 (1988).

[21] T. Takahashi, H. Tokailin and T. Sagawa, Phys. Rev. B**32**, 8317 (1985).

[22] N.A.W. Holzwarth, S.G. Louie and S. Rabii, Phys. Rev. B**26**, 5382 (1982).

[23] M. Posternak, A. Baldereschi, A.J. Freeman and E. Wimmer, Phys. Rev. Lett. **52**, 863 (1984).

[24] G.B. Bachelet, D.R. Hamann and M. Schlüter, Phys. Rev. B**26**, 4199 (1982).

[25] A. Catellani, M. Posternak, A. Baldereschi, H.J.F. Jansen and A.J. Freeman, Phys. Rev.B **32**, 6997 (1985).

[26] R.W.G. Wyckoff, *Crystal Structures* (Interscience, New York, 1965), Vol. 1.

[27] P. Villars and L.D. Calvert, *Pearson's Handbook of Crystallographic Data for Intermetallic Phases* (American Society for Metals, Metals Park, 1985), Vol. 2.

[28] J. Robertson, Phys. Rev. B**29**, 2131 (1984).

[29] A. Catellani, M. Posternak and A. Baldereschi, Phys. Rev. B**36**, 6105 (1987).

[30] W.P. Pearson, *A Handbook of Lattice Spacing and Structures of Metals*, Vol. 2, (Pergamon Press, Oxford, 1967), p. 35.

[31] B. Post, *Boron, Metallo-Boron Compounds and Boranes*, edited by R.M. Adams (Wiley, New York, 1964).

[32] F. Hulliger and E. Mooser, Prog. Solid State Chem. **2**, 330 (1965).

[33] H.J. Juretschke and R. Steinitz, J. Phys. Chem. Solids **4**, 118 (1958).

[34] P. Pfuger and H.J. Guntherodt, *Festkörperprobleme XXI*, ed. J. Treusch (Vieweg, Braunschweig, 1981), p. 271.

[35] D.M. Adams, *Inorganic Solids* (John Wiley, New York, 74) p. 223.

[36] H. Ihara, M. Hirabayashi, and H. Nakagawa, Phys. Rev. B**16**, 726 (1977).

[37] T. Tanaka, Y. Ishizawa, E. Bannai, and S. Kawai, Sol. State Commun. **26**, 879 (1978).

[38] R.C. Linton, Thin Solid Film **20**, 17 (1974).

[39] D.W. Fisher and W.L. Baun, J. Appl. Phys. **39**, 4757 (1968).

[40] L. Ramqvist, B. Ekstig, E. Köllne, E. Noreland, and R. Manne, J Phys. Chem. Solids **30**, 1849 (1969).

[41] I.I. Lyakovskaya, T.M. Zimkina and V.A. Formichev, Soviet Phys.: Solid State **12**, 138 (1970).

[42] I.I. Tupitsyn, I.I. Lyakovskaya, M.S. Nakhmanson, and A.S. Sukhikh, Sov. Phys.: Solid State **16**, 2015 (1975).

[43] Y.S. Tyan, L.E. Toth and Y.A. Chang, J. Phys. Chem. Solids **30**, 785 (1969).

[44] J. Castaing, R. Caudraun, G. Toupance, P. Costa, Solid State Commun. **7**, 1453 (1969).

[45] E. Dempsey, Philos. Mag. **8**, 285, (1963).

[46] A.H. Silver and P.J. Bray, J. Chem Phys. **32**, 288 (1960).

[47] I.I. Tupitsyn, Sov. Phys. Solid State **18**, 1688 (1976).

[48] D.R. Armstrong, A. Breeze, and P.G. Perkins, J. Chem. Soc. Faraday Trans I,I **73**, 952 (1977).

[49] S.H. Liu, L. Kopp, W.B. England, and H.W. Myron, Phys. Rev. **B11**, 3463 (1975).

[50] D.L. Johnson, B.N. Harmon and S.H. Liu, J. Chem. Phys. **73**, 1898 (1980) and references therein.

[51] D.R. Armstrong, Theoret. Chim. Acta **64**, 137 (1983).

[52] S. Massidda and A. Baldereschi, Solid State Commun. **66**, 855 (1988).

[53] S. Massidda, Ph.D. Thesis, ISAS Trieste (Italy) 1985, unpublished.

[54] C. Froese Fisher, Atomic Data and Nuclear Data **4**, 301 (1972).

[55] S. Massidda, B.I. Min and A.J. Freeman, Phys. Rev. **B35**, 9871 (1987).

[56] S.H. Wei and A. Zunger, Phys. Rev. Lett. **59**, 144 (1987).

[57] A. Continenza, S. Massidda and A.J. Freeman, J. Magn. Magn. Mat. **78**, 195 (1989).

[58] A. Continenza, S. Massidda and A.J. Freeman Phys. Rev. **B42**, 2904 (1990).

[59] G.A. Prinz, B.T. Jonker, J.J. Krebs, J.M. Ferrari and F. Kovanic. Appl. Phys. Lett. **48**, 1756 (1986).

[60] G.A. Prinz and J. Krebs, Appl. Phys. Lett. **39**, 397 (1981).

[61] G.A. Prinz, Phys. Rev. Lett. **54**, 1051 (1985).

[62] G.A. Prinz, B.T. Jonker and J.J. Krebs, Bull. of Am. Phys. Soc. **33**, 562 (1988).

[63] B.T. Jonker, K.H. Walker, E. Kisker, G.A. Prinz, and C. Carbone, Phys. Rev. Lett. **57**, 142 (1986).

[64] J.J. Krebs, B.T. Jonker and G.A. Prinz, J. Appl. Phys. **61**, 374 (1987).

[65] S. Ohnishi, A.J. Freeman and M. Weinert, Phys. Rev. B**28**, 6741 (1983).

[66] A.J. Freeman and R. E. Watson in *Magnetism*, Vol. IIA, edited by G. T. Rado and H. Suhl (Academic Press, New York 1966).

[67] L. Pauling in *The Chemical Bond* (Cornell Univ. Press, New York, 1967).

[68] A.R. Miedema, F.R. de Boer and P.F. de Chatel, J. Phys. **F3**, 1558 (1973).

[69] C.A. Mead, Solid State Electr. **9**, 1023 (1966); C.A. Mead, Phys. Lett. **18**, 218 (1966).

[70] F. Xu, M. Vos, J.H. Weaver and H. Cheng, Phys. Rev. B**38**, 13418 (1988).

[71] S.G. Anderson, F. Xu, M. Vos,J. H. Weaver and H. Cheng, Phys. Rev. B**39**, 5079 (1989).

[72] G.P. Das, P. Blochl, O.K. Andersen, N.E. Christensen and O. Gunnarsson, Phys. Rev. Lett. **63**, 1168 (1989).

[73] A. Continenza, S. Massidda and A.J. Freeman, Phys. Rev B**41**, 12013 (1990).

[74] A. Continenza, S. Massidda and A. J. Freeman, Phys. Rev. B**42**, 3469 (1990).

[75] C.G. Van de Walle and R. Martin, Phys. Rev. B**37**, 4801 (1988) and references therein.

[76] W.R.L. Lambrecht and B. Segall, Phys. Rev. B**41**, 2832 (1990).

[77] A. Baldereschi, S. Baroni, and R. Resta, Phys. Rev. Lett. **61**, 734 (1988).

[78] R.P. Schneider, Jr. and B.W. Wessels, Appl. Phys. Lett. **54**, 1142 (1989).

[79] K. Huang and B.W. Wessels, J. Appl. Phys. **64**, 6770 (1988).

[80] D.V. Lang, M.B. Panish, F. Capasso, J. Allam, R.A. Hamm, A.M. Sergent and W.T. Tsang, J. Vac. Sci. Technol. B**5**, 1215 (1987).

[81] J.R. Waldrop, R.W. Grant and E.A. Kraut, Appl. Phys. Lett. **54**, 1878 (1989).

[82] S. Massidda, A. Continenza, A.J. Freeman, T.M. De Pascale, F. Meloni and M. Serra, Phys. Rev B**41**, 12079 (1990).

HELIUM ATOM SCATTERING STUDIES OF THE SURFACE PHONONS OF LAYERED MATERIALS

by

James G. Skofronick
Department of Physics,
The Florida State University,
Tallahassee, Florida, USA 32306

and

J. Peter Toennies
Max Planck Institut für Strömungsforschung
Bunsenstr. 10
W-3400 Göttingen, Germany

ABSTRACT. The scattering of He atoms provides a nearly ideal probe of the structure and vibrations of crystal surfaces because the energy and momentum of the neutral He atoms match closely those of the surface phonons. Moreover He atoms are by virtue of their low energy only sensitive to the outer layer and are non-destructive and chemically inert. In this review we describe He scattering studies in which the surface phonon dispersion curves of the layered compounds of graphite, 2H-$TaSe_2$, 1T-TaS_2, GaSe and the high T_c superconductor $Bi_2Sr_2CaCu_2O_8$ were measured. Despite the small coupling between the two-dimensional layers in the bulk of these systems, only the graphite surface behaves as predicted by theoretical modeling based on its bulk properties. The other surfaces show special phenomena such as phase changes, critical phenomena, charge-density-waves and commensurate-incommensurate behavior which are different than similar phenomena observed in the bulk.

G. Benedek (ed.), Surface Properties of Layered Structures, 151–218.

1. Introduction

There is an ever increasing basic research effort to develop a better understanding of the chemical, physical and engineering properties of low-dimensional materials [1,2]. The surfaces of naturally layered materials provide information and guidance toward the understanding of low-dimensional systems. This information will also be useful for understanding superlattices prepared by molecular beam epitaxy.

This review is devoted to He atom scattering (HAS) studies on the phonon dispersion curves from the surfaces of naturally layered crystals. The emphasis is given to graphite [3-5], $2H\text{-}TaSe_2$ [6-12], $1T\text{-}TaS_2$ [13-15] and GaSe [16,17]. Some preliminary work on the layered high-T_c superconductor $Bi_2Sr_2CaCu_2O_8$ is included because of the intense interest in this material [18]. The results are mainly in the form of surface structure and phonon dispersion relations as well as studies of phase transitions and related phenomena. Experimental surface information obtained by other techniques is included when appropriate, but generally these alternate means are not thoroughly reviewed.

As a surface science technique, HAS can be compared with Low Energy Electron Diffraction (LEED), Electron Energy Loss Spectroscopy (EELS) for the impact regime ($E \geq 100eV$) and Scanning Tunneling Microscopy (STM). Although HAS dates back to the 1930's, its modern development goes back only about 10 years [19,20]. As with electron scattering, HAS is a diffractive technique in which angular and energy distributions of the particles scattered from a clean single crystal surface are measured [19,21-24]. In order to have the same de Broglie wavelengths, the energy of the He atom beams is typically about 20 to 60 meV compared to 100-200 eV for an electron beam. Helium atoms interact strongly with the electronic charge distribution of the crystal, about 3Å from the surface ion cores, and are therefore sensitive only to the first layer; further, at these low incident energies they are non-destructive. Electrons, on the other hand, penetrate several layers and can also lead to a certain amount of radiation damage. Thus HAS does have unique advantages with respect to surface sensitivity and non-destructiveness, as compared to electron probe surface techniques.

In this review, the crystal surface is considered at two levels of complexity. The simpler approach regards the surface as rigid with an interaction with He atoms described by a potential which has a corrugation with the same periodicity as the surface lattice. The chemically inert He atoms are elastically scattered in the form of a specular (mirror) reflection peak and diffraction (Bragg) peaks caused by the corrugation. The intensity of the peaks depends upon the static corrugation amplitude of the surface potential. Further, the intensities may be modified by interference effects caused by the He being temporarily trapped by resonances with bound states of the He-surface potential [23-25].

The next level of complexity treats the surface atoms as having collective vibrations about their equilibrium positions. The collective vibrations are called surface phonons and for a simple crystal with one atom per unit cell there are three (acoustic) modes, two having polarizations transverse and one with a polarization longitudinal with respect to the direction of propagation. The surface phonons are further characterized by dispersion curves which relate their frequencies and wavevectors. He atoms scatter inelastically from the vibrating surface with the exchange of energy and momentum. Experimentally, inelastic scattering is studied using time-of-flight (TOF) techniques: the nearly monoenergetic He beam produced in a nozzle beam expansion is first chopped into short pulses after which the atoms collide with the surface where they either annihilate or create single phonons. The scattered He atoms then travel a flight path long enough for the atoms with different velocities to spread out spatially and be resolved as a function of time as they pass through the detector. The TOF information and the scattering angles determine the energy and momentum of particular phonons. By changing scattering angles the entire surface dispersion curve is mapped out [19,21].

To a large degree, HAS parallels thermal energy neutron scattering. Both probes use neutral particles with energies and momenta closely matched to those of the phonons. Their interactions are, however, entirely different; whereas the neutrons penetrate the crystal and interact only with the point nuclei, the He atoms scatter from the electron density distribution on the crystal surface at distances of 3-4Å away from the topmost atomic layer [21,26]. The neutrons are thus sensitive to the bulk vibrations while the He atoms are only sensitive to the surface vibrations. In general, the surface vibrations can be attributed to: 1) localized surface modes, 2) the projection of the bulk modes onto the surface, which appear as bands, and 3) surface resonances, which are surface modes which lie in and are mixed with the bulk bands [27]. The surface localized Rayleigh mode usually couples strongest to the He atoms because it is predominantly polarized normal to the surface plane and is invariably observed. Additional higher energy modes are also frequently observed. The experimental surface mode dispersion curves usually differ from the behavior predicted by assuming the same force constants at the surface as in the bulk lattice. Finally because of the extreme surface sensitivity of HAS, this technique has an advantage over neutrons in that only microscopically thin samples are needed. For neutron scattering, single crystal samples with a volume of typically about $1cm^3$ are required depending upon the size of the neutron cross sections.

In the layered crystals under consideration the forces between the atoms are very anisotropic. They are characterized by a strong intralayer bonding with a much weaker interlayer coupling, the latter usually being of the van der Waals type. The weak interlayer bonding suggests that the bulk behaves dynamically as a nearly two-dimensional material and that the bulk and surface vibrational properties should be very similar. For graphite (Section 3), this appears to be the case, but for the other materials considered in this review, the vibrations differ by various degrees from the bulk results. In particular, mode softening, phase changes and the charge-density-waves (CDW) are different at the surface to those found in the bulk.

Some of the present surface dynamical studies follow the early pioneering structural studies using He scattering reported by the Genoa group around 1980 [11-14]. For 2H-$TaSe_2$ and 1T-TaS_2 they found the same CDW structures that had been observed in the bulk. For graphite, they were able to obtain some dynamical information through the use of an inelastic resonance effect. At that time they did not have the capability for energy analysis (TOF) of the scattered He beam and they were unable to measure surface phonon dispersion curves out to the zone boundary [4]. The inelastic results, which comprise the main part of this review, were nearly all obtained at the Max Planck Institut für Strömungsforschung (MPI) in Göttingen (Gö) as part of a focussed research effort on layered materials [3,4,6-10,15,16,18,28,29].

The experimental procedures are discussed in the next Section. This is then followed by five Sections in which the results for each of the layered crystals studied by HAS are presented. The review concludes with a Summary Section. In a companion chapter of this book, a more detailed theoretical interpretation of these results is presented.

2. Helium Atom Scattering Experimental Techniques

Two schematic views of the scattering geometry and of the instrument used for HAS surface vibrational studies are shown in Figs. 1a and 1b [19]. The sagittal scattering plane is defined

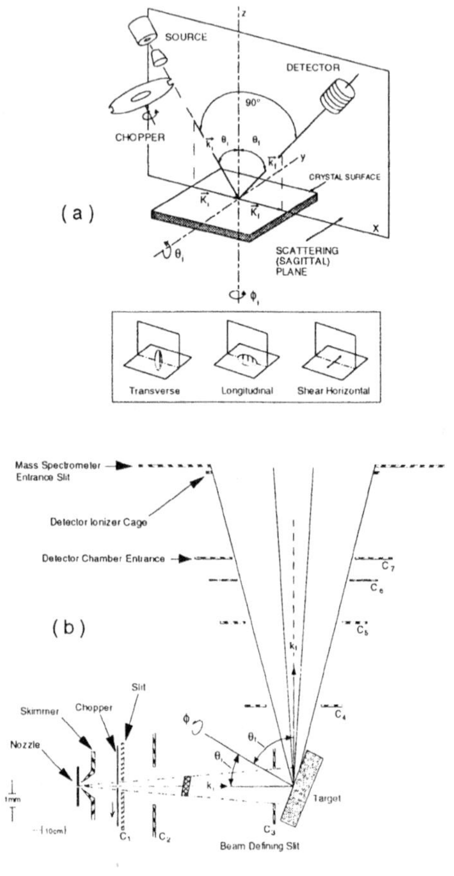

Figure 1. (a) Schematic view of the scattering geometry showing the incident and final scattering wavevectors, $\mathbf{k}_i$ and $\mathbf{k}_f$, their projections onto the crystal surface, $\mathbf{K}_i$ and $\mathbf{K}_f$ and the incident and scattered angles, θ_i and θ_f. The lower panel inset shows schematically the three types of surface polarizations with respect to the sagittal plane. (b) Schematic view of the scattering apparatus. The beam is formed in a nozzle source and chopped to form narrow pulses for use in time-of-flight measurements. The pulses pass through at least two additional stages of differential pumping before they collide with the crystal target. He atoms which are scattered at 90° to the incident beam pass through three stages of differential pumping before reaching the detector chamber where they are ionized, mass selected and individually detected. The crystal-to-detector distance is about one meter. The angular distribution is measured with the chopper moved out of the beam path. The C_i's refer to collimator and slit locations in the different vacuum chambers which separate the differentially pumped sections of the scattering apparatus (not shown on the sketch) [19].

by the incident wavevector $\mathbf{k}_i$ and the surface normal z direction as shown in Fig. 1a. The three polarizations of the surface phonons, shown in the inset of Fig. 1a, are defined with respect to this plane. The transverse and longitudinal modes lie in the sagittal plane and the shear horizontal mode is polarized perpendicular to it. Whereas the latter has a well-defined polarization normal to the sagittal plane, the transverse and longitudinal modes nearly always form coherent mixtures of both polarizations. When scattering in the sagittal plane, only the transverse and longitudinal modes couple to the He atoms.

The He beam is produced by expanding He gas at a pressure of typically 200 bar through an aperture about 10 μm in diameter into vacuum [19,30]. This produces a beam of atoms with a velocity spread of only $\Delta v/v \approx 1\%$ where the most probable beam velocity v can be varied by heating and cooling the source. For example, at liquid nitrogen source temperatures, the velocity v is about 900 m/s (energy $\approx$ 19meV) with a de Broglie wavelength of about 1Å and a wavevector of about 6Å^{-1}. The beam passes through the skimmer into a second differentially pumped chamber where for the inelastic experiments the beam is chopped into 5mm long 5μs pulses with a repetition rate of typically 10^3Hz. The atoms then pass through at least two additional stages of differential pumping before reaching the target chamber where they collide with the crystal. For all of the experimental scattering work presented here the detector axis was fixed at 90° to the incoming beam and the crystal was rotated to obtain different incident angles, θ_i, measured with respect to the surface normal. The scattered atoms then pass through typically three more stages of differential pumping before reaching the detector chamber where they are ionized, mass selected and the individuals ions are detected by an open electron multiplier. Angular distributions of the total intensity as a function of incident angle are measured with the chopper removed from the beam path, with an overall angular resolution (FWHM) of 0.1-0.3°. The TOF measurements are then carried out for fixed incident wavevector and different incident angles to map out the surface phonon dispersion curves.

It has been found to be satisfactory to use normal vacuum technology in the beam source chamber and the chopper chamber with base pressure of about 10^{-7} mbar. These chambers are sealed with O rings, but the rest of the system is an ultrahigh-vacuum apparatus with bakeable copper gaskets [19,21]. The background pressures in the target and detector chambers are typically 10^{-10} mbar and 5×10^{-11} mbar, respectively. During operation with the He beam on, the pressure in most of the rest of the beam path is in the 10^{-9} mbar range except for the source and chopper chamber where the pressures are approximately 10^{-4} and 10^{-6} mbar, respectively. Special additional details will be provided in the sections on each crystal.

The crystals were prepared in several different ways. Some were simply mounted onto the crystal holder on the manipulator by clamps and sticky tape was used to remove a layer to provide a fresh surface in air or in an argon atmosphere. Following this, the sample and manipulator were installed in the target chamber and the chamber was evacuated. In other cases, the sample was glued onto a target holder with a high temperature conducting epoxy [31] and a lever glued to the top of the target. By pushing on the lever, either in air or vacuum, a clean surface was exposed. In all cases, a bakeout of the target chamber at about 150° C insured the vacuum discussed above.

The crystal temperatures during operation were mostly below room temperature, cooling being achieved with liquid nitrogen or liquid helium. The temperature was stabilized with a computer controlled resistive heat source, achieving an uncertainty in absolute temperature of $\pm$0.5 K at 300 K and $\pm$2 K at 60 K and a relative temperature stabilization of $\pm$0.3 K.

2.1. ANGULAR DISTRIBUTIONS

The diffraction peaks observed in the angular distributions are analyzed with the aid of the Bragg equation

$$\Delta \boldsymbol{K} = \boldsymbol{K}_f - \boldsymbol{K}_i = \boldsymbol{G} \tag{1}$$

where $\Delta\mathbf{K}$ is the surface parallel momentum transfer ($\hbar = 1$), and $\mathbf{K}_i$ and $\mathbf{K}_f$ (see Fig. 1a) are the projections onto the surface of the incident and final He-atom wavevectors $\mathbf{k}_i$ and $\mathbf{k}_f$, respectively. ($K_i = k_i \sin\theta_i$ and $K_f = k_f \sin\theta_f$.) The vector $\mathbf{G}$ is a reciprocal surface lattice vector of magnitude $G = 2\pi m/a$, where a is the spacing between the diffracting lattice rows and m is a positive or negative integer. Equation 1 is frequently referred to as describing the conservation of parallel quasi-momentum. From the angular positions of the Bragg peaks the size of the surface unit cell is determined and the Bragg intensities give information on the amplitude of the corrugation of the He-atom surface potential. The intensities are plotted either as a function of incident angle θ_i or of surface momentum transfer ΔK as in Eq. 1.

The theory relating the Bragg peak intensity to the corrugation amplitude has been considered by a number of investigators [22,32,33]. In this review it will be sufficient to work with a simple approximate eikonal form which assumes the He-surface interaction can be described by a hard wall potential which is sinusoidal in shape and has an amplitude which is small compared to the distance between the atoms. The theory is analogous to that describing the scattering of sound waves from a corrugated wall as first described by Lord Rayleigh in his pioneer studies of sound [34]. The height of the hard-wall repulsion $z = \zeta(\mathbf{R})$ is a function of $\mathbf{R}$, the two-dimensional position coordinate on the crystal surface. For a one-dimensional crystal the following (simple) corrugation profile assumes a sinusoidal corrugation with maxima at the locations of the atoms and minima between the atom cores.

$$\zeta(x) = \zeta_1 \cos(2\pi x/a). \tag{2}$$

Other corrugation functions for the investigated two dimensional surfaces are provided in the corresponding Sections.

The scattered intensity P_G, depends on the corrugation function through the scattering amplitude A_G^0 as given by

$$P_G = |A_G^0|^2 \cos\theta_f / \cos\theta_i, \tag{3}$$

where in the eikonal approximation, A_G^0 can be expressed as an integral over the surface unit cell with an area $\mathscr{F}$

$$A_G^0 = \frac{1}{\mathscr{F}} \int \exp\{i[\boldsymbol{G}\cdot\boldsymbol{R} + (k_{iz} - k_{fz})\zeta(\boldsymbol{R})]\}\, d\boldsymbol{R} \ . \tag{4}$$

Here k_{iz} and k_{fz} are the components of the wavevectors normal to the surface for the incident and scattered waves, respectively. By comparing the experimental diffraction intensities with those calculated from Eqs. 2 through 4 both the form and the amplitude of the corrugation can be determined.

2.2. TIME-OF-FLIGHT SPECTRA

The angular distributions reveal a total residual intensity between the diffraction peaks with typically 10^{-3} to 10^{-4} of the specular intensity. This intensity between the Bragg peaks is due to inelastic scattering from one-phonon and multiphonon surface collisions as well as elastic scattering from surface defects [19]. In this work, the one-phonon inelastic collisions are of primary interest, and the beam energy and surface temperature are optimized to enhance their relative intensities [35-37].

To describe the inelastic scattering from single phonon-creation or -annihilation events, the conservation of parallel momentum has to be modified to include the surface projection of the phonon wavevector **Q**. Thus Eq. 1 becomes

$$\Delta K \equiv K_f - K_i = G + Q. \tag{5}$$

In addition, conservation of energy in a one-phonon He atom collision requires that

$$\hbar\omega = \frac{\hbar^2}{2m}(k_f^2 - k_i^2), \tag{6}$$

where $\hbar\omega$ is the phonon energy and ω is the phonon angular frequency which is positive for energy gain by the He atoms (annihilation event) and negative for energy loss by the He atom (creation event) and m is the mass of the He atom. From the time of flight of the peaks in the measured TOF spectra and the chosen incident and final angles, Eqs. (5) and (6) can be used to evaluate the surface phonon frequency as a function of **Q**, thus determining points on the surface dispersion curve, which is written as $\omega(\mathbf{Q})$. Invariably the lowest energy transverse Rayleigh mode and additionally often the longitudinal and higher energy sagittal plane modes or resonances are observed. A coupling to the shear horizontal modes is forbidden in the sagittal plane arrangement. (See inset in Fig. 1a.)

Equations (5) and (6) can be combined to provide a relationship between surface phonon energy $\hbar\omega$ and $\mathbf{\Delta K}$ for given incident and final angles and fixed incident energy $E_i = \hbar^2 k_i^2/2m$. The resulting expression is useful for interpreting the experimental data and describes the apparatus scan curve. For an apparatus with $\theta_i + \theta_f = 90°$, the scan curve equation is

$$\hbar\omega / E_i = [(\Delta K + k_i \sin\theta_i)^2 / k_i^2 \cos^2\theta_i] - 1. \tag{7}$$

When Eq. 7 is plotted on a theoretical surface dispersion curve in an extended zone scheme, the intersections of the scan curves and the dispersion curves identify the peaks which can be seen in TOF measurements. From the experimental point of view, the energy transfer corresponding to the peaks in the TOF spectra are projected on to the scan curves to determine the dispersion curves [19]. As will be seen later in this review, the scan curves are often very helpful in the interpretation of TOF spectra.

The raw data are taken as TOF spectra and often they are presented in that form. To facilitate their interpretation, they are frequently transformed to energy distributions where the elastically scattered signal corresponds to zero energy loss and positive (negative) energy transfers are associated with phonon-annihilation (-creation) events, respectively.

3. Graphite

3.1. INTRODUCTION

Graphite is probably the best known of all of the layered materials discussed in this article. The large surface to volume ratio in finely divided graphite [38] as is present in exfoliated pyrolytic graphite makes this system accessible to a wide variety of techniques. Since it has been used in many chemical and physical studies it is an important prototypical layered compound. For example, the adsorption of various gases and the intercalation of many substances have been extensively studied [39,40]. Neutron scattering measurements of Nicklow et al. [41, called NWS] and Dolling and Brockhouse [42] provided much of the early information on the bulk dispersion modes and several models were used for an interpretation of the lattice dynamics of the clean surface [41-43]. Early measurements also included ir reflectance [44-46] and Raman spectroscopy [44,45,47]. Starting in 1981 efforts were undertaken to evaluate theoretically the surface lattice dynamics of graphite [40,48,49]. In 1986 the first surface HAS time of flight measurements were reported [3]. Since then there have been further studies by HAS [28], by Electron Energy Loss Spectroscopy (EELS) [50,51] and by Scanning Tunneling Microscopy (STM) [52]. With the exception of the HAS experiments and the EELS results of Oshima et al. [51] all the work has been on exfoliated graphite in which the crystallites have random azimuthal orientations. In this Section mostly HAS studies on clean graphite single crystal surfaces will be reviewed. Structural information and results on the bound states for the He-Graphite potential will be presented first. Although not of direct relevance to the surface dynamics this discussion serves to illustrate other information which can be obtained by the HAS method. Next the surface lattice dynamics results will be presented. Finally, a recent helium scattering study of xenon adsorbed on graphite will also be included as another illustration of a potentially fruitful area of future research also for the other layered compounds.

3.2. EXPERIMENTAL

Graphite is a layered material consisting of sheets of carbon atoms which are bound together in the layers by strong covalent bonds to form a hexagonal two-dimensional array. The sheets are held together by weak van der Waals forces. Figure 2a shows the crystal structure for graphite and the unit cell which contains four atoms. Figure 2b shows the surface Brillouin zone for graphite and the two principal high symmetry directions.

There have been three HAS studies carried out on graphite, the initial one by the Genoa group [4] the other two from the MPIGö [3,28]. The first of the MPIGö results, has already been published [3] but has since been superceeded by a more recent unpublished study [28]. All three HAS studies used natural graphite single crystals [53] but in the latest MPIGö work a synthetically grown Kish sample [54] was also studied [55].

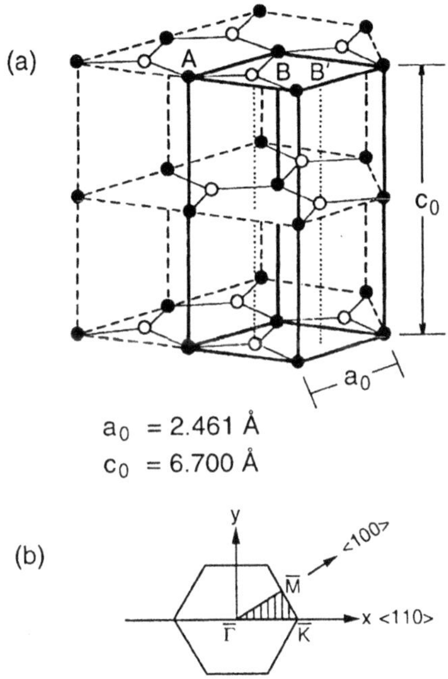

Figure 2. (a) Crystal structure for graphite. The solid lines show the bulk graphite unit cell. (b) The lower sketch shows the surface Brillouin zone for graphite and the two principal directions. After Ref. 41.

The pioneering Genoa research was primarily directed toward the surface structure and bound states for the He-graphite interaction [4]. Although they did not have TOF capabilities they were able to use information from inelastic bound state resonances to obtain a part of the graphite surface dispersion curve for the Rayleigh mode. Both of the MPIGö efforts concentrated on inelastic studies in an apparatus specifically designed for such measurements [19]. LEED studies were done on both the Kish and the natural crystals in the latest work and the crystals were found to consist of several subcrystallites each rotated and tilted slightly with respect to each other. On the Kish crystal, a single crystal region of about $1.5 x 2 mm^2$ could be identified. Thus in the HAS measurements, it was possible to probe only the single crystal region by reducing the collimator in front of the Kish target to 1mm diameter (target spot size 1.6mm). The LEED peaks from the natural crystal were much broader and a scan of the crystal revealed a mosaic structure over the entire surface of the sample (about $7 x 10 mm^2$) with single crystallites of about $1 mm^2$ or smaller.

Both crystals were cleaved in air prior to insertion into the target chamber in which the

LEED and the HAS measurements were done. After the normal bakeout of the vacuum chamber, the crystals were typically flashed for cleaning to approximately 700 K.

3.3. SURFACE STRUCTURE AND HE-SURFACE POTENTIAL

Figure 3 shows the specular and first-order Bragg peaks seen in an angular distribution for He on the natural graphite crystal in the <100> direction [28]. The natural crystal had broader diffraction peaks than the Kish crystal (not shown) primarily because of the mosaic spread described above. Several small broad maxima and minima are seen between the diffraction peaks in the angular distribution of Fig. 3. They are attributed to enhanced phonon intensities as a result of selective adsorption resonance capture into bound states and have been used to determine the bound state energies [56-58]. For the He-graphite system these bound state energies are compared in Table I with those observed in several other previous measurements.

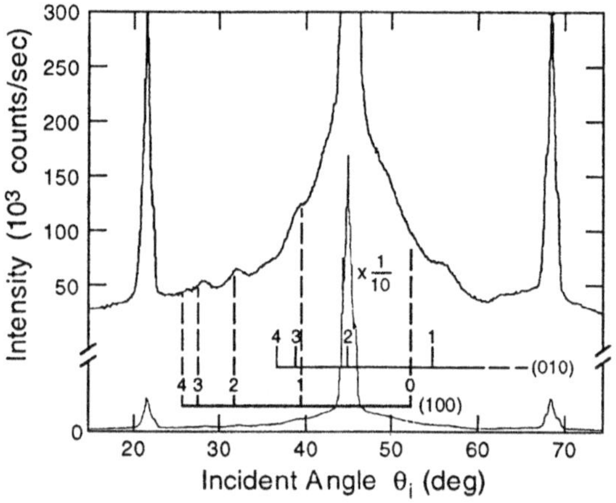

Figure 3. Measured angular distribution (full scale and a reduced scale version) for He scattered from natural graphite for incident wavevector $k_i = 5.24 Å^{-1}$, in the <100> direction and for a target temperature of 110 K. The angle between the incident and outgoing beams was fixed at 90°. The large peaks at 21° and 69° are the first-order Bragg peaks while the smaller peaks, which are labeled according to their selective adsorption bound state and reciprocal lattice vector, are due to resonance enhanced scattering.

With the exception of the n=4 and 5 states, most of the bound state energy levels shown in Table I are in reasonable agreement. The missing value for the ground state (n=0) in the work of Ref 28 is due to the fact that the corresponding peak may be distorted by another resonant channel. The n=5 level measured both by the Penn State group, [57] and the Princeton group, [58] is not observed in Ref. 28 and in the results of Ref. 56. In the recent MPIGö investigation considerable effort was made to confirm this additional state. Because of the fixed angle arrangement, Fig. 1, the more customary measurements of specular intensity as a function of incident angle were not possible. Besides the angular distribution measurements, a measurement of the specularly scattered intensity as a function of incident wavevector was carried out. In such measurements the resonances show up as small minima appearing in the specular intensity at the resonance incident wavevector. In the range of incident wavevectors investigated, no minimum corresponding to the extra state, n=5, could be observed. For the n=4 levels the results of Ref. 28 come closest to those of Ref. 56. The other results of Refs. 57 and 58 also agree fairly well and predict deeper bound states as well as the n=5 bound state. Probably the measurements of Ref. 57, being the most extensive, are the most accurate. The well depth D for the He-graphite potential deduced from these measurements lies between 14.9 - 15.2 meV [59].

Table I. Bound State Energy Levels (meV) for He on Graphite

Level n	MPIGö Ref. 28	Genoa Ref. 56	Penn State Ref. 57	Princeton Ref. 58
5	---	---	-0.16	-0.13
4	-0.23	-0.17	-0.47	-0.49
3	-1.15	-0.99	-1.24	-1.20
2	-2.95	-2.85	-3.05	-3.08
1	-6.45	-6.33	-6.59	-6.56
0	---	-11.98	-12.42	-12.27

The diffraction peak intensities provide information on the surface potential corrugation as described in Section 2.1. The corrugation function $\zeta(R)$ used to evaluate the data was first given by Boato et al. [60].

$$\zeta(R)=2\zeta_{10}\{\cos(2\pi x/a)+\cos(2\pi y/a)+\cos[2\pi(x-y)/a]\}, \tag{8}$$

where only one corrugation amplitude ζ_{10} need be determined. The Genoa group reported a peak to peak corrugation of ζ_{10} = 0.21Å [60] while the result of Ref. 28 was 0.15Å±0.01Å for both the Kish and natural crystals. As with $TaSe_2$ (Section 4), the differences are well within the expected errors. Thus it appears that the corrugation and the bound state energy levels of the He-graphite potential are reasonably well understood.

3.4. SURFACE VIBRATIONS

The bulk lattice dynamics of graphite have been extensively studied both experimentally [41,42] and theoretically [40,41,43]. Since each unit cell contains four atoms, two from each layer there are twelve bulk phonon modes. At the zone origin these modes can be labelled according to the symmetry of the atomic displacements as shown in Fig.4. The Raman and infra-red active modes are indicated [46]. All of the modes involve weakly dispersive interplaner motions and thus form fairly narrow bands of states at the surface.

The surface dynamics of graphite was first modeled theoretically by de Rouffignac et al. (RAW) [40] and Alldredge et al. [48]. They performed a 13-layer slab dynamics calculation using carbon-carbon interactions carried out to 2nd and 3rd nearest neighbors for the interplanar and intraplanar interactions, respectively [40]. The calculated surface phonon dispersion curves are shown in Fig. 5 for the $\bar{\Gamma}\bar{M}$ surface direction. The bands labeled 2 (E_{1ua}/E_{2g1}) and 5 (E_{2g2}/E_{1u}) are shear horizontal modes which have polarizations in the surface plane and are perpendicular to the sagittal plane, whereas the modes labeled 1 (A_{2ua}/B_{1g1}), 3 (E_{1ua}/E_{2g1}), 4 (B_{1g2}/A_{2u}) and 6 (E_{2g2}/E_{1u}) lie in the sagittal plane. Of the latter four modes, the modes 3 and 6 are longitudinally polarized in the direction of propagation, while modes 1 and 4 are transverse modes with a polarization normal to the surface and to the direction of propagation (See inset in Fig. 1a). Note that for wavevector values above the middle of the surface Brillouin zone there are band crossings which depend upon the dynamical character of the modes. The crossings are indicated in Fig. 5 and discussed thoroughly in Ref. 40.

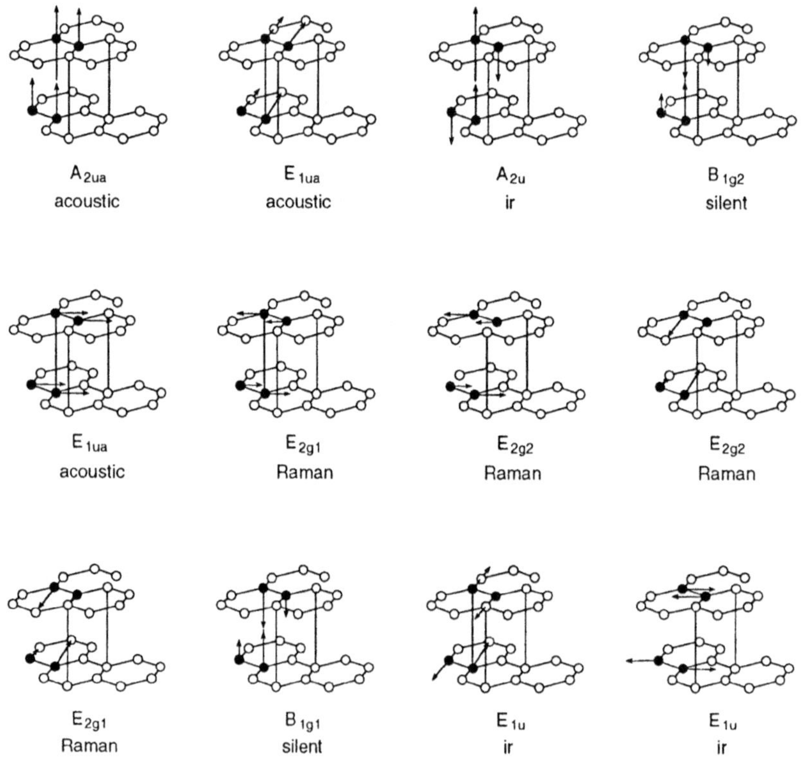

Figure 4. Atomic displacement pattern of the 12 normal modes at the zone origin of bulk graphite. The infrared and Raman activity is indicated (after Ref. 46).

The energy of the lowest mode of the surface dispersion curves shown in Fig. 5 is slightly below the bottom of the bulk band. The bulk modes can be used to approximate the bulk bands; they are not plotted in Fig. 5 because they would obscure the other curves. A later slab dynamics calculation removed some inadvertent symmetries which were in the NWS model, but it did not substantially change the previous results [40,48]. In what follows, the theoretical slab dynamics of Fig. 5 will be used for comparison with the experiments. Results for the $\overline{\Gamma K}$ direction were not published, but it was stated that it had features similar to those of the $\overline{\Gamma M}$ direction and for equivalent wavevectors, the surface dispersion curves for the two directions were very nearly the same [40].

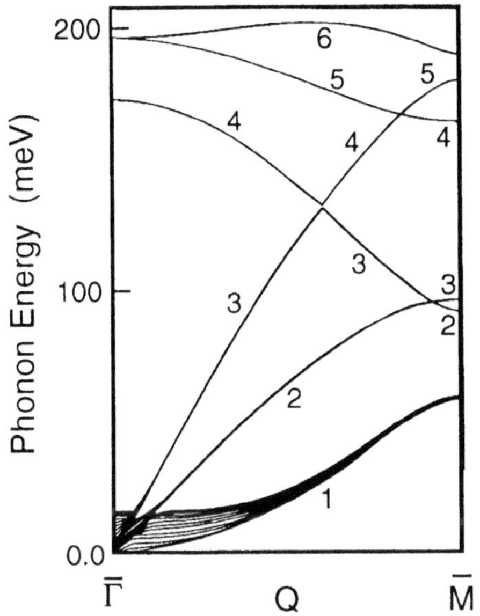

Figure 5. Theoretical surface phonon dispersion curves in the $\bar{\Gamma}\bar{M}$ direction for graphite as obtained from the slab dynamics calculation (After Ref. 40). See text for meaning of indices in the figure.

The important predictions of the RAW theory for HAS measurements are that the graphite surface dynamical behavior is similar to that which would be predicted from the bulk modes with the localized surface modes lying only slightly below the bulk band edges. Since graphite has strong covalent intraplanar bonding and weak interplanar bonding, it was believed that the same behavior should occur for other layered materials described in this chapter which have similar bonding characteristics.

In a remarkable achievement the Genoa group [4] was already able to determine the first surface phonon dispersion curves for graphite in 1972 from observations of the angles at which inelastic bound state resonances affected the intensities in the angular distributions. The dispersion curve of the Rayleigh mode determined in this way is shown in Fig. 6 for the $\bar{\Gamma}\bar{M}$ direction up to parallel wavevectors of 0.3 G_{10}, where G_{10} = 2.95 Å^{-1}, and for mean phonon energies up to 20 meV. As seen from the Figure the results are in reasonable agreement with both the theoretical surface dispersion curve from RAW for the Rayleigh mode as well as the bottom of the bulk acoustic band as determined from the model calculations of NWS [40,41].

More detailed and extensive results were made available by the more recent high resolution TOF HAS studies of Refs. 3 and 28. Two particularly interesting examples out of about 100 measured TOF spectra are shown in Figs. 7a and 7b for the <110> direction of the Kish crystal and the <100> direction for the natural graphite crystal, respectively [28]. The spectra exhibit several peaks of which the one labeled (E) is a diffuse elastic peak, caused by defects on the surface. There are one or two additional peaks associated with the Rayleigh mode (R) and sometimes a higher energy mode attributed to a sagittal plane longitudinal mode (L). The results from the Kish crystal are clearly better resolved as compared to the spectrum from the natural crystal. In Fig. 7c, an additional spectrum from the Kish crystal is shown which includes a higher energy peak from the longitudinal mode.

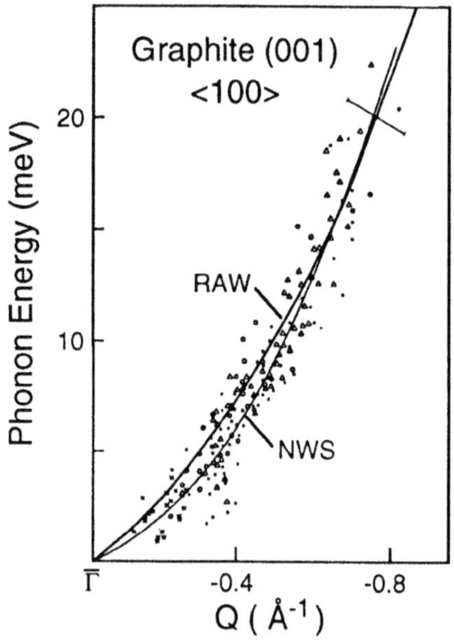

Figure 6. Surface phonon dispersion curves in the $\bar{\Gamma}\bar{M}$ direction for graphite as obtained from phonon-assisted resonances in the He atom scattering from graphite [Modified from Ref. 4]. The different symbols refer to different experimental conditions and the error bar represents the angular uncertainty in the position of the resonance. The two solid lines are from neutron scattering experiments (NWS refers to Ref. 41) and from slab dynamical calculations (RAW refers to Ref. 40).

The corresponding dispersion curves are shown for the Kish crystal in Fig. 8 for the <110> direction and two representative error bars are included for two of the measurements made at an incident wavevector of 7.6 Å^{-1}. There is good agreement with the theoretical Rayleigh and longitudinal modes as was also observed in the case of the experimental Rayleigh mode in the initial study [3,28]. An additional mode at higher energies labeled S_7 in the first study was not observed in the more recent work [3,28] and therefore does not appear in Fig. 8. The longitudinal S_6 mode is only observed for the Kish crystal over a narrow energy range. For the natural graphite crystal, there were additional peaks in the <100> direction corresponding to surface phonon energies of about 33 meV and phonon wavevectors $Q \approx 1$Å^{-1} which extended and agreed with the results in Fig. 8. They are not shown separately, but are included in Figs 9 and 10. The more recent work of Ref. 28 is now considered to be more reliable than the earlier study [3].

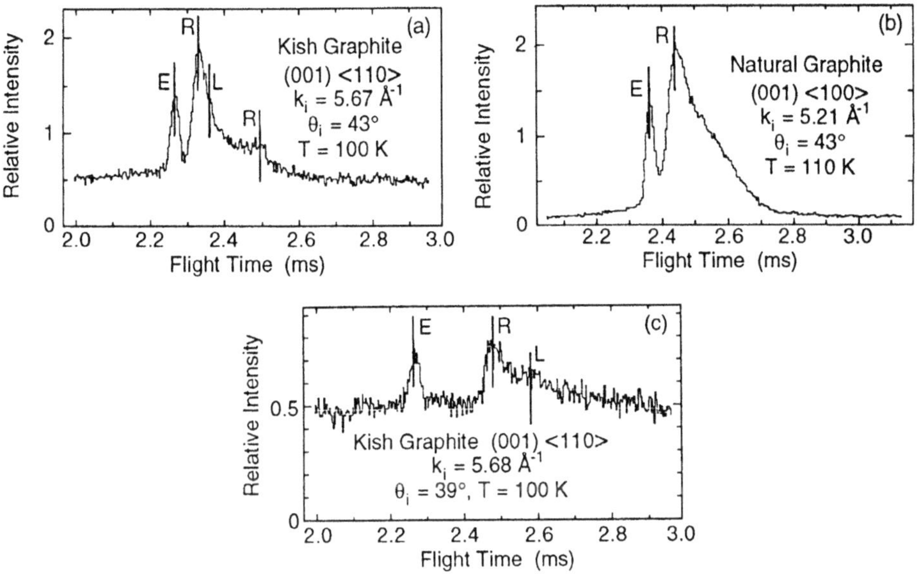

Figure 7. Several TOF spectra taken for He atoms scattering from graphite. The incident wavevectors, angles and target temperatures are given in the figures along with the interpretation of the modes. E indicates the diffuse elastic peak, R is a Rayleigh mode and L is a longitudinal mode. Peaks in the spectra which occur after the diffuse peak are from surface phonon creation events, while those which occur at times shorter than the diffuse peak come from surface phonon annihilation collisions. The spectrum labeled (a) taken from the Kish crystal should be compared to (b) which was taken from the natural graphite crystal under nearly the same experimental conditions. The peak labeled L in (c) is that for the highest energy longitudinal mode measured in this work.

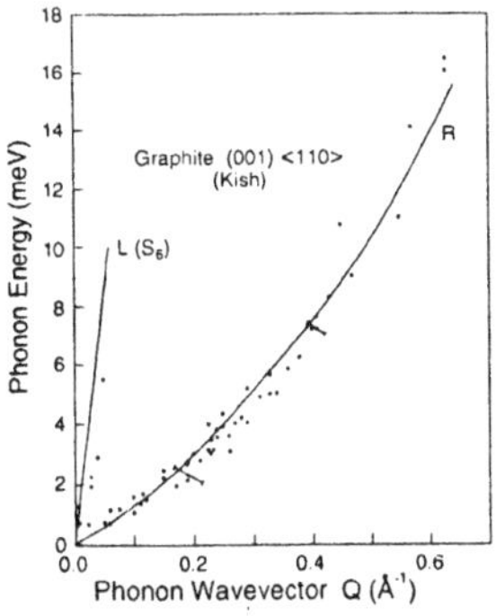

Figure 8. Comparison of the HAS experimental Rayleigh and longitudinal surface phonon dispersion results for the Kish graphite crystal in the ΓK direction with the theoretical model of Ref. 40 (solid lines). The longitudinal and Rayleigh modes are labeled by L (S_6) and R, respectively.

By using a highly ordered pyrolytic graphite sample, Wilkes et al. [50] in an EELS study were able to obtain dispersion results over a considerably expanded range. Because of the broad mosaic spread of such crystals and resolution limitations of their instrument, the results are however an average over all azimuths. Their data and some representative values of the HAS data for both directions are compiled in Fig. 9. Also included in this figure are the bulk zone origin data from ir reflectance [44-46] and Raman spectroscopy [44,45,47] as well as bulk dispersion curves from neutron scattering [41,42]. The theoretical phonon band edge is from the calculations of Al-Jishi and Dresselhaus which gave the best fit to the EELS data [43]. In the range of the HAS measurements it is nearly the same as the RAW results [40].

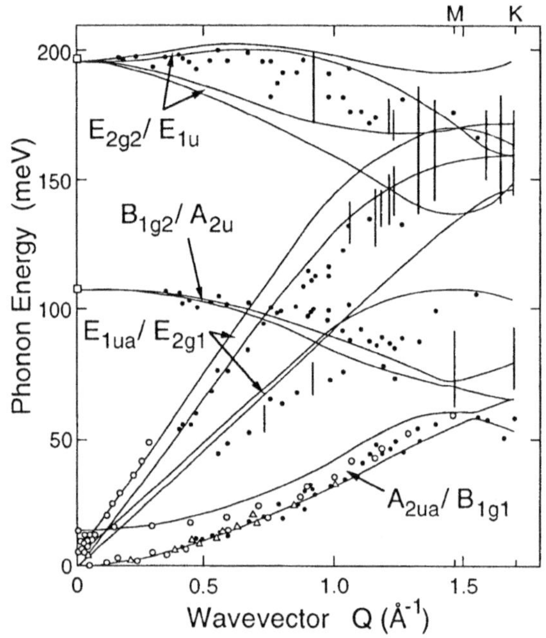

Figure 9. A compilation of graphite surface phonon data from several different sources. Some selected results from the HAS work for the <110> (Fig. 8) and the <100> directions are shown as triangles, the neutron scattering results as open circles [41], the EELS as solid points and vertical lines [50], and the ir measurements [44] are shown as large squares. The vertical lines correspond to features in the EELS measurements which were much broader than the resolution of the spectrometer and the length of the lines corresponds to the half-maximum widths of each of the features. Adapted from Wilkes et al.[50] with the theory curves shown as solid lines taken from Ref. 43.

The most recent EELS work by Oshima et al.[51] who used a flux grown Kish type graphite crystal now represents the most definitive results on the graphite surface. They were able to reduce considerably the uncertainties in the measurements by Wilkes et al. [50]. For the ΓM direction, Fig. 10a, they found the six modes which were theoretically predicted by RAW as shown in Fig. 5. An expanded scale of their lower energy results, Fig. 10b, (open circles) are compared to the HAS measurements (crosses) of Ref. 28. The solid lines are from the RAW calculations [40].

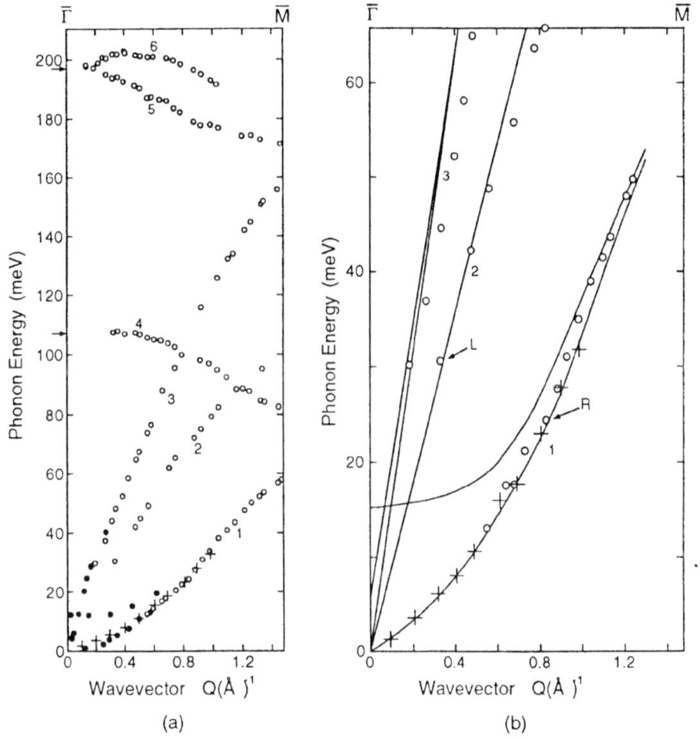

Figure 10. A comparison of surface phonon dispersion curves for graphite for the entire $\bar{\Gamma}\bar{M}$ direction. In (a) the open circles are from Oshima et al. taken by EELS, Ref. 51, the closed circles from neutron data, Ref. 41 and the crosses from several representative points from the HAS measurements on the natural crystal in the <100> direction. The indices labeling the curves follows that of Ref. 40. (b) An expanded energy scale for (a) showing the EELS measurements of Oshima et al. Ref. 51 as open circles and several representative HAS Kish results as crosses. The solid theoretical curves are from Ref. 40. The L and R label the longitudinal and Rayleigh modes, respectively. The labels 1, 2 and 3 are those of RAW [40].

3.5 XENON ADSORBATE VIBRATIONS

The rare gas physisorbed layers on graphite are probably the most extensively studied two dimensional layered systems. As pointed out in the introduction of this section, graphite has the advantage of providing a large surface area thus facilitating the application of bulk diffraction techniques involving x-rays, neutrons, high energy electrons as well as thermodynamic techniques. Of special importance for the rare gas studies is the smoothness, inertness and inherent cleanliness of the graphite surface. Very little is known concerning the dynamics of the rare gas overlayers. The frequencies are too low for resolution with EELS and moreover EELS would lead to collision induced desorption. The heavy rare gas atoms have very large neutron capture cross sections and so far only ^{36}Ar on graphite has been studied by Taub et al. [61]. Interesting results are becoming available for adsorbates of ^{4}He [62] and D_2 [63] which are more favorable for neutron studies. All these measurements have been performed on pyrolytic graphite so that no azimuthal dependencies could be studied.

For the experimental adsorbate studies, the Kish sample was used and the surface preparation was similar to that described above [64]. The Xe deposition was monitored via the specular HAS signal while the Kish surface was exposed to a continuous flux of Xe gas. In the temperature range between 65 and 75 K, the specular peak drops sharply by about a factor of 10 to 30 in intensity corresponding mainly to an increase in the Debye-Waller factor during formation of a single monolayer; between 50 and 60 K, another drop in intensity takes place indicating the formation of a second layer. For all of the adsorbate measurements, the graphite temperature was maintained constant and a few degrees above the bilayer formation temperature so that only a monolayer would be present and a steady Xe flux was used to maintain a constant effusive pressure over the monolayer.

The monolayer angular distributions showed a $\sqrt{3}x\sqrt{3}R30°$ Xe monolayer which depending on the conditions can be slightly incommensurate with the underlying graphite surface in agreement with previous work [64,65]. Figure 11 shows a series of TOF spectra, converted

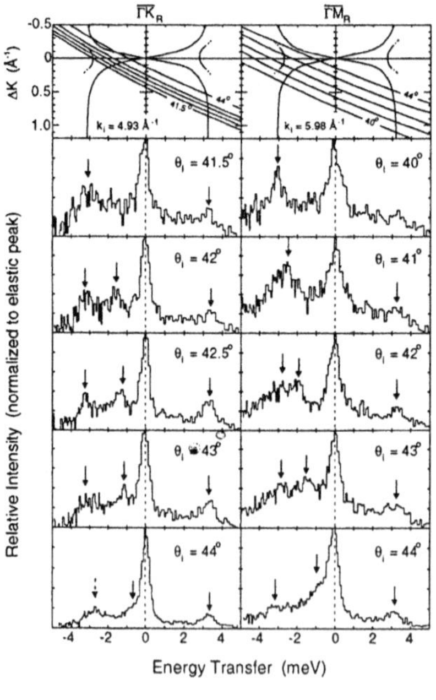

Figure 11. Helium atom TOF spectra, converted into energy transfer distributions, measured from a monolayer Xe covered graphite substrate along both high symmetry directions and for different incident angles and different incident wavevectors [64]. The arrows on the phonon peaks correspond to crossing positions of the scan curves and the dispersion curves as shown at the top of each column.

into energy transfer distributions, for the two high symmetry directions of the overlayer at several different incident angles and incident wavevectors. Note that the two high symmetry directions $\overline{\Gamma K}_R$ and $\overline{\Gamma M}_R$ of the overlayer are rotated with respect to the substrate as indicated in the inset of Fig. 12. The peaks in the TOF spectra in Fig. 11, including those with zero energy loss corresponding to elastic scattering from defects on the Xe surface, can be interpreted by referring to the scan curves plotted at the top of each of the respective columns. Over the temperature range of 55 to 65 K, the inelastic peaks did not change position as long as a full monolayer coverage is maintained.

Figure 12 summarizes the experimental dispersion curves for the Xe covered crystal for the two high symmetry directions. The solid and dashed lines are from a slab model calculation while the shaded line marks the edge of the transverse bulk band. Two modes are clearly visible in the data. The relatively thin slab of 13 layers in the theoretical calculation leads to an interference between the modes of the top and bottom surfaces causing them to split apart much more than they would for an infinitely thick slab [66]. However, for a thick slab, a broad resonance near 2.3 meV and another mode at the zone center are expected [66]. The data for small Q indicate that the lower mode coincides with the bulk band edge, while the upper mode is near 3.0 meV in some disagreement with the prediction of 2.3 meV for the vertically polarized adsorbate-substrate mode. The avoided crossing of the two modes at $Q \approx 0.2$ Å^{-1} in the $\overline{\Gamma K}_R$ direction is similar to that predicted theoretically by Hall, Mills and Black [67] for rare-gas overlayers on smooth metal surfaces as was observed for Kr on Pt(111) [68]. The difference in frequencies at the avoided crossing is a sensitive measure of the coupling of the surface and substrate vibrations.

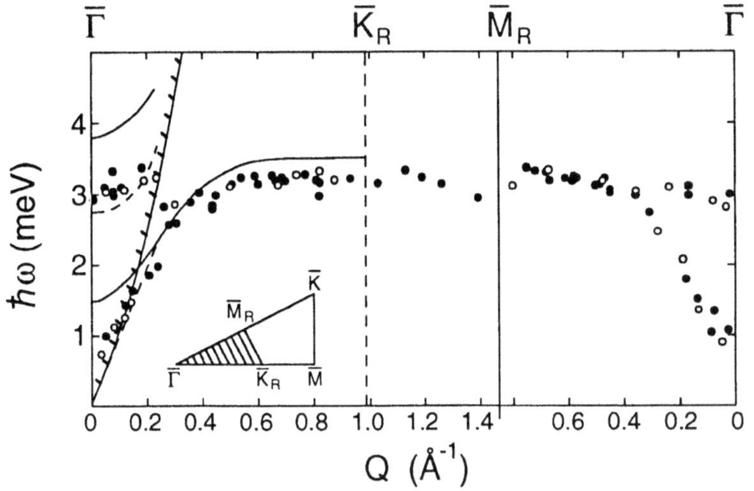

Figure 12. The experimental surface phonon dispersion curves of phonon energy versus wavevector for a monolayer Xe coverage on graphite [64]. The various lines are explained in the text. The data are shown by open and closed circles, the former from the previous figure. The inset shows the irreducible part of the clean graphite surface and the $\sqrt{3}\times\sqrt{3}R30^\circ$ Xe overlayer zone (shaded).

3.6. DISCUSSION

The agreement between the HAS and EELS data appears to be very good and within experimental error as shown by the comparison in Figs. 9 and 10. Indeed, both sets of measurements follow rather closely the bulk band edges as determined from neutron scattering.

Thus all the graphite surface dynamical measurements are in good agreement with those predicted by slab dynamics calculations as shown in Figs. 6,8,9 and 10. The experimental data show the quadratic functional form for the Rayleigh mode at small wavevector as predicted for a nearly two-dimensional crystal system [40,41]. Some experimental HAS results were obtained for the longitudinal mode, but these are only available for the Kish graphite crystal and are very weak. Thus we conclude that the surface dynamics of graphite are very well explored experimentally and good agreement exists between the theory and experiments at the low energies of the HAS measurements. The higher energy dispersion results of Wilkes et al. [50] were best fit by the theory of Al-Jishi and Dresselhaus [43]. The results of Oshima et al. [51] in Fig. 10a provide nearly complete surface dispersion results for the $\overline{\Gamma M}$ direction. The agreement with neutron scattering measurements [41] and HAS results in the low-energy range is very satisfying. The HAS measurements agrees with the EELS results and the theoretical curves (solid lines, Ref. 40) at energies below 30 meV. Above about 40 meV, the theory and EELS results disagree which suggests that a new surface analysis would be desirable in order to further refine the force constants for graphite. However, it is clear that graphite is one of the most studied and better understood layered materials. Graphite thus is an ideal benchmark two-dimensional layered material.

The adsorbate studies indicate that there is a strong hybridization-induced splitting between the localized adsorbate mode and the clean graphite surface Rayleigh mode. This splitting provides important information on the adsorbate-substrate interaction.

4. 2H-$TaSe_2$

4.1. INTRODUCTION

In the layered dichalcogenides 2H-$TaSe_2$ and 1T-TaS_2 (the latter discussed in the next section) there is a significant electron-phonon interaction with charge-density-waves (CDW) leading to structural phase transitions [6-15]. Several mechanisms are involved. In the simplest case Fermi surface nesting leads to divergencies in the dielectric response function at certain wavevectors Q. These divergencies affect the phonon dispersion curves via the electron-phonon interaction leading to a frozen-in soft phonon. At the same time a gap appears in the Fermi surface leading to a modulation of the electronic charge density in real space which can be either commensurate or incommensurate with the crystal lattice geometry. With HAS it has become possible to not only observe these structural phase transitions but also to study the phonon softening.

The crystal structure for the 2H polytype of $TaSe_2$ is shown in Figure 13. The unit cell contains two Ta ions and four Se ions [10,69,70]. The Se-Ta-Se subunits form strongly bonded layers which are weakly bonded to another similar layer. The weak bond connects the cleavage planes (001) and because of the strong anisotropy in the bond structure, the material is expected to behave as a nearly two-dimensional substance, similar to graphite. Different stacking arrangements between the layers lead to different polytypes with different unit cells and different electronic and structural properties. For the 2H case, the unit cell spans two layers [71]. Figure 13c shows the surface Brillouin zone and the high symmetry directions.

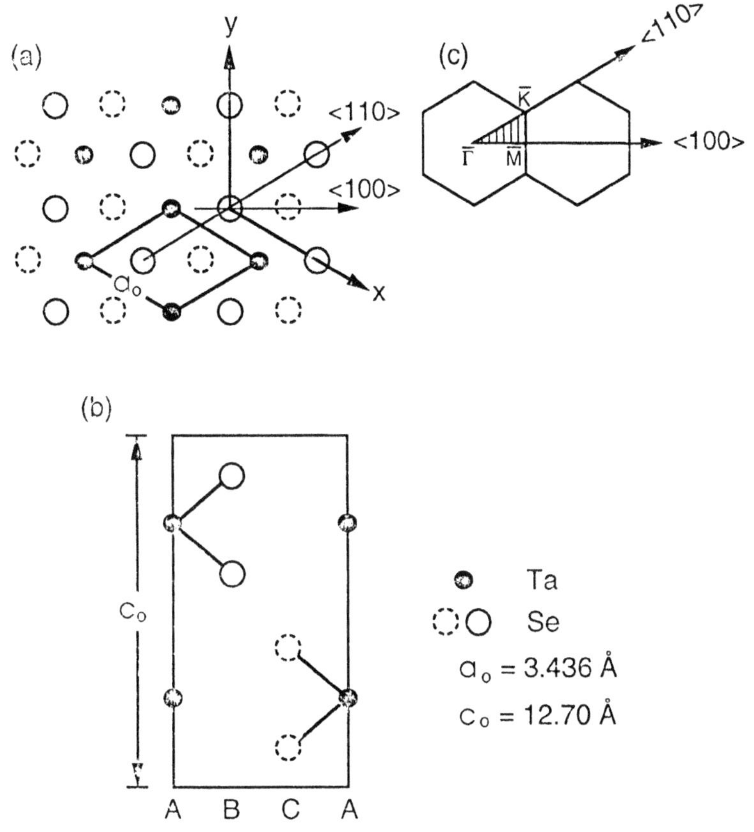

Figure 13. (a) The crystal structure of $2H\text{-}TaSe_2$ showing the atomic arrangement as seen from the surface. In (b) a side view of the unit cell is shown, while (c) shows the high symmetry directions and the irreducible segment of the surface Brillouin zone [After Ref. 77].

The phase diagram for the temperature dependence of $2H\text{-}TaSe_2$ is shown in Fig. 14. The normal phase exists above the transition temperature for the onset of the CDW which occurs at T_o= 122 K. The transition temperatures and the occurrence of certain phases depend upon whether the crystal is being cooled or heated. On cooling the charge density wave transforms the normal phase indicated by an N, into a fully incommensurate phase (I) at 122 K. In the incommensurate phase the three CDW wave vectors are slightly different than those for the commensurate (3x3) superlattice lowest temperature C phase. With increasing crystal temperature, the C phase transforms to a new phase not observed on cooling, called the S striped incommensurate phase at a significantly higher temperature of 92K. In the S phase two of the wave vectors are incommensurate, while the third is commensurate [71-87]. This leads to a striped real space pattern which was first observed by high resolution electron microscopy [80]. Upon further warming, the S phase goes into the same incommensurate phase found on cooling, but at 113K. Then finally the system returns to the N phase at the same temperature of 122K as found on cooling.

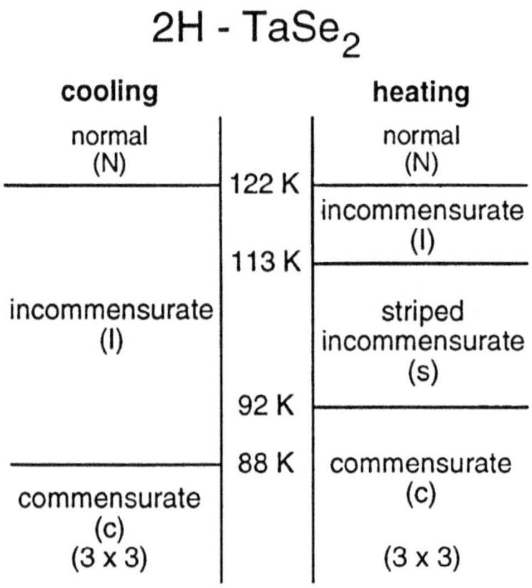

Figure 14. The phase diagram for 2H-$TaSe_2$ for cooling and heating [After Ref. 78].

The N-I phase transition is of considerable interest and the HAS results presented below have been mainly devoted to this transition. The I phase has been discussed theoretically with two different models. One model uses a network of parallel discommensurate lines which lower the hexagonal symmetry to orthorhombic and results in a second-order phase transition. These results agree with several experiments including the extended neutron work of Moncton et al. [73] and the electron microscope results of Chen et al. [80]. The HAS experiments also indicate that the phase transition is second order. The other model assumes a honeycomb network which maintains the hexagonal symmetry of the normal phase [79]. From this model, a six component order parameter results, along with a first-order phase transition for the N-I phase change. Some of the available data also support this model [83]. The other phase transitions S-C, I-S and the C-I are all believed to be first-order.

4.2. EXPERIMENTAL

Over 25 different samples were tried in the course of the HAS-TOF experiments [10]. The sizes were typically several mm^2 in surface area with thicknesses of approximately 1mm. An earlier study was done on samples cleaved in air [6], but most of the temperature dependent dynamical work was done on one sample cleaved in vacuum [10]. The other samples had either marginal scattering intensity or broad diffraction peaks, or exhibited considerable mosaic structure so that they did not produce useful results. There was no observable difference between samples cleaved in air or in vacuum.

The air cleaved samples were quickly transferred to the vacuum chamber and then gently baked at approximately 100°C for about 20 hrs. The crystals were then additionally flashed to 150°C for a short period, after which measurements were started. These relatively low temperatures were not exceeded to avoid decompositional damage to the crystal. The vacuum cleaved samples were cleaved after an initial 100°C bakeout, the crystal then underwent the same procedure as above.

Even though the background gas pressure in the scattering chamber was in the mid 10^{-10} mbar range, some residual gas adsorption took place during the measurements, especially when the crystal was kept at temperatures near 60 K. Typically this leads to a decrease in the scattering intensity by 10 to 15 percent over ten hours, but was restored by a bakeout to approximately 150°C. The crystal was baked before each set of measurements, and whenever the signal intensity decreased during a long-duration low-temperature measurement.

4.3. SURFACE STRUCTURE AND He-SURFACE POTENTIAL

Figure 15 shows four different angular distributions for the two high-symmetry directions and at temperatures corresponding to the normal and CDW phases. In the $<100>$ direction shown in Fig. 15 b) and d) the CDW peaks are clearly in evidence at T=50K. In the $<110>$ direction no superlattice CDW induced peaks are observed probably because they are below the signal-to-noise limit of this instrument. They have, however, been observed by neutron scattering [73].

4.3.1. Corrugation Amplitude of the Atom Lattice. At temperatures below the onset of the CDW the He atom-surface interaction can be characterized by two corrugation amplitudes which can be extracted from the diffraction intensities. The simplest corrugation function, as first proposed by Cantini [11], is $\zeta(\mathbf{R})=\zeta^{lat}(\mathbf{R})+\zeta^{CDW}(\mathbf{R})$. The first term, which is largest, describes the temperature independent corrugation of the normal surface lattice, while the second term accounts for the temperature dependent amplitude of the CDW periodic lattice distortion. Both are determined from the corresponding diffraction peaks via the eikonal approximation discussed in Section 2.1. The spatial dependence of the normal lattice corrugation $\zeta^{lat}(\mathbf{R})$ was assumed to have the same simple functional form as for graphite (See Eq. 8).

$$\zeta^{lat}(\boldsymbol{R})=2\zeta_0^{lat}\{\cos(2\pi x/a_o)+\cos(2\pi y/a_o)+\cos[2\pi(y-x)/a_o]\} \qquad (9)$$

where a_o=3.44 Å (Fig. 13), is the lattice spacing for 2H-$TaSe_2$ and ζ_0^{lat} is the corrugation amplitude. The directions x and y are shown in Fig 13a. Since there was no evidence for significant deviations from hexagonal symmetry which had been taken into account by Cantini [11], this was not included. The value of ζ_0^{lat} was adjusted until a best fit of the ratios of the intensities of the Bragg peaks to the specular peak was achieved. The value of $\zeta_0^{lat}=0.04\pm0.01$Å obtained from data including that shown in Fig. 15, is in rather good agreement with the earlier determination of $\zeta_o=0.06$Å±0.005Å [11]. The differences between these two values is not too surprising in view of the different apparatus geometries and related limitations imposed by the eikonal approximation. The small value of the corrugation ζ_0^{lat} is qualitatively consistent with recent quantitative STM and AFM studies of the IT polytype [88].

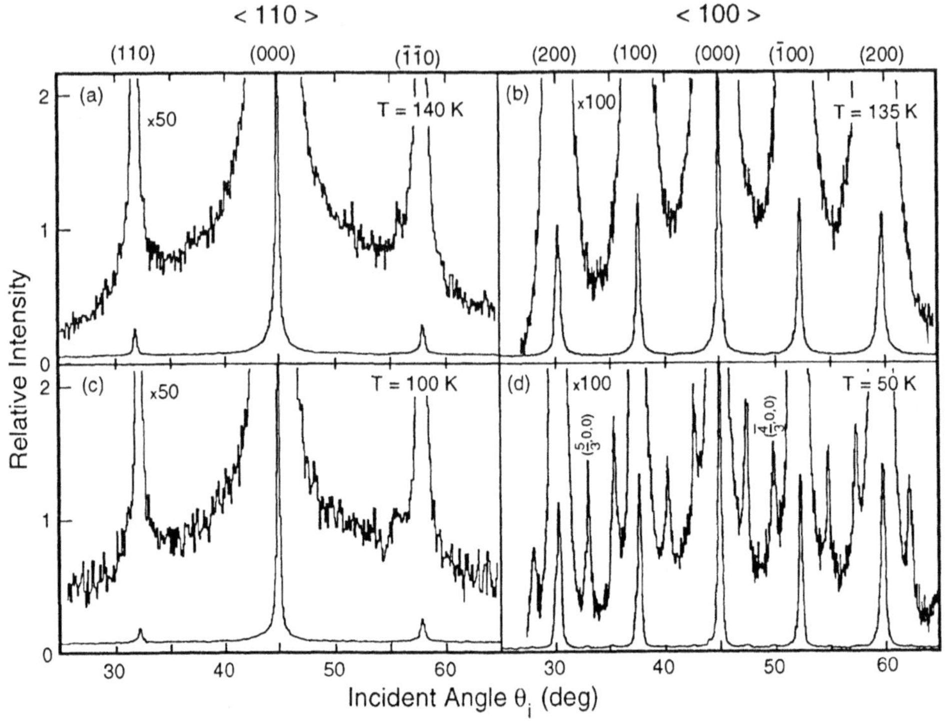

Figure 15. Measured angular distributions for He scattered from the 2H-$TaSe_2$ crystal surface for the <110> direction (a and c) and the <100> direction (b and d) for different temperatures [After Ref. 10]. Panels (a) and (b) are for temperatures above the N-I transition and show only the normal peaks. A magnified version of each is included. Panels (c) and (d) lie below the N-I transition and in (d) the additional CDW peaks are clearly visible, but not for (c) where they are too small to be measured. The Bragg peaks and two of the CDW peaks are labeled. The incident wavevector for the <110> measurements was $k_i = 11.5Å^{-1}$ and for the <100> direction was $k_i = 11.7Å^{-1}$. On (a) and (c), the small additional satellite peaks next to the high angle Bragg peaks are likely caused by a small flake of material on the crystal surface.

4.3.2. Corrugation Amplitude of the Charge-Density-Wave. The corrugation $\zeta^{CDW}(\mathbf{R})$ due to the CDW distortion is of considerable interest as a temperature dependent order parameter which can be described in terms of a critical exponent [9]. The CDW corrugation is obtained from the intensities of the small CDW induced diffraction peaks in the angular distributions as a function of the temperature below the transition temperature T_o. Figure 16 shows a series of uncorrected angular distributions for different target temperatures. The increase in intensity with decreasing temperature is due both to the Debye-Waller effect and the induced corrugation from the CDW.

The Debye-Waller effect is approximately accounted for by normalizing to the intensity from the nearest Bragg peak, the results then being sensitive primarily to the temperature dependent CDW corrugation term. The intensity ratios of the (5/3,0,0) and the ($\bar{4}$/3,0,0) CDW peaks, labeled in Fig. 16, are plotted in Fig. 17 as a function of temperature for the N-I transition. Note that there is some hysteresis in the data depending upon whether the intensities were measured during heating or cooling. This can be regarded as a signature of a first-order phase transition. Looking back at the phase diagram, Fig. 14, the difference in behavior is understandable. In particular, for increasing temperature the relative intensities are affected by the hysteresis of the first-order phase transitions of the S-I phase transition at 113 K and the C-S transition at 92 K and for decreasing temperature by the C-I transition at 88K [73,79,83].

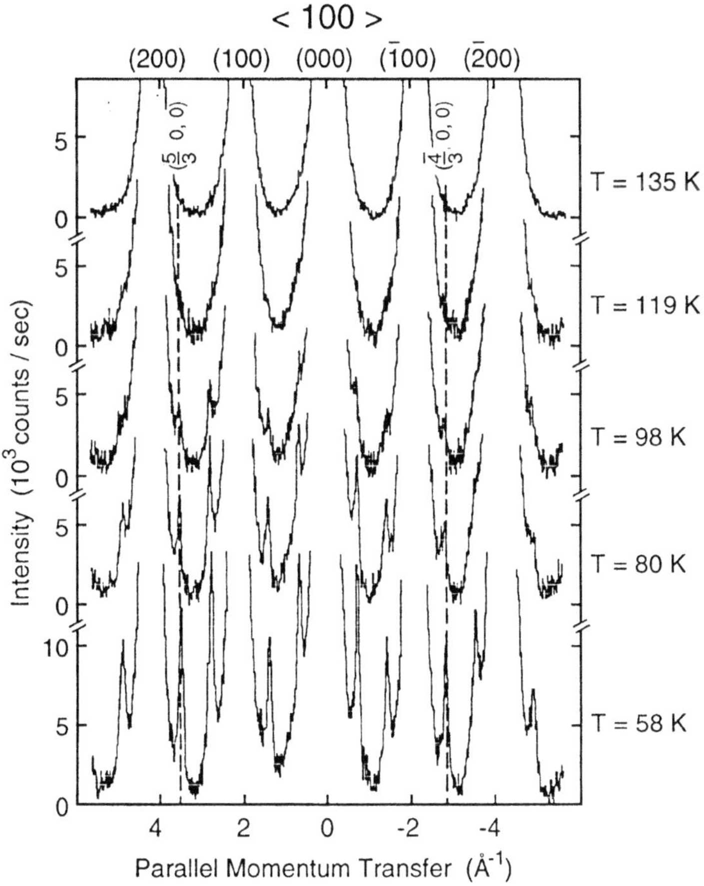

Figure 16. A series of HAS angular distributions in the < 100 > direction taken as a function of crystal temperature [After Ref. 10]. The angle between the incident and outgoing beams was fixed at 90°. The distributions were taken for increasing temperature and k_i=11.7Å^{-1}. All of the Bragg peaks are labeled along with two of the CDW peaks.

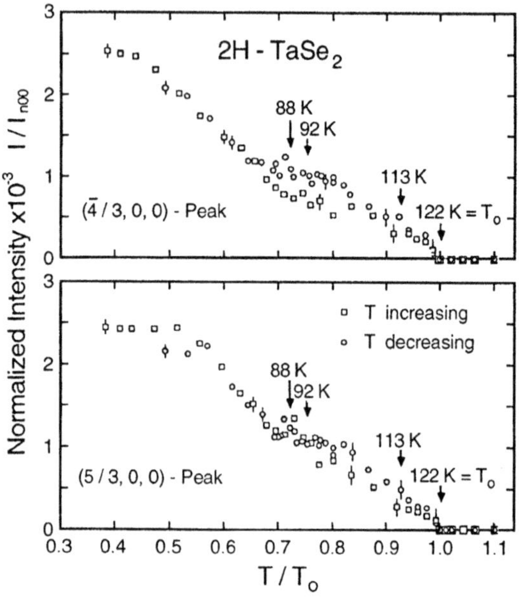

Figure 17. The normalized ratio of the scattered intensities for the $(\bar{4}/3,0,0)$ and (5/3,0,0) CDW induced diffraction peaks as a function of temperature [9,10]. The normalization is with respect to the nearest Bragg peak and the squares correspond to data taken for increasing temperature and the circles for decreasing temperature. Some small hysteresis is seen around the S-I transition temperature (113K) and in the region of the I-C (88K) and the C-S (92K) transitions.

With the eikonal approximation the normalized CDW diffraction intensities of Fig. 17 can be converted to a CDW corrugation amplitude ζ_0^{CDW} with the corrugation function for the (3x3) structure

$$\zeta^{CDW}(R) = 2\zeta_0^{CDW}\{\cos(2\pi x/3a_o)+\cos(2\pi y/3a_o)+\cos[2\pi(y-x)/3a_o]\} \tag{10}$$

where ζ_0^{CDW} is shown as a function of $1-T/T_0$ in Fig. 18 both for heating and cooling. The slope gives the critical parameter β. For cooling the best fit of all of the data gives $\beta=1/3$ which is smaller than the mean field value of $\beta=1/2$. For increasing temperature the hysteresis from the S-I phase transition significantly affects the results so that a critical parameter can not be reliably inferred.

Since the Se atoms are in the surface layer whereas the Ta atoms are in the second layer the most direct explanation of the increase in CDW corrugation accompanying the N-I phase transition is to attribute it to shifts in the z direction associated with a complicated periodic lattice distortion in the position of the Se atoms. By extrapolating the temperature dependence of the CDW corrugation to 0 K a value of $4\zeta_0^{CDW}=0.025$ Å is obtained in direct agreement with Cantini [11]. It is also in reasonable agreement with a value of 0.017 Å deduced from neutron data by Moncton et al. [73]. It is interesting to compare this corrugation with STM and AFM measurements of the 1T polytype which yield a CDW corrugation amplitudes of 12Å and 1Å respectively at room temperature [88]. Only qualitative STM results are available for 2H-$TaSe_2$ at 77K [89]. They indicate that the atomic corrugation is observable and that the CDW corrugation is smaller.

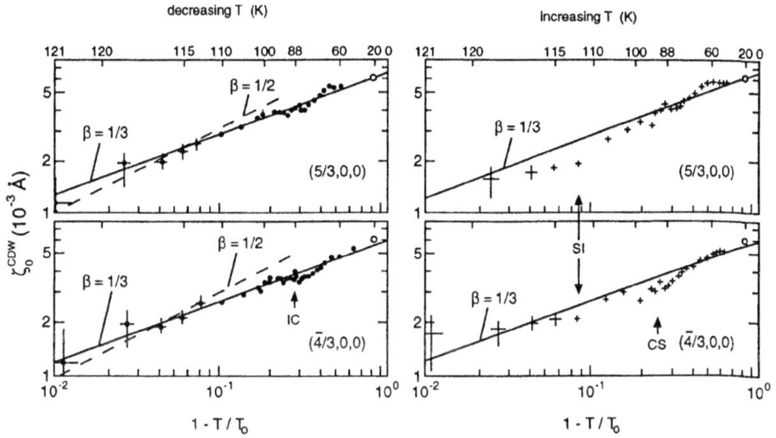

Figure 18. Log-log plot of the temperature dependence of the CDW induced corrugation ζ_o^{CDW} for the (5/3,0,0) and (4/3,0,0) CDW induced diffraction peaks [After Refs. 9,10]. The results have been plotted for both decreasing crystal temperature (left) and increasing temperature (right). The dashed lines correspond to a slope $\beta = ½$ which is the mean-field behavior and the solid line has a $\beta = ⅓$. The corrugation value obtained by Cantini (Ref. 11) is shown as an open circle.

In the incommensurate temperature region, small angular shifts in the (3x3) CDW superlattice peaks of the order of 0.5% have been observed by neutron scattering[73]. These shifts are attributed to the incommensurability of the CDW wavevector. However, the width of the HAS peaks and the limited angular resolution of the instrument has impaired their measurement in the HAS experiments.

4.4. SURFACE VIBRATIONS

HAS TOF measurements were used to investigate the surface phonon dispersion curves both above and below the CDW phase transition temperature in a search for dynamical manifestations of the CDW at the surface. The neutron dispersion curves of Moncton et al. [73] had already provided some information on the CDW induced softening of the longitudinal acoustic mode and the results had been modeled by Feldman [90].

Examples of TOF spectra transformed to an energy abscissa for both the <100> and the <110> directions for different incident angles, θ_i, are shown in Fig. 19. The target temperature was at 140 K, well within the normal phase region, and the incident energy for the <100> direction was E_i=71.5 meV and for the <110> direction, E_i=53.4 meV. All the spectra show a weak diffuse elastic peak at zero energy transfer which is due to surface imperfections. One or two low-energy inelastic peaks are observed on both sides of the diffuse peak, which are attributed to the transverse-polarized Rayleigh mode. In addition, on the creation side many of the spectra reveal high energy-transfer peaks from optical phonons. The dispersion curves from all of the spectra are shown in Fig. 20. For the <100> azimuth they are compared to the surface adapted bulk calculations of Feldman [90] who used a force-constant model based on bulk modes from neutron measurements [73] and Raman data [91-93]. The overall agreement is good except for: 1) the experimental Rayleigh frequency near the zone boundary that lies below the corresponding Σ_3 acoustic bulk band edge and 2) the experimental Σ_3 optical mode that lies above Feldman's prediction. Since no optical dispersion results were available to Feldman from neutron scattering when he formulated his model, his calculations are expected to be less accurate for the optical modes. The lowering in the Rayleigh mode frequency at the zone boundary and in the optical mode provided an impetus to improve the existing model. Thus a new fit to these data using the Dispersive-Linear-Chain Model was carried out by Miglio et al. [94].

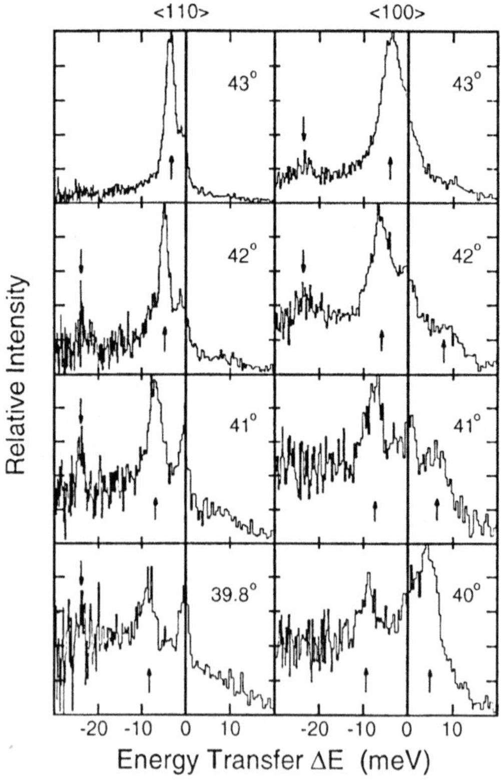

Figure 19. A series of TOF spectra for the 2H-$TaSe_2$ crystal, converted into energy transfer distributions, for several different angles and a target temperature of 140 K [10]. The incident energy E_i for the <110> spectra was 53.4 meV and 71.5 meV for the <100> direction. The phonon peaks are marked by arrows and the peak at $\Delta E=0$ corresponds to diffuse elastic scattering.

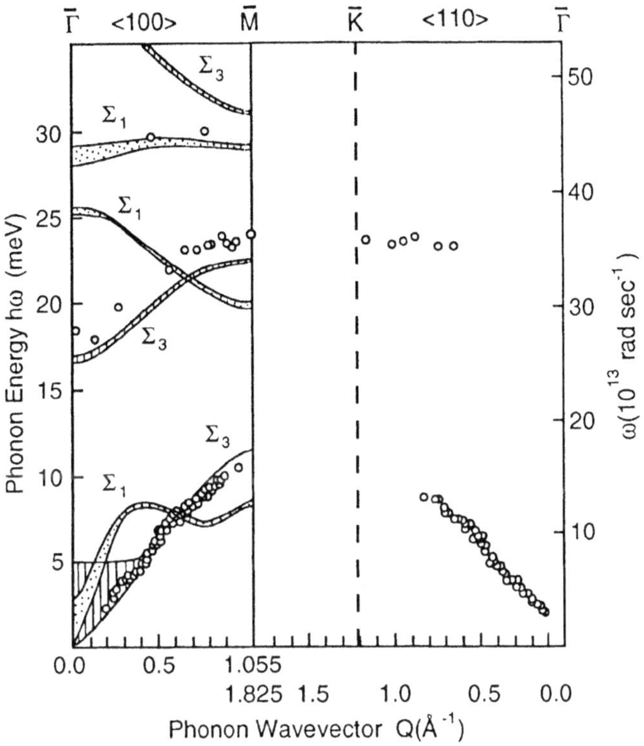

Figure 20. A comparison between the experimental HAS surface dispersion curves (open circles) and the surface adapted (shaded) calculations of Feldman [90] which are available only for the < 100 > direction. All of the data were taken at 140 K and only the theoretical sagittal plane modes, Σ_3 and Σ_1 are included [10].

In a search for CDW induced anomalies the measurements were repeated for a range of temperatures between 50 and 300K. A careful examination of the data revealed a softening of the Rayleigh mode located at $Q_S = 0.53 Å^{-1}$ with a maximum downward shift of about 2 meV at T=110K. An expanded view of the data is shown in Fig. 21a. In Fig. 21b these observations along with the softening of the bulk longitudinal mode with its estimated shape at 120K is shown [8,73]. It is located at a wavevector of $Q_B \approx 0.7 Å^{-1}$ which is considerably larger than that for the surface softening. Further TOF measurements were carried out in the < 100 > direction for the target at 60 K, well below the CDW commensurate transition. A comparison of this new data with the previous 140 K experimental Rayleigh wave dispersion curves (Fig. 20) shows that the two sets of data nearly overlap. A careful examination of the 60 K TOF spectra shown in Fig. 22 suggests moreover that there is a weak hybridization over a narrow range of incident angles θ_i (corresponding to $Q \approx 0.6 Å^{-1}$) between the Rayleigh and another surface mode which is thought to be longitudinally polarized. In order to explain the relative intensities it was assumed that the cross section for the longitudinal mode is normally small and was therefore not seen previously. It is observable only because of enhancement by the hybridization. The 60K experimental results

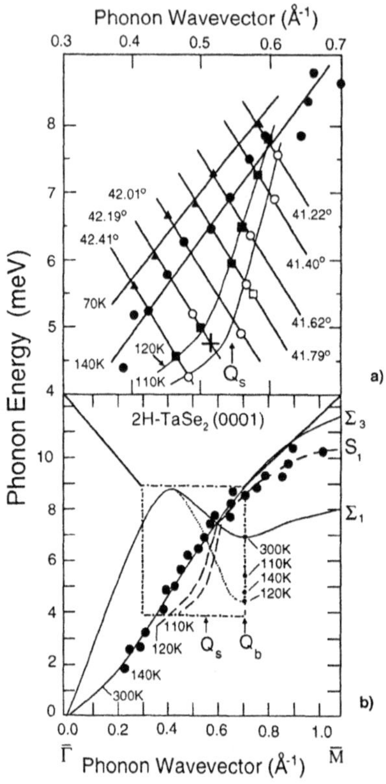

Figure 21. Temperature dependence of the Rayleigh mode dispersion curve in the <100> direction [After Refs. 8,10] (a) for a limited wavevector range which shows a softening near $Q_s \approx 0.53\text{Å}^{-1}$. The filled circles were taken at 140 K, the filled squares at 120 K, the open squares at 115 K, open circles at 110 K, a cross at 100 K and the triangles at 70 K. The solid thick lines are to guide the eye and the thin lines are the scan curves for different incident angles and $k_i = 11.7\text{Å}^{-1}$. In part (b) the experimental points correspond to the data at 140 K. The solid lines are from the bulk transverse (Σ_3) and longitudinal (Σ_1) acoustic modes [73]. The dashed lines show the Rayleigh-mode (S_1) dispersion curve and its temperature-dependent anomaly for 110 K and 120 K. They can be compared to the temperature dependent softening in the Σ_1 bulk mode near $Q_b \approx 0.7\text{Å}^{-1}$ [73]. The dotted line shows a possible dependence of the bulk mode on temperature.

showing the longitudinal mode are summarized in Fig. 23. Recent calculations [94], based on a Dispersive-Linear-Chain fit, are shown as a solid line (Rayleigh mode). The shaded and hatched regions shown in Fig. 23 are from a slab calculation of the bulk bands at the surface based on a new fit to the existing bulk measurements [94]. It is gratifying that the new theory agrees better with experiment than the previous theory by Feldman. The comparison with the new theory suggests that both the Rayleigh and the longitudinal modes are slightly softened by about one meV at the zone boundary.

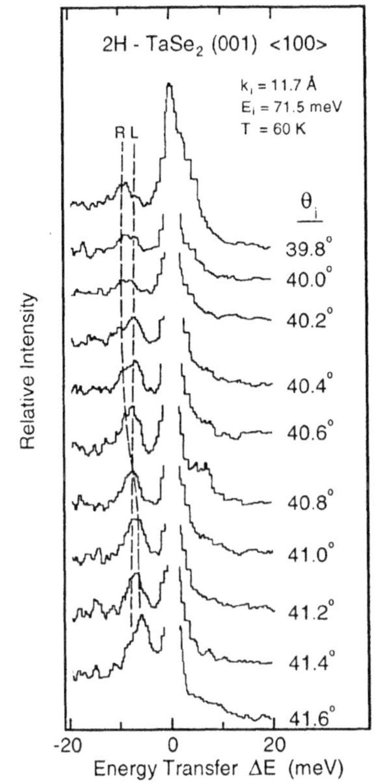

Figure 22. A series of TOF spectra, converted into energy transfer distributions, showing a weak hybridization between a longitudinal (L) acoustic and a Rayleigh (R) mode for different incident angles at a surface temperature of 60 K. The dashed lines suggest that the peak is really a double peak due to a hybridization of a Rayleigh mode and a longitudinal mode [10].

4.5. DISCUSSION

The HAS experiments on the 2H-$TaSe_2$ surface reveal CDW induced diffraction peaks which increase in intensity with decreasing temperature below the N-I transition at 122 K where the onset of the second-order CDW phase appears. From the CDW peak intensities, the temperature dependence of the corrugation amplitude was determined and found to give a critical exponent $\beta = 1/3$ in agreement with the 3d-xy model. The observed temperature dependence of the corrugation amplitude (Fig. 18) is indicative of a low-symmetry incommensurate CDW and thus disagrees with an earlier proposal of hexagonal symmetry for the incommensurate phase which was based on an x-ray analysis [79]. The HAS data is in better agreement with the more recent observations by microscopic techniques [75,80,84,85]. For example, Chen et al.[80] have found that the hexagonal symmetry for the incommensurate phase is not seen on the cooling cycle, as had been observed in earlier x-ray diffraction results. Instead they find on the microscopic scale an orthorhombic symmetry (point group D_{2h}).

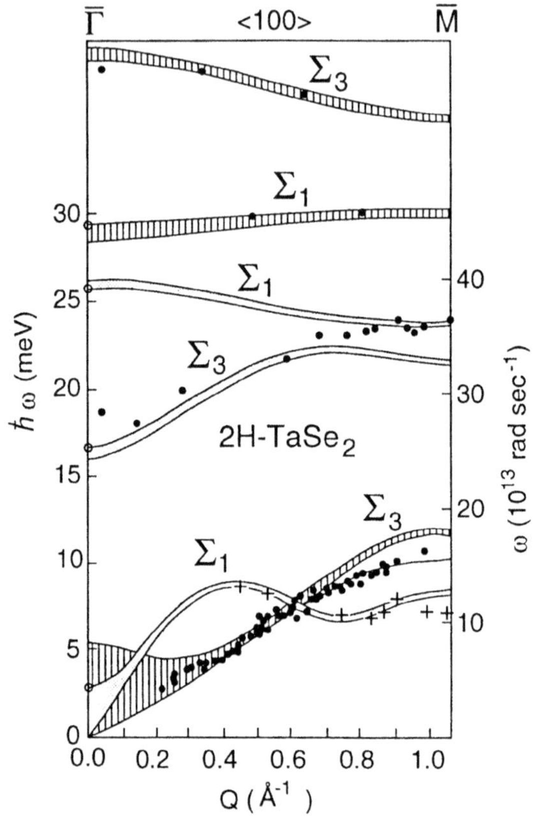

Figure 23. The HAS data in the <100> direction taken at 60 K along with the longitudinal data from the hybridization case of the previous figure also taken at 60 K (crosses) are compared to slab-adapted bulk phonon bands (vertical striped for transverse and shaded for longitudinal polarization) and the calculated Rayleigh branch based on the Dispersive-Linear-Chain method (solid line) [94]. The four open circles at Q=0 are from Raman measurements [10].

Overall the 2H-$TaSe_2$ surface dispersion curves follow closely those of the bulk. Surface optical modes, which were not seen by neutron scattering, were observed for the first time by HAS. Furthermore, a slight softening of the Rayleigh mode is seen near the zone boundary. Of considerable interest are the HAS measurements which indicate that the transverse Rayleigh mode softens at a temperature of about 110 K and at a wavevector of about 0.53$Å^{-1}$ whereas in the bulk only the longitudinal mode softens near a wavevector of 0.7$Å^{-1}$ at a transition temperature near 120K [8,73]. If we assume that the anomalies are due to a Kohn-type mechanism, the different wavevector for the surface anomaly suggests that the surface electronic states at the Fermi energy have a wavevector nesting at a different wavevector than the bulk electronic states. Another possible explanation of the anomaly at Q=0.53$Å^{-1}$ is a hybridization of the Rayleigh mode with a temperature-dependent surface longitudinal mode.

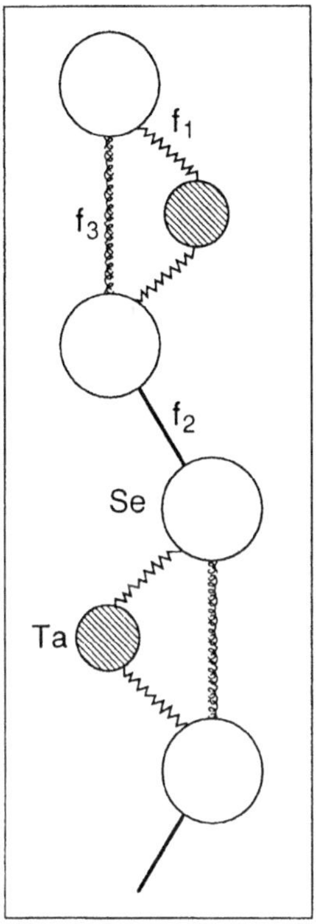

Figure 24. The spring model used for $TaSe_2$ in the dispersive-linear-chain method [94]. The coupling between the subunits in the layers is not shown. See the text and Ref. 94 for further details.

The force constants arrangement used in the Dispersive-Linear-Chain model [94] which is able to fit the $TaSe_2$ surface dispersion curves is shown in Fig. 24. The model offers a convenient parameterization method for these nearly two-dimensional crystal systems. For $TaSe_2$, the components of motion normal and parallel to the surface are almost completely decoupled and can be treated separately. f_1 is a wavevector dependent force constant which represents the covalent binding between the Ta atom and the Se shells within the layers, whereas the term f_2 accounts for the weak van der Waals coupling between the layers and the term f_3 accounts for the electrostatic interaction between the Se atoms within the layers. As can be seen in Fig. 23, the model is able to nicely fit the HAS results. From the model it can be concluded that the longitudinal Coulomb interactions in the Ta plane causes the anomalous behavior in 2H-$TaSe_2$ and that these interactions are related to the CDW instability, in agreement with recent calculations [74]. The anomalies in the bulk longitudinal mode and the surface Rayleigh mode may well be due to the same mechanisms. At the surface there is a mixing not present in the bulk of the longitudinal mode with the shear vertical polarized Rayleigh mode. This probably explains the observed hybridization at $Q=0.6Å^{-1}$ and may be responsible for the Rayleigh mode anomaly at $Q=0.53Å^{-1}$.

5. 1T-TaS_2

5.1 Introduction

There have been many experimental investigations of the various phases of the 1T polytype of bulk TaS_2, including their two-dimensional nature as well as the structural phase transitions, the nature of the charge density wave and the commensurate/incommensurate behavior as a function of temperature [11,13-15,69,75,95-104]. This review concentrates mostly on recent HAS studies [13-15,29] of the low temperature commensurate $1T_3$ phase of 1T-TaS_2. Fig. 25 shows the crystal structure for the high temperature 1T polytype, which is that of CdI_2.

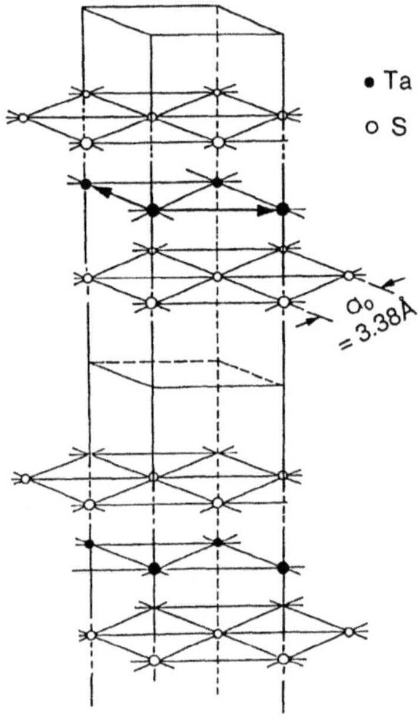

Figure 25. Schematic view of the TaS_2 crystal in the 1T normal phase which is the CdI_2 structure.

The various structural phases which are accessible by cooling or heating the sample are shown in the phase diagram Fig. 26 [69, 96, 98]. Above 543 K, the normal 1T phase is metastable and slowly transforms into the 2H polytype. Below 543 K there are several different 1T polytype phases which are distinguished by numbered subscripts. On cooling, the $1T_1$ hexagonal incommensurate (I) phase exists between 543 and 351 K [104]. On further cooling the $1T_{2,1}$ nearly commensurate (NC) rotated phase appears between 351 and 183 K. Finally, at temperatures below 183 K the $1T_3$ low-temperature commensurate C $\sqrt{13}x\sqrt{13}$ R13.9° phase appears. On warming from the C phase, a $1T_{2,2}$ triclinic incommensurate (I) phase [93] exists between 223 and 283K followed by the other phases described for the cooling cycle and shown in Fig. 26. A similar behavior was seen in the heating cycle for 2H-$TaSe_2$, Fig. 14, which also had one more phase than the cooling cycle. The normal lattice and the CDW reconstructed $1T_3$

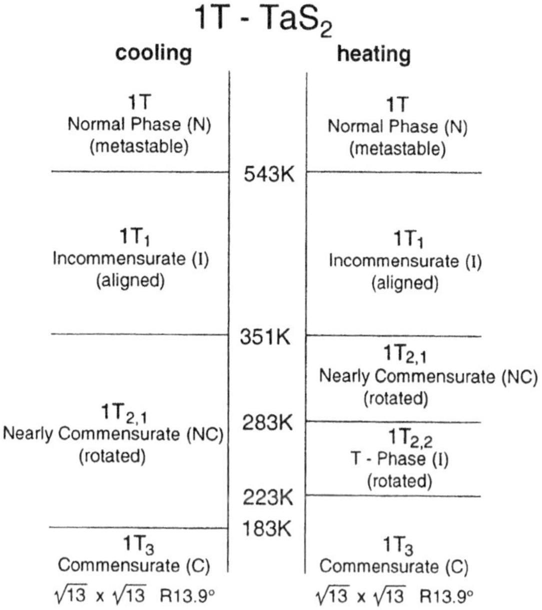

Figure 26. Phase diagram of the 1T-TaS_2 layered material showing the different phases encountered for decreasing and increasing temperatures. The $1T_3$ is emphasized in this review.

real space surface structure are shown in Fig. 27a. Brower et al. [105] explained how the Ta atoms at room temperature have moved in the plane of the metal to form a 13 atom star-of-David cluster, which leads to the $1T_3$ phase. They further argued that the S atoms were squeezed out of the Ta_{13} cluster. The corresponding reciprocal space structure is shown in Fig. 27b.

The bulk dynamical information on the various phases is rather limited. Apparently it is difficult to obtain sufficiently thick samples needed for neutron measurements [100]. As pointed out earlier, thick samples are not required for the HAS work. The first HAS angular distributions were measured by Cantini et al. who observed diffraction from the normal and CDW induced lattice in the $1T_3$ phase [11,13,14]. From the diffraction intensities they determined the corrugation amplitudes for both the normal lattice and the CDW reconstructed $1T_3$ surface. The substantial new information resulting from the more recent MPIGö studies of both the surface structure and phonon dispersion curves will be reviewed below [15,29].

5.2 EXPERIMENTAL

In the MPIGö experiments, Laue diffraction patterns were first obtained [29] to check the quality and confirm the orientation of the crystals. Three thin samples of cross sectional area of approximately 3x3 mm^2 were used [106]. After mounting inside of the vacuum chamber, a 10 hr bakeout at 450 K was used to achieve UHV conditions. The crystal was then cleaved in vacuum following the technique discussed in Section 2 to obtain a clean basal plane surface and LEED measurements were made to check the quality of the surface. The $1T_3$ phase was studied in the temperature range from 223 down to 50 K.

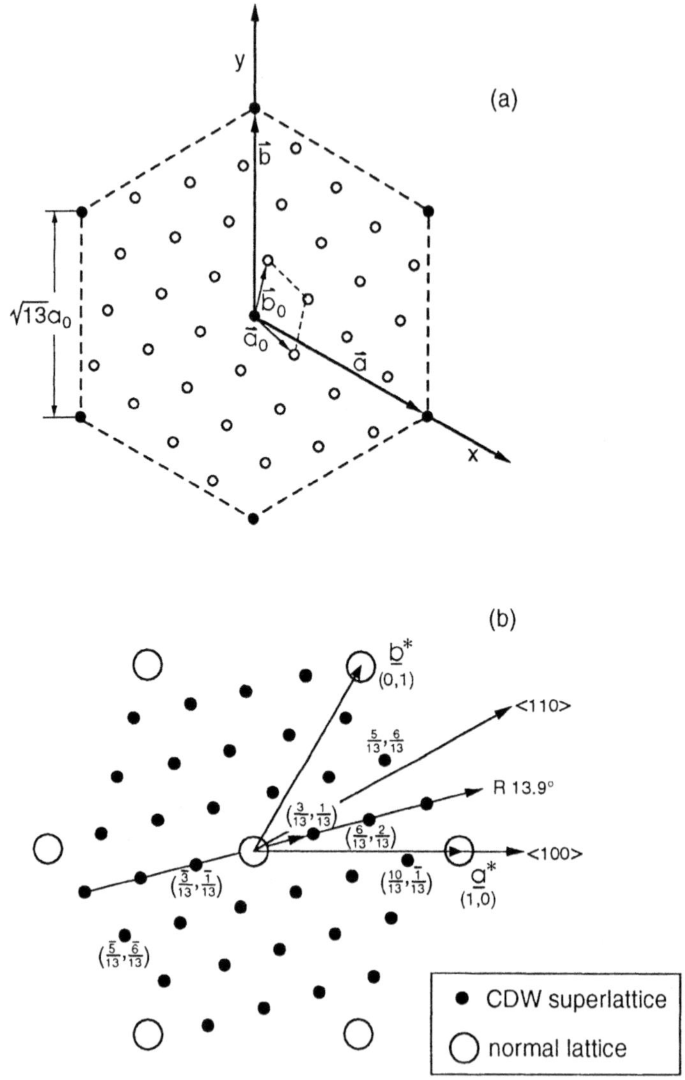

Figure 27. a) Schematic view in real space [13] of the normal and the CDW lattice structure for $1T_3$-TaS_2. Included are the coordinate systems for the normal and the CDW structures. The magnitude of a_o is 3.38Å. b) Schematic view of the TaS_2 surface in reciprocal space showing the reconstructed $1T_3$ phase. The two high symmetry directions are indicated along with the direction of the CDW superlattice which is rotated 13.9° from the <100> direction. Several of the superlattice peaks are labeled.

5.3 SURFACE STRUCTURE AND He - SURFACE POTENTIAL

Figure 28 shows measured angular distributions of total intensity for the low temperature commensurate IT_3 - phase at 120 K measured with an incident been wavevector k_i = 8.56Å^{-1}. The three different azimuths shown are along the <100> and the <110> high symmetry directions and also in the 13.9° rotated direction of the commensurate CDW superlattice. These three panels illustrate the striking difference between the 1T - TaS_2 and 2H - $TaSe_2$ layered

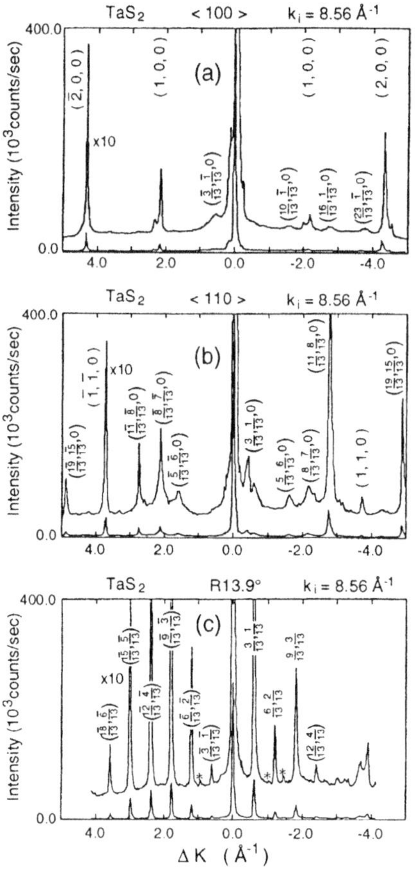

Figure 28. Angular distributions for the $1T_3$ commensurate phase of TaS_2 taken for an incident wavevector k_i=8.56Å^{-1} and the crystal at 120 K [29]. For the <100> (a) and <110> (b) directions there are peaks corresponding to those from the normal lattice as well as some components from the CDW superlattice. The intensities of the superlattice peaks (c) are comparable to those for the other two directions (a,b) indicating that the corrugations have nearly the same amplitude. Each of the curves show an expanded scale (x10) which aids in identifying the less intense peaks. Several of the peaks are labeled and those marked by stars are of unknown origin.

compounds. Whereas in the latter case the additional CDW diffraction peaks intensities were very small compared to the normal lattice peaks, they are comparable or even larger in the case of TaS_2. The CDW superstructure diffraction peaks are especially prominent in Figs. 28c. Their intensities are within 30% of that of the specular peak. The asymmetry on the intensity with respect to the specular peak is consistent with the trigonal symmetry of the CDW structure. The relatively large angular widths of the peaks are attributed to the mosaic spread and finite size of the crystallites on the surface. The labeling of the peaks is the same as was employed in Fig. 15, for 2H-$TaSe_2$ but differs from that used by Cantini [13].

The positions of the integral order peaks in the $<100>$ and the $<110>$ directions establish the surface structure and symmetry of the normal lattice. The measured lattice spacing is in good agreement with the established bulk value of a_o = 3.38Å [100]. Likewise in the R13.9° direction a reciprocal lattice vector of G = 0.60 $Å^{-1}$ for the CDW superlattice is obtained in agreement with the measured bulk value of G = 0.595$Å^{-1}$ [15,98,106-108]. The angle of rotation of 13.9° for the CDW superlattice with respect to the $<100>$ direction of the normal lattice is also the same as in the bulk (See Figure 27). This indicates, as has also been verified by LEED [109], scanning tunnel microscopy (STM) [88,89,110] and atomic force microscopy (AFM) [88,111] measurements on several of the layered dichalcogenides, that the surface and bulk structure of the various phases are identical.

As with the other layered compounds the peak intensities in the angular distributions are fitted by assuming a hard-wall model surface corrugation. Again it is assumed that the corrugation amplitudes can be added $\zeta(R) = \zeta^{lat}(R) + \zeta^{cdw}(R)$, where the terms have the same meanings as for $TaSe_2$.

$$\zeta^{lat}(\boldsymbol{R}) = 2\zeta_0^{lat}\{\cos[a^*(4x-y)]+\cos[a^*(x+3y)]+\cos[a^*(3x-4y)]\}. \tag{11}$$

This spatial dependence of the corrugation follows that of the Genoa group [11,13] for the normal structure which has maxima on the sulfur atoms. Note that R = (x,y) as defined in Fig. 27a. For the CDW structure the spatial dependence is given by

$$\zeta^{cdw}(\boldsymbol{R}) = 2\zeta_0^{CDW}\{\cos(a^*x)+\cos(a^*y)+\cos[a^*(x-y)]\}, \tag{12}$$

where the maxima are shifted away from the sulfur atoms to the Ta positions by the transformation $\mathbf{R}\rightarrow\mathbf{R}+\mathbf{R}_0$ with $\mathbf{R}_0=(5a^+/39, 7a^+/39)$, where $a^+ = 2\pi/[a_o\mathrm{sqrt}(13)]$. The two corrugation amplitudes ζ_o^{lat} and ζ_o^{CDW} are determined from a best fit of the corresponding peak intensities. The overall corrugation predicted by Eqs. (11) and (12) produces the observed trigonal symmetry. In the fitting procedure [97] the well depth of the He-TaS_2 potential (D=8.7meV) has also been included.

The values of the corrugation amplitudes first determined by the Genoa group for an incident wavevector k_i=11.05$Å^{-1}$ and at a surface temperature of 80 K are ζ_o^{lat} = 0.055±0.010Å and ζ_o^{CDW} = 0.040±0.005Å. They are in good agreement with the more recent MPIGö measurements at k_i = 8.56$Å^{-1}$ and 120 K [97]. It is interesting to observe that the corrugation amplitudes for the normal and the CDW superlattice are nearly the same in absolute size.

The same surface has recently been studied by STM and AFM by several groups at room temperature. The periodic modulation in the height of the structures corresponding to the lattice atoms due to the CDW is clearly visible [110]. Barrett et al. [111] reported a CDW amplitude between 0.1 - 0.3Å which is comparable with the HAS results. Slough et al. [88] reported larger values of 6Å (STM) and 0.3 - 1.0Å (AFM). The only measurements at low temperatures are the 77K STM studies which show a strong modulation at the wavelength of the CDW [89].

It is particularly interesting to note as first pointed out by Boato et al. [14] that in X-ray, neutron or high energy electron diffraction studies the CDW superlattice peaks are 20 - 50 times weaker than the normal lattice peaks whereas with HAS they are comparable in intensity. Low energy election diffraction (LEED) patterns of 1T - TaS_2 at 130K [109] also show comparatively weak superlattice peaks. To understand this difference with respect to HAS we recall that the other diffraction probes are mainly sensitive to the high density of electrons close to the ion cores and the ion core atomic positions. HAS on the other hand is sensitive to the weak electronic densities of the order of 10^{-3} e(a.u.)$^{-3}$ at distances of about 3Å, relatively far from the cores of the outer plane of atoms. Thus it is conceivable that whereas the other probes see only the periodic lattice distortion that HAS is sensitive also to the modulation of the electronic density associated with the CDW itself [14,112,113].

It should be pointed out that several additional peaks of unknown origin remain unaccounted for in the R13.9° HAS angular distribution. They have been seen in the three crystal surfaces tried and, in one case, they were also observed in an angular distribution with an incident wavevector of only k_i=4.77Å^{-1}. Three of these peaks are marked by stars in Fig. 28c. Similar, unexplained peaks were also reported by the Genoa group [14].

The diffraction studies help to understand the detailed nature of the phase transitions. This is illustrated in Fig. 29 which shows a series of angular distributions in the R13.9° direction for k_i=6.13Å^{-1} as a function of temperature during heating from 150 to 385 K. The fact that the superlattice peaks lie at nearly identical positions in the four different phases is an artifact of the measurements. As a result of the limited angular resolution the shoulders of peaks in the rotated frame of the nearly commensurate intermediate phases are also detected. Figure 30 shows the dependence of the intensities of the specular peak and the superlattice peak labeled "L" in Fig. 29. The temperature of the crystal was raised and lowered between approximately 150 and 370 K. At the phase transition a much more erratic behavior is observed instead of the relatively smooth change in intensity found with 2H-$TaSe_2$. On cooling, the specular peak shows a step at the 351 K phase transition whereas the L-peak shows only a small change [97]. Both peak intensities show significant oscillations on cooling through the 183 K phase transition. On heating both peaks show significant oscillations and a sharp step at the first phase transition near 223 K. For the superlattice peak, a hysteresis region covering nearly 60 K is seen from approximately 223 to 280 K. In addition, both peaks show steps on heating for the phase change near 350 K. These structures are not fully understood but appear to indicate that the phase transitions are accompanied by several intermediate steps. Related structures have also been seen in ultrasonic pulse studies [114], in He^+ ion back-scattering experiments [115] and by the resonant flexural vibration technique [116].

5.4 SURFACE VIBRATIONS

Extensive measurements of TOF spectra were taken at 120 K in the 1T_3 commensurate phase for the <100>, the <110> and the R13.9° directions as studied in the angular distributions. The lower panels in Fig. 31 show a typical series of TOF spectra converted to an energy transfer distribution for the <110> direction. The peaks at zero-energy are a result of elastic scattering from defects while the positive (negative) energy peaks are from single surface phonon annihilation (creation) inelastic collisions with the He atoms. At the top part of Fig 31

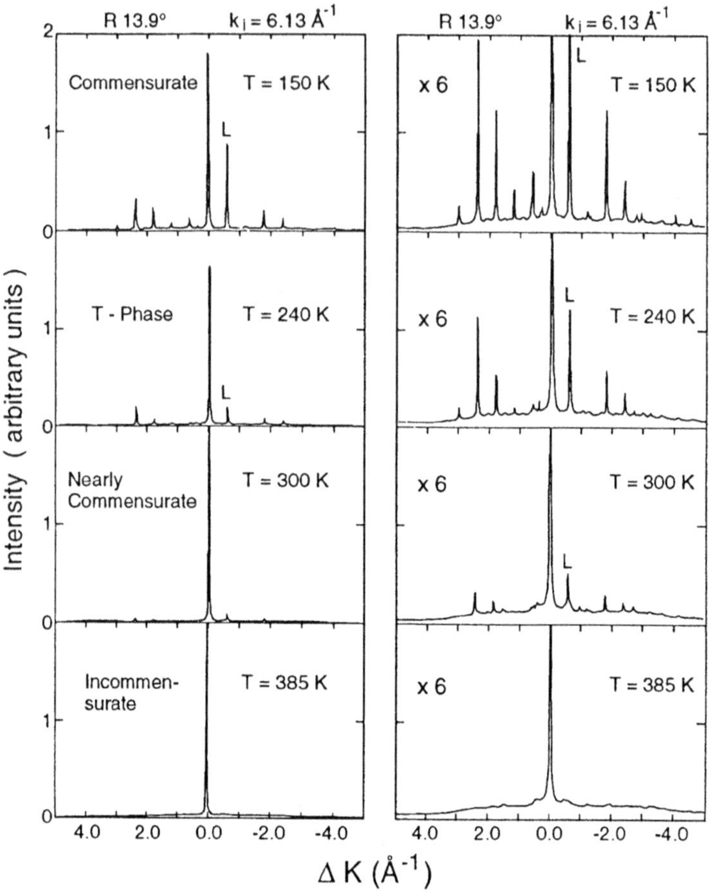

Figure 29. Series of angular distributions in the R13.9° direction for various crystal temperatures taken as the crystal temperature was increased [29]. The helium atom beam wavevector was k_i=6.13Å^{-1} in all measurements. Also shown on the right of each panel is a factor six enlargement of the distributions. The letter "L" marks one of the superlattice peaks.

the peak locations indicated by solid square points can be identified with the aid of the scan curves plotted as dashed lines for the angles and energies of the TOF spectra shown in the figure below. All of the data for the dispersion curves for the three directions are summarized in Figures 32a-c. Surprisingly the dispersion curves show an unusual behavior with a larger number of curves than expected for the normal lattice and there are minima between the maxima. This unusual behavior, which is explained in detail below, is attributed to the CDW induced surface reconstruction. The heavy line curves are the theoretical transverse Rayleigh modes calculated with account taken of the reconstruction. The open square points are from the HAS measurements. Room temperature neutron measurements for the <100> direction are shown in Fig. 32b as filled circles [100].

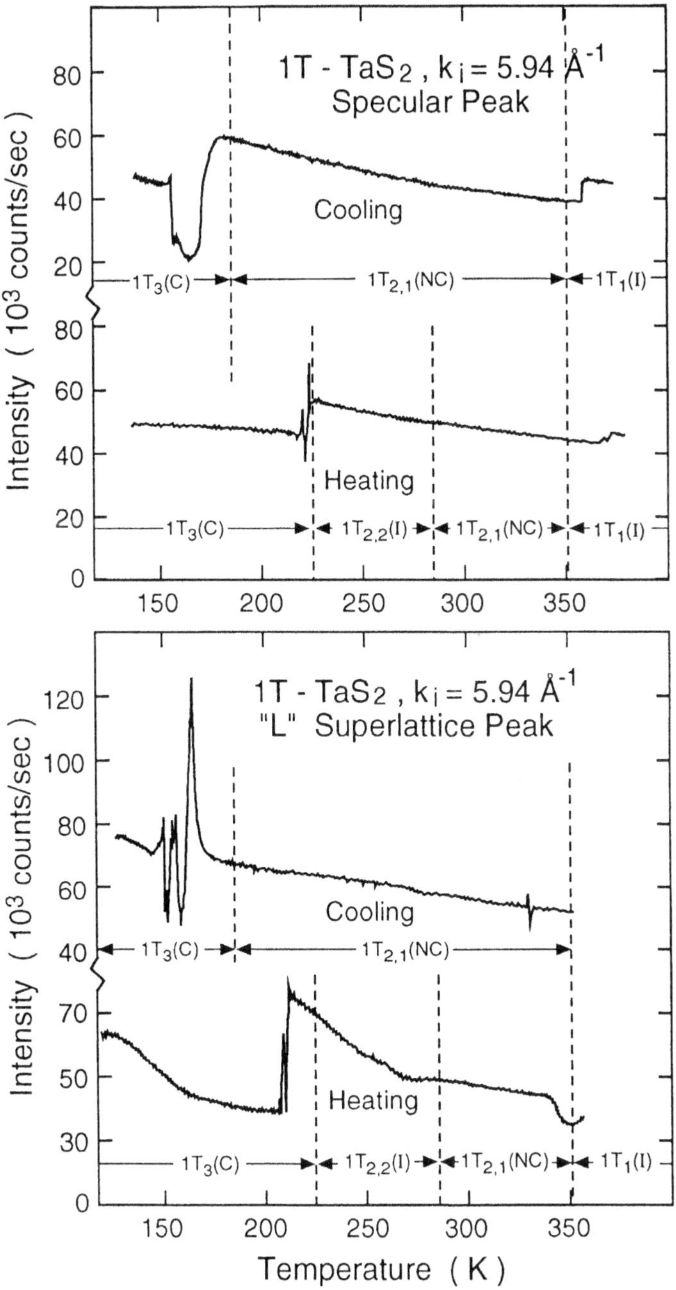

Figure 30. The behavior of the intensity in the specular peak and the superlattice peak (marked "L") as a function of temperature for both cooling and heating of the crystal [29].

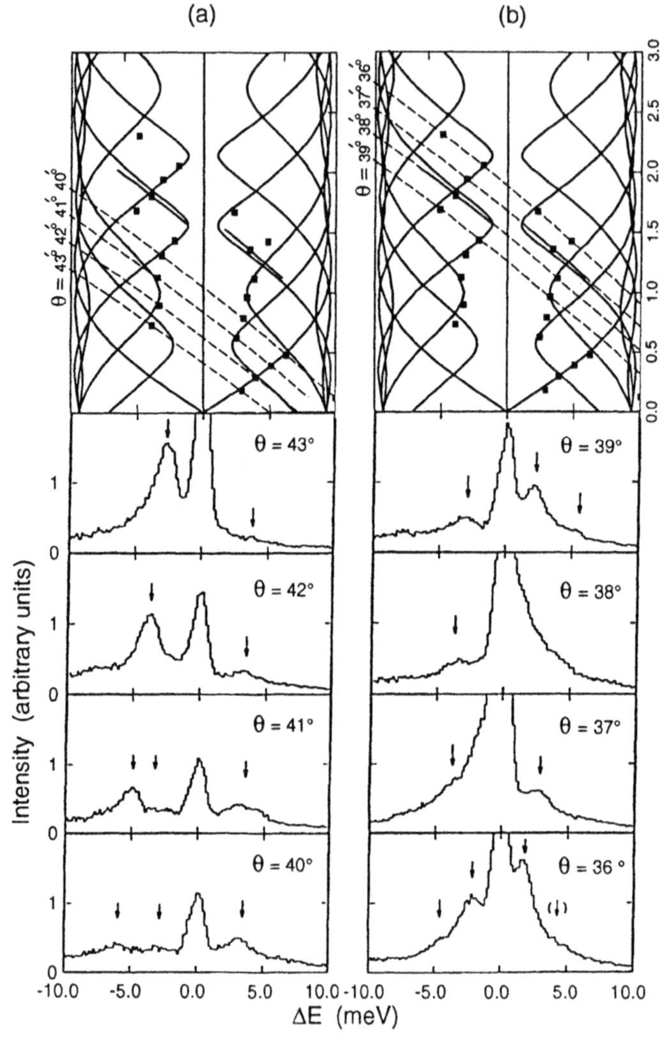

Figure 31. The lower panels in a,b show a series of time-of-flight spectra, converted to energy transfer distributions, in the <110> direction for incident wavevector $k_i = 8.56 Å^{-1}$, for a 120 K target temperature and for various incident angles between 40° to 43° and 36° to 39°, respectively. The spectra have been converted to energy gains and losses of the surface phonons, the energy gains corresponding to annihilation collision events, while the losses are from phonon creation collisions. The experimental data from the peaks marked by arrows are converted into phonon energy and wavevector values and they are plotted in the top panel which shows the phonon energy as a function of phonon wavevector including the scan curves (dashed lines) and the Rayleigh modes (heavy lines), the latter is based on a model interpretation of the dynamics [29].

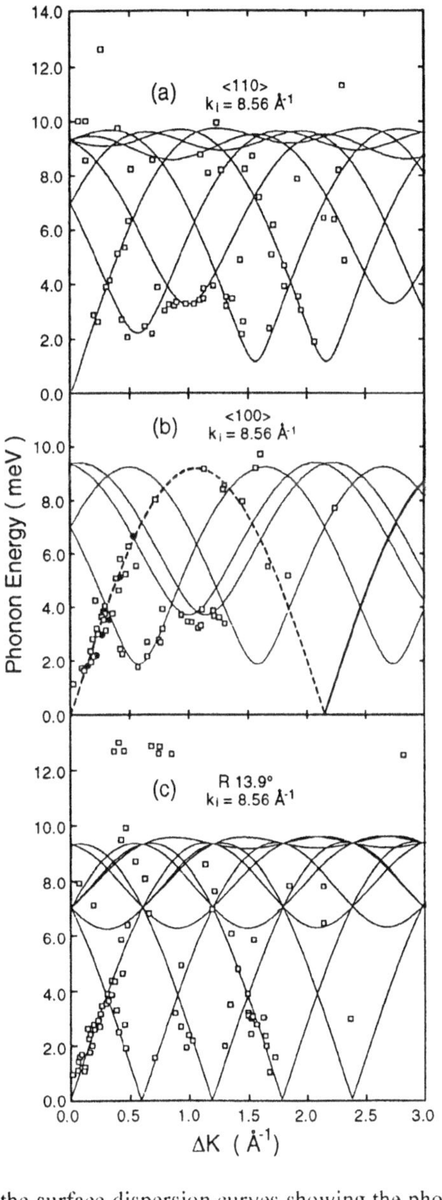

Figure 32. a-c) Plots of the surface dispersion curves showing the phonon energy as a function of phonon wavevector for the Rayleigh modes (solid and dashed lines) and the TOF measurements taken at $k_i = 8.56Å^{-1}$ and target temperature of 120 K for the three directions, <110>, <100> and R13.9° [15,29]. The experimental data are shown by open squares. In the <100> direction the filled circles are from neutron measurements [100]. The dashed line sine curve in the <100> direction is discussed in the text.

A nearest-neighbor force constant model is able to provide a simple physical interpretation of the transverse acoustic Rayleigh mode for all three directions as seen in Figure 32 [15]. In the model shown schematically in Fig. 33 only the motion of a rigid TaS_2 molecule is considered. The TaS_2 molecules lie in the basal x,y plane and are constrained to have motion only in the z direction. The forces connecting the central molecule with the six nearest neighbors control the transverse z-directed vibrational properties through an effective force constant. The equations for the z motion (Rayleigh mode) for rigid molecules of mass m is then

$$m\frac{\partial^2 z_{i,j}}{\partial t^2} = k[6z_{i,j} - z_{i+1,j} - z_{i,j+1} - z_{i+1,j+1} - z_{i-1,j} - z_{i,j-1} - z_{i-1,j-1}]. \tag{13}$$

The equations of motion are solved with the usual ansatz,

$$z_{i,j}(t) = z\exp[i(\omega t - \boldsymbol{K}\cdot\boldsymbol{R}_{i,j})], \tag{14}$$

where ω is the vibrational frequency, z, the amplitude and $K=(K_x,K_y)$, the surface wavevector of the phonon. Solving the dynamical matrix for the dispersion relation gives

$$\begin{aligned}\omega^2/A^2 = {} & \sin^2(K_x/G') + \sin^2(K_x/2G' + \sqrt{3}\,K_y/2G') \\ & + \sin^2(-K_x/2G' + \sqrt{3}\,K_y/2G'),\end{aligned} \tag{15}$$

where A is the amplitude of the dispersion curve and $G = [\sqrt{3}G/2\pi]Å^{-1}$, with $G=G_{10}=2.146Å^{-1}$. These formula, which assume that the phonon dispersion curves have their origin at the reciprocal lattice vectors of the normal atomic lattice, are unable to explain the measurements, especially the observation that many of the modes pass through minima between intervening maxima. In order to explain these features it is necessary also to take into account the large diffraction intensity produced by the CDW superlattice. In this connection it is important to realize that as implied by the conservation Eqns. (5) and (6), the He wave packets are, in fact, inelastically diffracted from the surface. As a result the phonon intensities are directly proportional to the intensities of the nearest diffraction peaks, taking into account Umklapp processes. In the present system the CDW diffraction peak intensities are unusually large and it was found that the observed dispersion curves could only be explained by relating them via sagittal plane umklapp processes to the reciprocal lattice of the CDW superstructure $g=(g_x,g_y)$. These umklapp processes can rather simply be accounted for by introducing the transformation $(K_x \to K_x + g_x)$ and $(K_y \to K_y + g_y)$ into Eqn. 15 above to get

$$\begin{aligned}\frac{\omega_g^2}{A^2} = {} & \sin^2[\frac{(K_x+g_x)}{G'}] + \sin^2[\frac{(K_x+g_x)}{2G'} + \sqrt{3}\,\frac{(K_y+g_y)}{2G'}] \\ & + \sin^2[-\frac{(K_x+g_x)}{2G'} + \sqrt{3}\,\frac{(K_y+g_y)}{2G'}]\end{aligned} \tag{16}$$

The solid lines in Figs. 31 and 32 are based on this expression. It is gratifying to see that they provide an excellent fit to all of the data by adjusting only one parameter, A = 6.5 meV.

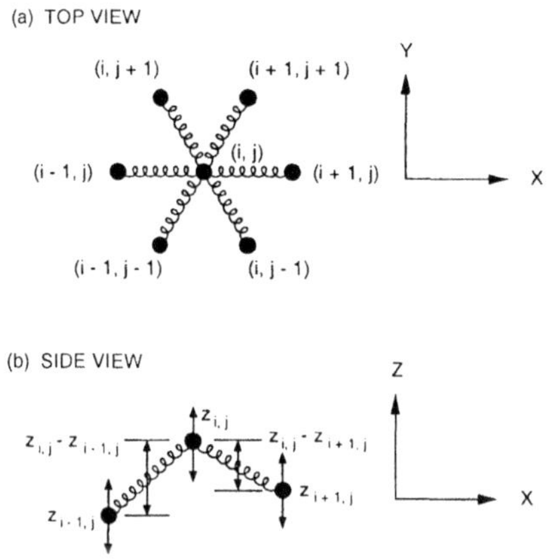

Figure 33. A simple, real-space model used to calculate the Rayleigh mode (transverse acoustic wave) for the surface dispersion curves. The rigid TaS_2 molecules located at (i,j) in the basal plane are all constrained to have motion only in the z direction. The only effect of the forces with the six nearest neighbor molecules is to provide for a restoring force in the z direction.. Only the z direction motion is considered as shown in the lower cross sectional view which shows the stretching of the springs from two of the neighbors acting on the (i,j) particle.

In other words, if the reconstruction were not present then the points would be expected to lie on a simple sine-shaped Rayleigh mode dispersion curve along the undistorted lattice direction. This is shown by a dashed line sine-curve in Fig. 32b for the <100> direction. In Fig. 32b it is seen that the room temperature neutron data lie as expected on the Rayleigh mode dispersion curve of the undistorted normal lattice. In the HAS measurements the large amplitude of the charge density wave provides a new set of CDW reciprocal lattice points with respect to which the phases are observed. The umklapp processes associated with the CDW lattice modify this simple form to give the other dispersion curves which do not touch the abscissa. For example, the dispersion curve with the minimum at $\Delta K \approx 0.6 Å^{-1}$ is associated with the g=(3/13,1/13) peak. The other curves can be explained in an analogous way. This system provides a striking example of how a relatively small CDW induced 10% lattice distortion, possible magnified by a direct interaction with the CDW electrons, can drastically change the shape of Rayleigh dispersion curves.

5.5 DISCUSSION

As we have just seen, in comparison to 2H- $TaSe_2$, the CDW corrugation of the 1T-TaS_2 commensurate phase has a dramatic effect on the surface phonon dispersion curves. When this is accounted for, many of the features of the 1T-TaS_2 surface phonon dispersion curves can be explained by a simple spring model devised to describe the acoustic Rayleigh dispersion curves. The agreement with the experimental results is good considering the simplicity of the model.

The fit is best for the lower part of the energy range shown in Fig 32. The simple dynamical model is, however, limited in several ways. First it does not predict the upward parabolic behavior expected for small values of the wavevector for the two-dimensional crystal [40]. This disagreement was expected a priori as it is a direct consequence of the sine ansatz (Eqn. 14) used in the model. Moreover, the model does not adequately account for the measured higher-lying phonon points which are seen in Fig. 32 near 13meV. These points may be the result of the opening-up of gaps in the mode structure. Further the model has hexagonal symmetry whereas the actual surface has a near trigonal symmetry. A more sophisticated theoretical approach, which addresses many of the above shortcomings, and which leads to a more detailed interpretation of the data is given in [154].

Presently there are only very few experimental results at temperatures other than those corresponding to the $1T_3$ region. Thus the other phases offer additional opportunities for further research. The specular intensity oscillations seen in Fig. 30 suggest that measurements in these temperature regions would be very interesting. Thus by a careful temperature stabilization of the intermediate peaks seen during cooling it might be possible to explore the structure and dynamics of the intermediate phases. These may well correspond to the steps in the "devils staircase" as discussed by Bak [117]. Furthermore it would be most interesting to look for a softening of the Rayleigh mode at higher temperatures just above the phase transition.

Intercalation represents a further direction for HAS measurements. For example, recent work on evaporation of Cu on 1T-TaS_2 shows that the Cu does not remain on the surface, but intercalates instead [118]. This causes the nearly commensurate $\sqrt{13}\times\sqrt{13}$ superstructure to transform into a 3x3 superstructure. Ultraviolet photoelectron spectroscopy showed that the Peierls gap opens near 300K during the course of the Cu deposition indicating the formation of a charge density wave. However, the phase transition at the lowest temperature is lost for the intercalated case [118]. Further HAS investigations on the Cu intercalates and how they modify this material represents a promising direction for future structural and dynamical studies of the layered compounds.

6. GaSe

6.1 INTRODUCTION

GaSe was the first layered crystal for which the surface dynamics was studied extensively by inelastic He atom scattering [16]. It appeared to be an ideal candidate for a pilot study, because the bulk properties indicated that it was free of phase transitions [119-123]. Also at that time the available calculations on graphite, showed that the surface modes for graphite could be predicted from the bulk mode behavior (see Section 3). Using the same arguments for GaSe, it was believed that there would be little difference between the bulk phonons and those on the surface. However, the inelastic helium atom scattering studies indicated just the opposite. While the very strong localization of the surface modes was found as expected, the surface modes softened considerably from the projected bulk band edges. Moreover the dispersion curves revealed considerably more modes than had been measured before by neutron scattering [119,120]. Of all the materials discussed up to this Section, GaSe differs the most from the ideal behavior as exemplified by graphite.

The structure of GaSe is similar to the layered dichalcogenides. Sheets of Ga dimers are sandwiched between two similar sheets of Se atoms to form a layer consisting of $(GaSe)_2$ dimers held together by strong Ga-Ga bonds lined up perpendicular to the planes [119]. The Se atoms and the Ga dimers are hexagonally close-packed in their respective sheets with the same periodicity. The bonding between the Ga and Se atoms between the sheets within the layers is of the strong covalent type, whereas, the bonding between the layers is due solely to weak van der Waals forces [119-123]. Different stacking of the layers with respect to each other gives different unit cells or polytypes [119-123]. So far only the ϵ polytype shown in Fig. 34a has been studied by HAS. A top view of the crystal surface structure and the high symmetry directions for the surface Brillouin zone are also shown in Fig. 34b.

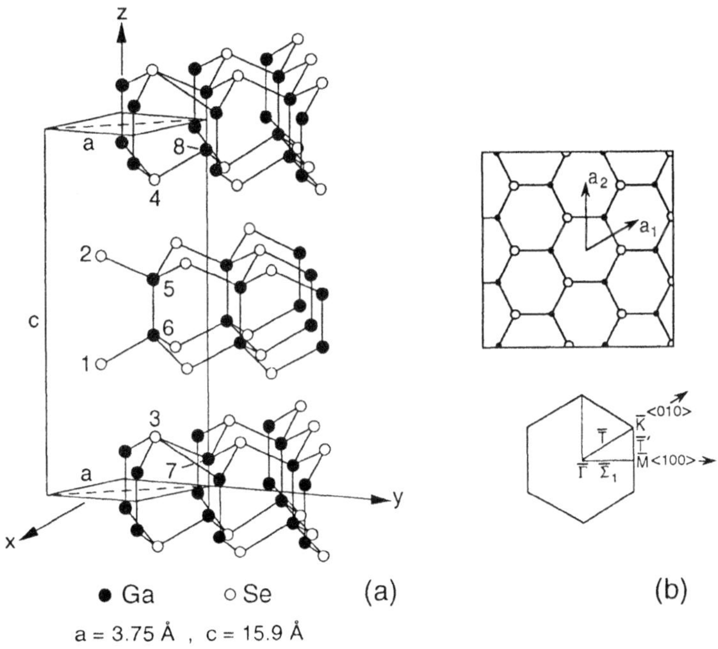

Figure 34. a) The unit cell for the ϵ polytype for GaSe showing the interlayer bonding. The unit cell contains the atoms labeled 1-8 (from Ref. 119). b) Top view of the lattice showing the lattice vectors. The surface Brillouin zone and the high symmetry directions are indicated below.

Presumably because it is difficult to prepare large samples, the neutron data on the bulk phonon dispersion curves is rather limited [119,120] and only available along the Γ-A-direction (z-direction) and the Γ-M-direction (see inset in Fig. 35). This data was sufficient to determine force constants and predict the dispersion curves for the other symmetry direction in reciprocal lattice space [121]. Since this crystal has by far the most complicated structure and normal mode behavior of all the layered crystals studied by He scattering, we show in some detail the different types of motions involved at different points in the irreducible Brillouin zone in Fig. 35. First we discuss the motion along the Γ-A direction corresponding to phonon propagation normal to the planes. The lowest mode at the origin is the TA-mode in which the two layers of each unit cell vibrate in the x-y-direction. With increasing frequency the weak coupling between layers is broken and the resulting motion is a vibration of adjacent single layers with respect to each

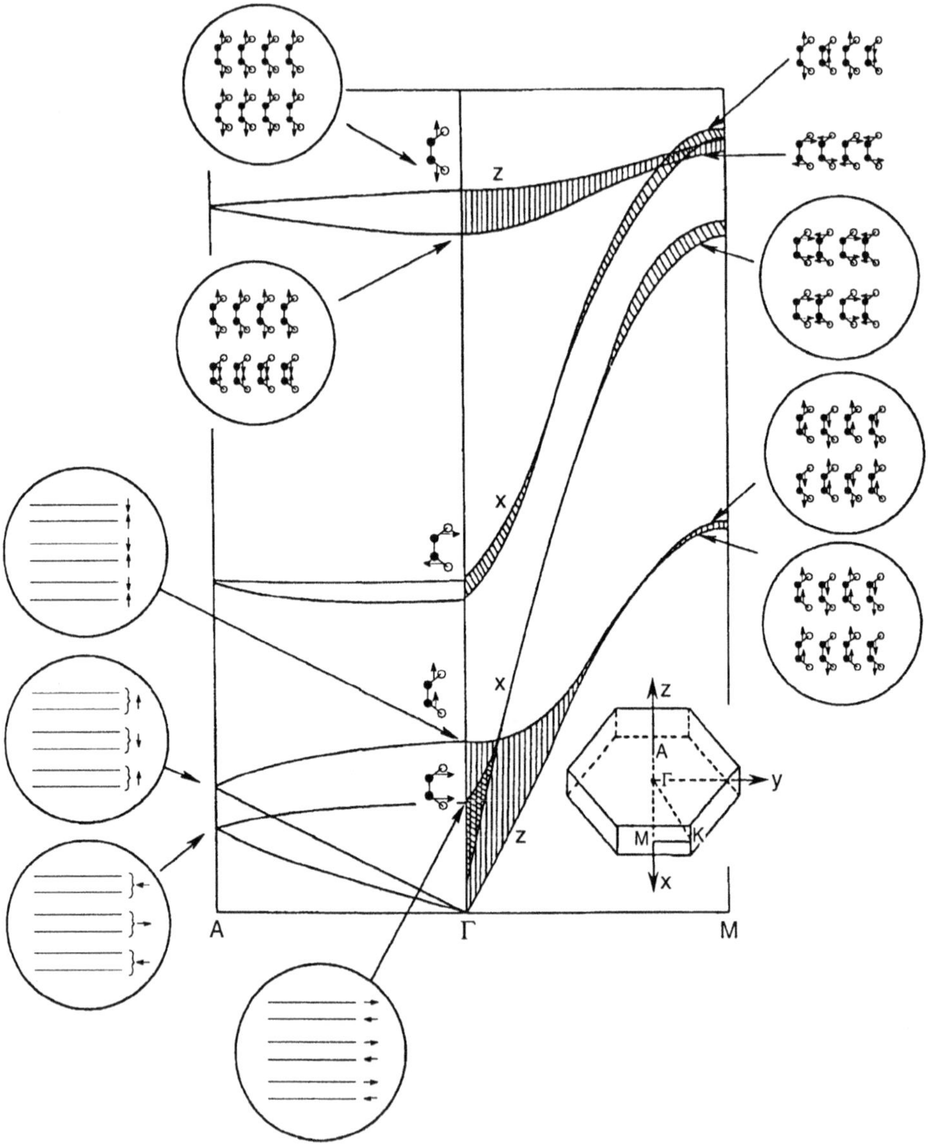

Figure 35. Some of the different motions of the GaSe units in the different regions of the bulk phonon dispersion diagrams. Along the Γ-A direction the phonons propagate perpendicular to the lattice planes. The horizontal lines in the inserts represents the individual layers. Along the Γ-M direction the shaded regions indicate the bands resulting from the projection of the perpendicular modes propagating in the Γ-A direction.

other. At the zone boundary the adjacent pairs of layers are moving in opposing directions. If the two layers of the unit cell are not coupled then the unit cell contains only a single layer. In this case the reciprocal lattice vector would extend twice as far. Thus the interlayer coupling leads to a folding back of the TA branch in the dispersion along the Γ-A direction. The same occurs for the LA mode propagating in the Γ-A direction, which has a slightly higher frequency than the TA mode and is also shown in Fig. 35. Next we note that at the zone origin the TA and LA vibrational motions along the Γ-A direction correspond to LA and TA vibrations, respectively, with infinite wavelength in the plane of the crystal. In the Γ-M direction we thus have two TA modes which differ in frequency corresponding to in-phase and out-of-phase motion of the two layers of the unit cell. Thus the interlayer coupling lifts the degeneracy and leads to a splitting in frequencies. Between these two extremes the modes correspond to different motions in the z-direction and thus contribute to the bulk bands projected on to the surface layer. Half way out to the zone boundary, where the wavelengths span several unit cells, these differences tend to disappear and the bulk bands become narrower. Then at the zone boundary where the motion is concentrated in single layers, the differences increase again. As expected the characteristic counter-phase motions of the dimers in the adjacent layers of the unit cells are found to have higher frequencies than the in-phase motion.

Much the same arguments hold at higher energy for the optical modes, which involve, just as in the spectroscopy of free molecules, distortions within the $(GaSe)_2$ dimer. In the optical modes shown in Fig. 35 only modes in which a GaSe monomer moves as a unit are involved. Distortions within the GaSe-parts have even higher frequencies and do not appear in the low energy part of the dispersion curves in Fig.35.

6.2 EXPERIMENTAL

The surface was cleaved from a single crystal (8x16 mm^2 by 2mm thick) by using sticky tape [124]. The cleavage, transport and mounting of the crystal into the scattering chamber took place in an argon atmosphere. Following this, the ultrahigh vacuum chamber was baked for 12 hrs at 100°C with the crystal baked at 150°C for an additional 12 hrs at a base pressure in the chamber of approximately 2×10^{-10} mbar.

6.3 SURFACE STRUCTURE AND He-SURFACE POTENTIAL

The angular distributions are shown in Fig. 36 for the <100> and the <010> directions [16]. They exhibit sharp diffraction peaks revealing the same lattice spacing as for the bulk and without any indication of a surface reconstruction. The <100> direction has much more intense diffraction peaks than the other direction indicating substantially more corrugation of the surface in that direction. The corrugation has been evaluated within the hard wall eikonal approximation, Section 2.1, in which the acceleration of the He atoms by the attractive well at the surface is approximately accounted for. For the <100> direction the corrugation is 0.7Å compared to only 0.12Å for the <110> direction [125]. These corrugations are quite large and comparable to those for He scattering from the alkali-halide (001) surfaces [32]. Iannotta has remeasured the structure of this material with He atom scattering and used a different form for the corrugation which accounted for asymmetries in the scattering [17]. His value for the peak-to-peak corrugation was 0.6Å which compares well with the value above.

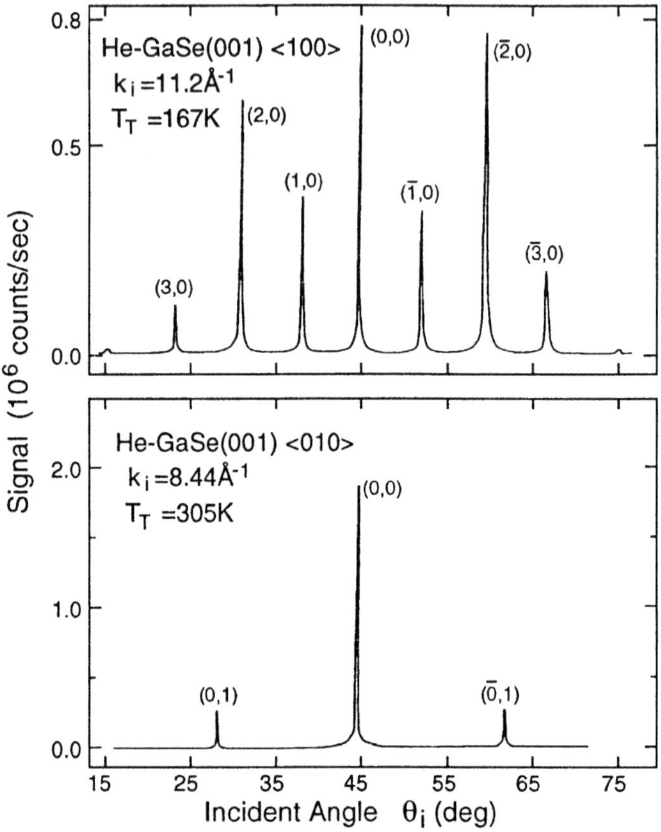

Figure 36. Scattered He intensity from the GaSe(001) surface in the <100> direction (upper panel) and in the <110> direction (lower panel) as a function of incident angle [16]. (0,0) marks the specular peak and the others labeled by (m,n) integer values are Bragg diffraction peaks.

6.4 SURFACE VIBRATIONS

Of the four layered materials considered in this paper up to now, the spectra from GaSe are by far the richest in detail revealing at least five different surface modes. Several examples of spectra, converted to surface phonon energy, for the two high symmetry directions are shown in Figure 37 for the <100> direction. In the spectra of Fig. 37a five different modes can be seen with the relative intensities changing in a remarkable way for the four closely spaced scattering angles. At low frequencies the Rayleigh mode S_1 is especially strong on the annihilation side. The other annihilation peak is attributed to the S_2 optical mode. On the creation side the S_1 mode coalesces with the S_8 mode and another optical mode S_3 is clearly resolved in two of the spectra. Figure 37b shows some spectra in which yet another mode, called the S_6 longitudinal mode, is visible between the S_2 and $S_{1,8}$ modes.

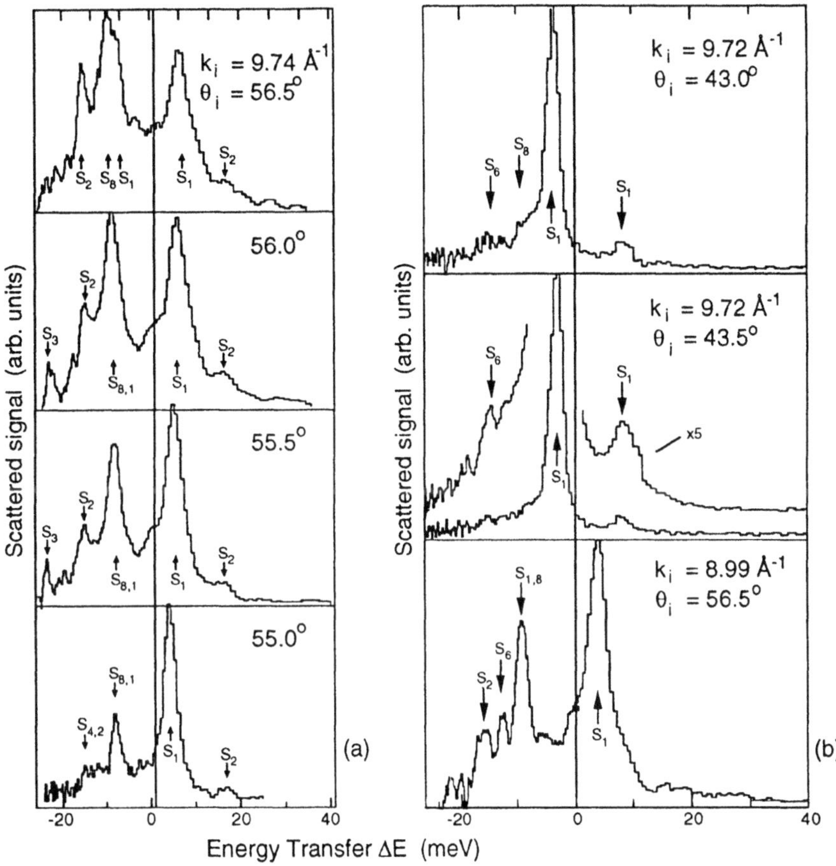

Figure 37. Sequence of time of flight spectra converted into energy transfer distributions for GaSe(001) in the < 100 > direction as a function of incidence angles [16]. (a) Spectra taken between the ($\bar{1}$0) and the ($\bar{2}$0) Bragg peaks. (b) Spectra taken under conditions where the S_6 longitudinal acoustic mode is in evidence. The spectra at 43.5° has an expanded scale. Note that the 56.5° spectra is at the same angle as one of the spectra in the other panel. Because the incident wavevector is different, the two spectra map out different modes of the surface dispersion curves. The surface modes are labeled according to the scheme of Ref. 127.

The spectra in the < 110 > direction are less spectacular since for this direction the specular peak dominates the angular distribution and thus intense phonon peaks are only visible in the vicinity of the specular peak. In Fig. 38a the S_8 mode is seen to appear in going from the top panel to the next panel immediately below it. As the incident angles move closer to the specular peak the S_1 and S_8 mode split apart, while the S_2 mode gradually fades away. The region probed in Fig. 38b on the creation side is near the K-point. With increasing θ_i the S_1 mode increases in intensity on the creation side and the S_2, S'_8 and S_6 modes appear. Surprisingly, in none of the spectra is an appreciable incoherent elastic intensity seen which thus indicates a surface with very few defects [126] and it appears that the crystals of GaSe are remarkably perfect when compared to all other crystals including metals and the easily cleaved alkali halides studied so far by HAS. This may be the reason why so many different modes could be observed for this surface.

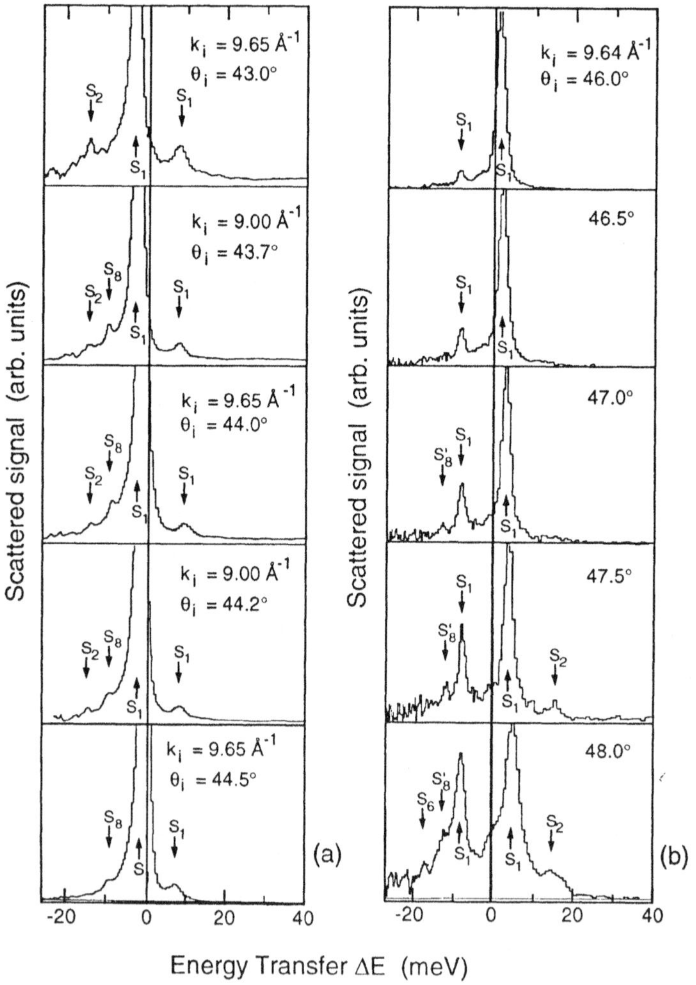

Figure 38. Time of flight spectra converted into energy transfer distributions for GaSe(001) for the <110> direction for various incidence angles [16]. (a) Sequence taken between the (01) diffraction and the specular peaks and (b) Sequence taken where the Rayleigh mode is near the K point on the Brillouin zone boundary. The weak S_6 and S'_8 peaks are observed in some of the spectra and the latter S'_8 points refer to the three isolated points in Fig. 39 for $1.3 < Q < 1.4$Å^{-1}. Note the appearance and disappearance of peaks with angle and incident wavevector conditions for both series.

All the experimental points giving the surface phonon dispersion curves are presented in a reduced zone diagram in Fig. 39. For comparison the surface projected bulk bands previously discussed in Fig. 35 are also shown. At long wavelengths the results for the Rayleigh S_1 mode are in reasonable agreement with the predictions. The S_2 and the S_3-modes, however, lie considerably below the bulk value as is best seen along the < 100 > directions. Comparison with Fig. 35 indicates that the S_2-mode corresponds to a transverse polarized motion of the two GaSe-molecules within the dimer with respect to each other. Since the He atoms impact with the stiffly bound GaSe molecules directly it is not surprising that in this case the optical mode, which otherwise is only weakly excited, has a fairly large intensity. At about half way out to the zone boundary a new mode called the S_8 mode appears and drops in frequency until it apparently coalesces with the S_1 mode near the zone boundary. Whereas the origin and nature of the S_1, S_2, S_3 and S_6 modes are reasonably clear from their proximity to the corresponding bulk modes, the S_8 mode is harder to explain. One explanation is that the S_8 mode derives from an x-polarized mode originating from the E' acoustic band which softens strongly halfway out to the zone boundary leading to its visibility in the time of flight spectra.

The unexpected deviations at the surface from the bulk behavior has lead to the development of the Dispersive-Linear Chain Model [94]. This force constant model is derived from the shell model approach developed for the alkali halides and is shown in Fig. 40a [127]. The interplanar force constants are approximated by first assuming that the Ga-Ga bond is so stiff that its vibrational frequencies lie well above the region of interest. The novel feature of this model is that the other force constants are assumed to be dependent on the wave vector Q. The term $f_1(Q)$ describes the nearest neighbor interactions between the Ga and Se atoms within the $(GaSe)_2$ units. The electrons of the negatively charged Se ions are accounted for by a separate shell which is coupled to the cores by a spring (with force constant g) and which thereby mediates the interactions to the other ions in the lattice. The term $f_2(Q)$ is the force constant for nearest neighbor Se sheets in two adjacent layers and $f_3(Q)$ describes the force between the Se sheets within the same layer (or dimer). With this model it was possible to fit the bulk phonon dispersion curves. The force constants then had to be significantly modified in order to fit the surface phonon dispersion curves, Fig. 40b. The largest change is observed for the f_3 vibrations which are appreciably softened.

6.5. DISCUSSION

The comparison of the rich experimental TOF spectra with the slab adapted surface projected bulk modes of Fig. 39 indicates that: 1) the Rayleigh S_1 mode follows the acoustic band edge over most of the Brillouin zone, 2) the S_2 surface mode is optical in character and related to the z polarized A'_1 bulk band mode, 3) the S_3 surface mode is also optical in character, and probably z polarized, 4) the mode S_6 is related to the longitudinal acoustic band edge and thus is a longitudinal acoustic (x polarized) surface mode. Finally we note that whereas these assignments are reasonably clear, the origin of the S_8 mode is still not completely understood. Additional questions raised by this work are mostly related to the upper branches. The optical modes are considerably softened, even more so than for 2H-$TaSe_2$, but the softening behavior is similar.

The Dispersive Linear Chain Model provides a reasonable fit to the experimental results and makes it possible to quantify the magnitude of the force constant reduction at the surface compared to the bulk as illustrated in the case of f_3. Unfortunately the model does not provide a simple physical mechanism to explain this large softening. Most likely a complex modification

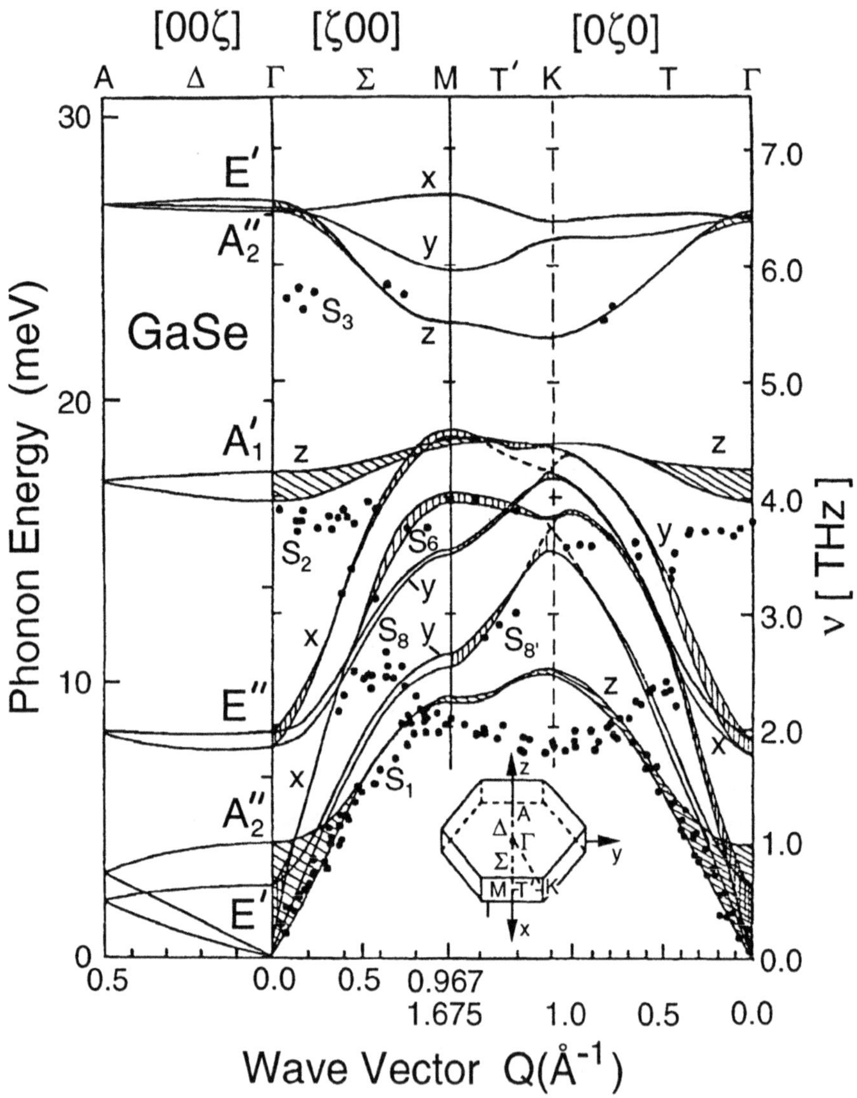

Figure 39. Dispersion relations for GaSe(001) as shown by the experimental points [16] which are labeled according to the surface notation of Ref. 127. S_1 refers to the Rayleigh mode, while S_2, S_3, and S_8 are optical modes. The S_6 mode is probably an acoustic longitudinal mode. The calculated dispersion curves and hatched bands are the slab-adapted forms from the bulk model of Ref. 119.

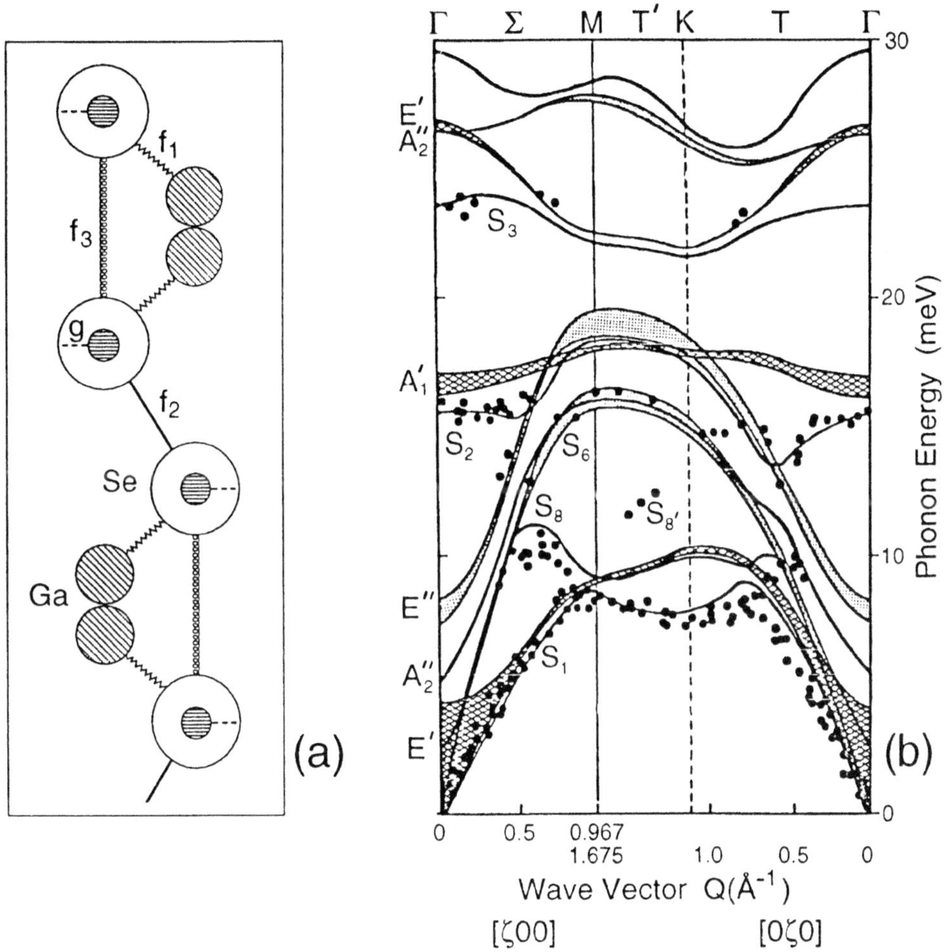

Figure 40. (a) Spring shell model of GaSe used for the Dispersive-Linear-Chain Model [94]. (b) The experimental dispersion data for ϵ-GaSe compared to the Dispersive-Linear-Chain Model results [94]. The latter has the bulk bands as shaded areas and the surface modes as solid lines. The S'_8 points in this figure were not fit in this interpretation.

of the bond-charge distribution within the surface layer is responsible for the dramatic changes [128,129]. Such changes in bond strength could be explained if localized excited electronic states were present at the surface. Unfortunately such states are not known for GaSe but have been observed in graphite [130]. We recall however that the phonons in graphite showed no irregular behavior.

Another possible explanation of the anomalous modes could be that intercalated impurities have a big effect on the intralayer bonding as for TaS_2 [118]. Substantial impurities do however appear unlikely on the basis of the He angular distributions and time of flight spectra. If impurities were intercalated between the layers in a regular periodic arrangement then we would expect these to disturb the corrugation of the surface layer to a small extent, which should be sufficient to produce weak superstructures in the angular distributions. Here it should be emphasized that He atom scattering is very sensitive to surface corrugations as small as 10^{-3} Å, which for example have been observed on low index metal surfaces [131]. Randomly distributed defects or impurities would lead to strong incoherent elastic signals [126] which also were not observed. Additional evidence against the impurity mechanism comes from the fact that the total density of phonons states was found to be compatible with inelastic electron tunneling spectroscopy signals measured prior to and independently of the surface phonons [132]. Finally it is intriguing to speculate about the possibility that the anomalies are due to a surface induced charge density wave.

This surface has been the most challenging for theory and should be reexamined experimentally with the aim of studying the temperature and incident wavevector dependence. Such experiments might, for example, reveal that this system reconstructs at higher temperatures and that the softening of the S_8 mode is a precursor to this behavior.

7. $Bi_2Sr_2Ca_1Cu_2O_8$ High Temperature Superconductor

7.1. INTRODUCTION

The discovery of ceramic-like materials which are superconducting at transition temperatures T_c above 25K has spurred an enormous effort to understand the fundamental behavior of these materials. Because of their importance, significant resources have gone into investigations aimed at gaining a firm understanding of their fundamental behavior so that informed technological developments could follow. The history of the high T_c materials goes back to 1986 with the discovery by Bednorz and Müller of the $(La,Ba)_2CuO_4$ crystal [133]. This was quickly followed in 1987 by the 95K $Y_1Ba_2Cu_3O_{7-x}$ material (abbreviated 123) discovered by Chu et al. [134]. Other materials were soon added, two families being the Bismuth series with T_c=105K and the Thallium series with T_c=125K. We will consider only $Bi_2Sr_2Ca_1Cu_2O_8$ (abbreviated Bi(2212)) with T_c=85 K in this section since it has been studied most extensively with HAS. The materials have very complex unit cells, but soon after their discovery the crystal structure was known [135,136].

The surface properties have been studied by LEED, STM and Auger electron spectroscopy [137-139] and recently by HAS [140]. Early work indicated that the Bi(2212) surface remains stable long enough in the vacuum chamber so that measurements can be made. The initial work on the 123 material indicated that the surfaces are only stable under vacuum at very low temperatures in order to prevent oxygen from diffusing out of the surface [138]. The Bi(2212) materials, on the other hand, do not have this limitation and since relatively large crystals can be produced [141] they offer the best opportunities for surface studies. The Bi(2212) crystal structure is shown in Fig. 41. In the direction labeled "a", the lattice distance is 5.4Å and at 90° to it is the "b" direction which is incommensurate with a spacing about 4.7 times the lattice constant or approximately 27.3Å [137,138]. Auger electron spectroscopy, LEED studies and STM measurements suggested that the cleavage plane is the surface between the two equivalent BiO planes as marked on Fig. 41 [137-139]. Here we review the first HAS angular distributions and TOF spectra on the Bi(2212) surface [18].

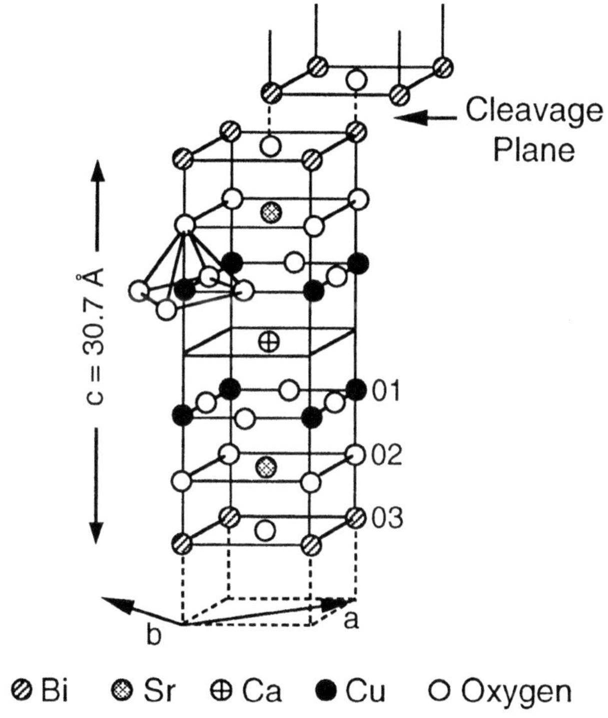

Figure 41. Schematic diagram showing the idealized crystal structure of $Bi_2Sr_2CaCu_2O_8$. Six layers containing oxygen atoms are shown. There are two Cu-01 layers and two Bi-03 layers per cell with the Cu-01 layers separated by Ca ions and the Cu-01 layers separated from the Bi-03 layers by a Sr-02 layer. The entire crystal unit cell consists of two of the displayed units shifted with respect to each other at the cleavage plane.

7.2 EXPERIMENTAL

Bi(2212) samples (about 3x3x0.1mm^3) were attached by epoxy to target holders so they could be cleaved in vacuum as described earlier [142]. The target was cleaved by the same procedure used for $TaSe_2$ and TaS_2 and measurements could be performed immediately. The scattering intensity for the TOF measurements were low and it took typically about 90 minutes to obtain a TOF spectrum. At low crystal temperatures (T = 100K) the crystal was cleaned every three hours by heating to about 350K. The Bi(2212) surface is equally inert as the other layered compounds discussed previously.

7.3 SURFACE STRUCTURE AND He-SURFACE POTENTIAL

A series of angular distributions in the <100>, the <110> and the <010> directions are presented in Fig. 42. The <100> direction shows very small first order diffraction peaks, but sizable second order peaks. The latter are attributed to dominant contributions to the form factor coming from the Bi atoms. The <110> direction has very weak diffraction intensities while the <010> direction has a very large number of closely spaced peaks corresponding to the superstructure.

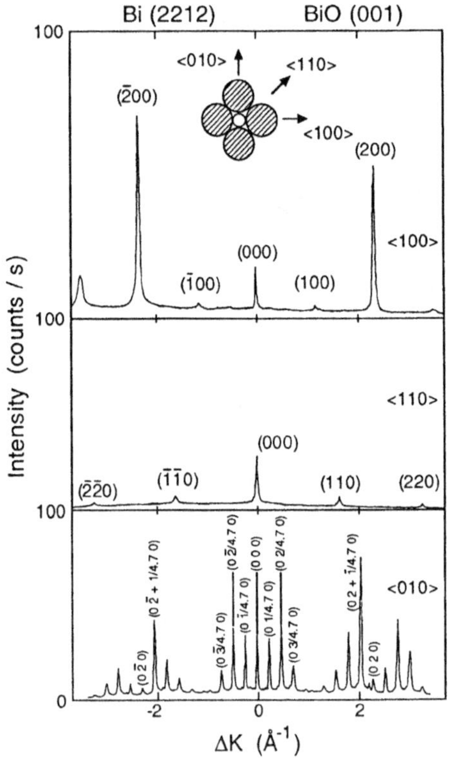

Figure 42. Angular distributions of the scattered He signal as a function of the parallel momentum transfer for the Bi(2212)(001) surface for a target temperature of 310K and incident wave vector $k_i = 6.4 A^{-1}$ [18]. Note the incommensurate superstructure in the <010> direction.

The intensities of the peaks are analyzed in the eikonal approximation as discussed in Section 2.2 by assuming the following corrugation function:

$$\zeta^{lat}(x,y) = \zeta_1^{lat}\cos(y/5) + \zeta_2^{lat}[\cos(x)+\cos(y)] + \zeta_3^{lat}[\cos(x+y)+\cos(x-y)] \qquad (17)$$

In this case the first term is used to approximately evaluate the incommensurate superstructure term in the <010> or "b" direction. A period of 5 times is used instead of the 4.7 to circumvent the necessity of a much larger surface unit cell. The optimum values obtained by the eikonal fit are $\zeta_1^{lat} = 0.4Å$, $\zeta_2^{lat} = 0.008Å$ and $\zeta_3^{lat} = 0.22Å$. This corrugation form reproduces the measured angular distributions and provides additional evidence that the predominant cleavage plane has the symmetry of the BiO surface and not that of the CuO_2 plane.

7.4 SURFACE VIBRATIONS

The HAS method is ideally suited for dynamical studies of the high T_C materials since so far it has been difficult to obtain sufficiently large samples for neutron scattering studies. In Fig. 43, a series of TOF spectra for the Bi(2212) <100> direction are shown for the sample

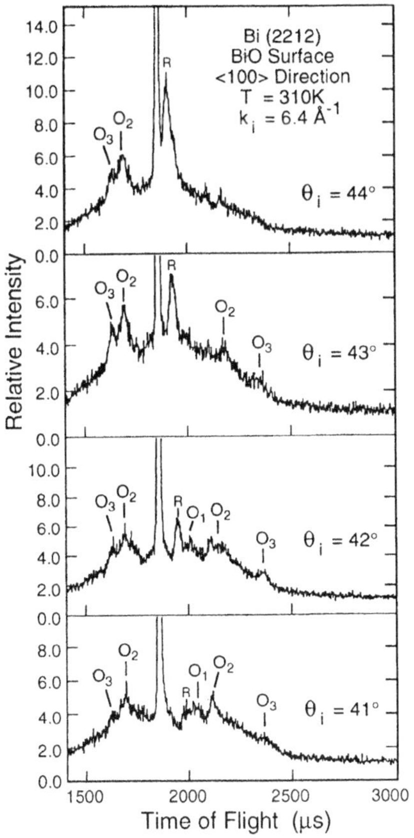

Figure 43. A series of TOF spectra for the Bi(2212)(001) surface in the <100> direction for incident wave vector $k_i = 6.42 Å^{-1}$ and for a target temperature of 310K [18]. The largest peaks are due to diffuse elastic scattering caused by the defects on the surface. The peaks labeled R are from a Rayleigh mode while those labeled O_1, O_2 and O_3 are from optical modes.

at 310K and for an incident wavevector at $k_i = 6.4 Å^{-1}$ for several different incident angles near the specular peak. The very large diffuse elastic peaks indicate that these surfaces have a very high defect concentration [126]. Only the peaks labeled by R which have been assigned to the Rayleigh mode show a distinct dispersion. The peaks from these and other TOF spectra lead to the dispersion curves shown in Fig. 44. Clearly there are three optical modes, O_1, O_2 and O_3, in addition to the Rayleigh mode. Fig. 44 also indicates the predicted energetic position of the ir (L) mode of 5.4 meV as well as the measured Raman E_g mode at 8.02 meV and the Raman A_{1g} mode at 10.7 meV [143-145]. These optically active modes are marked by arrows at the zone origin. For each of these bulk modes there appears to be a surface mode which is shifted downwards in frequency from 5.4 to 3.0 meV, 8.0 to 6.0 meV and from 10.7 to about 8.9 meV. This downward shift in frequencies is consistent with the dispersion curves measured for GaSe and $TaSe_2$. A slab-dynamics calculation to fit these results is in preparation. The preliminary result suggest that the surface modes are very sensitive to the actual surface plane exposed or to small amounts of reconstruction.

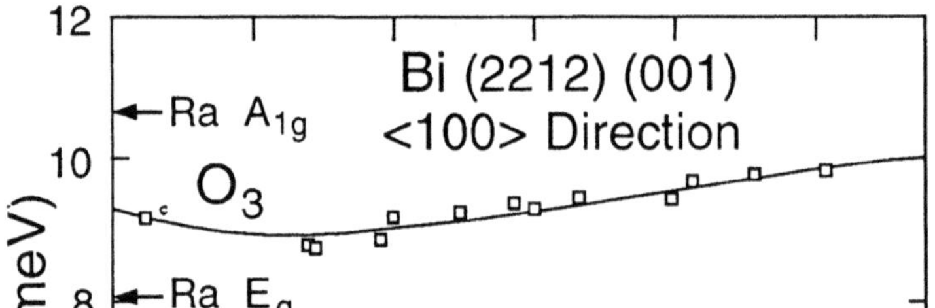

Figure 44. Measured dispersion curves for the Bi(2212)(001) surface for the <100> direction [18]. The lines have been drawn to guide the eye. The arrows at the zone origin indicate the energies of the known Raman (Ra) and infrared (ir) lines [143-145].

7.5. DISCUSSION

These initial exploratory studies on the high T_C superconductors illustrate the ability of the HAS technique to investigate systems for which only small samples can be produced. The structural studies confirm the reconstruction along the <010> direction found in previous studies [18,146]. The HAS technique reveals a considerable concentration of defects greater than that for the other layered compounds. The surface phonon dispersion curve presented here for Bi(2212) provide evidence for a considerable softening of the force constants at the surface. Additional work is needed to establish if this softening is due to relaxation or other structural changes at the surface or due to changes in the electronic surface states.

8. Summary and Conclusions

In this review we have brought together the results of recent helium atom scattering studies of the surface and lattice dynamics of five different layered compounds: graphite, 2H-$TaSe_2$, 1T-TaS_2, GaSe and the Bi(2212) high-T_c superconductor. Initially it was anticipated that because of the weak coupling between layers that the lattice dynamics at the surface would be nearly the same as in the bulk. On the contrary for all of the systems studied, except graphite, significant differences were observed in the surface phonon dispersion curves determined by helium atom scattering compared to those measured largely by neutron scattering in the bulk. Only the graphite surface shows the nearly quadratic functional form for the Rayleigh mode dispersion curve near the zone origin which had been predicted for two-dimensional systems [3,28,40]. Overall the bulk theory predicts very well the observed dynamical behavior of the graphite surface.

For 2H-$TaSe_2$ in the high temperature normal (N) phase, the surface dispersion curves also follow rather closely those of the bulk [10,73]. There are, however, some small differences in the details of the dispersion curves in the vicinity of anomalies. In the bulk only the longitudinal mode softens at the CDW phase transition temperature of 122 K at a wavevector of about 0.7 $Å^{-1}$. At the surface, however, the transverse Rayleigh mode is observed to soften at a somewhat lower temperature near 110 K and at a wavevector near 0.53 $Å^{-1}$. For the 2H-$TaSe_2$ in addition to the strong diffraction from the normal lattice a weak diffraction (about 1%) was also observed from a CDW induced additional periodicity. The weak additional corrugation of the interaction potential associated with the CDW increases with decreasing temperature below 122 K. Thus the corrugation amplitude could be used as an order parameter and the measured critical exponent $\beta = 1/3$ was found to be in agreement with the 3d-xy model. The helium TOF measurements also reveal for the first time several optical modes, not observed previously. These experiments clearly demonstrate the advantage of He scattering experiments over neutron scattering which require much larger samples.

The $1T_3$ phase of 1T-TaS_2, which is the only phase of this substance which has been extensively studied by helium atom scattering so far, reveals a CDW which has much greater diffraction intensities than in the case of 2H-$TaSe_2$. Although as mentioned above the periodic lattice distortion associated with the CDW on TaS_2 is weakly seen in electron and neutron diffraction, surprisingly the intensities of the same peaks in HAS are roughly 30-50 times greater [14]. The CDW induced diffraction peaks are often somewhat larger than those observed for the normal lattice. Because of the weak neutron intensities the CDW induced periodic distortion of the lattice, it is conjectured, may in itself be too weak to satisfactorily explain the HAS results. For this reason it has been proposed that the helium atoms may be diffracted directly from the small modulations in electron density produced by the CDW. In this connection it is interesting to recall that in this system the phonon dispersion curves are also strongly affected by the diffraction from the CDW induced structure. The observed unexpected shape of the surface phonon dispersion curves measured with respect to the normal lattice could be explained by a simple model which was developed to account for the additional diffraction from the CDW. This model could explain many of the major features of the surface dispersion curves.

Of all the surfaces studied the GaSe surface shows the richest structures in the inelastic TOF spectra and the most striking differences in the dispersion curves from those predicted from the bulk behavior [16]. Several soft modes were observed with a strong wavevector dependence and additional unexpected modes were found. A simple explanation for these intensity anomalies could so far not be found. For this reason it appears that the GaSe surface needs further study. Since neither the dependence of the dispersion curves on the surface temperature and on the incident wavevector have been reported these should be studied next.

Finally the preliminary results on the Bi(2212) high-T_c superconductor illustrate the continuing promise for HAS observations on layered materials. The structure of the lattice plane containing Bi and O atoms was studied at room temperature for all three directions. In the <010> direction, both the normal Bragg structure and the incommensurate lattice structure were seen. The surface phonons at the same temperature for the <100> direction were also examined and found to have four modes, three of which were slightly below the Raman and ir frequencies at the zone origin. These are the first measurements of phonon dispersion curves for this new material.

As mentioned above the observed differences between the surface and bulk behavior are surprising in view of the fact that the layers in all of the compounds investigated are bound together by weak van der Waals forces with force constants roughly one order of magnitude smaller than those binding the atoms and molecules within the layers. A completely satisfactory explanation for this phenomenon is still not available and at present we can only make some general remarks.

One possible effect may come from electronic surface states produced on exposure of the surface. This mechanism has been discussed in connection with the many anomalies found on GaSe [16,94]. Unfortunately, however, little is known about the surface electronic states of those layered compounds for which the surface phonon dispersion curves have been measured. Of the surfaces discussed in this review we are only aware of evidence for surface electronic states on graphite coming from inverse-photo-emission experiments [147] and calculations [130]. Ironically, this is just the surface for which no anomalies are observed. Thus the existence and nature of the surface electronic states on these layered compounds and their possible effect on the surface phonons remains an intriguing question.

From analogous work on metal surfaces, which have been more extensively studied, we are lead to expect that possibly two additional effects may be important. For one the observed structural changes of relaxation and reconstruction of metals are attributed to a redistribution of electrons at the surface [148]. Electronic redistribution has also been invoked to explain the unexpected appearance of longitudinal modes on the noble and transition metals [149-151]. The usually small changes in the Rayleigh phonon frequencies on metals appear to be related to surface stress [152] which also can be explained in terms of electronic readjustments at the surface [153]. Most of the electronic readjustment occurs within the topmost layers. For this reason analogous phenomena may also occur at the surfaces of the layered compounds, even though as mentioned above the coupling between layers is much weaker.

The other effect is related to the unique interaction of the He atoms with the surface. As is well known electrons, neutrons and X-rays, interact with the large electronic densities close to the ion cores. He atoms, on the other hand, by virtue of their own electronic shell, interact at large distances of about 3 Å or more from the cores, where the electron densities are only 10^{-3} e/Å^3. As a result, as discussed in connection with the TaS_2 results [154], helium atoms are more sensitive to minute additional changes in the electron densities at the surface. Confirming independent evidence for this comes from helium scattering measurements on the W(001) surface which exhibits a temperature induced reconstruction at 280K. This reconstruction has been extensively studied by LEED and X-ray scattering. However only He atoms are able to detect an incommensurate surface structure which precedes the reconstruction on cooling the sample. This observation has been attributed to a weak diffraction from a CDW induced additional corrugation at temperatures above the transition temperature for the well-known reconstruction [113]. Finally in this connection it may be of interest that very recent HAS experiments have revealed a large increase in the relative probability for exciting the longitudinal surface mode compared to the Rayleigh mode on the Cu(111) and Cu(100) surfaces [149]. This has very recently been explained in the framework of a pseudocharge model by an indirect excitation of the longitudinal vibrational motions via the collision induced distortion of the density distribution of electrons situated between the cores. A similar but simplified pseudocharge model has been found to be uniquely successful in explaining surface dispersion curves of TaS_2 and the observation by HAS of certain optical modes (see in [154]). Additional work is now needed to investigate the coupling of He atoms to the pseudocharges on this and the other layered metallic compounds.

The future of HAS for further studies of two dimensional materials looks very promising. Compared to neutron and X-ray scattering only very thin samples with surface areas of several mm^2 are required. Compared to low energy electron scattering (LEED and EELS) the neutral He atoms are a non-destructive, chemically inert, and a purely surface sensitive probe. A further advantage of HAS is its versatility. In addition to the studies of surface effects, discussed in this review, such as surface phonons, structural changes accompanying phase transitions, and the coupling of charge density waves to the surface phonons, other phenomena such as the periodic as well as the randomly distributed defect structures, the mechanisms of surface diffusion [155], surface melting [156], epitaxial growth [157-159] and quantum size effects [160] can all be studied.

9. Acknowledgements:

We acknowledge support from NATO Grant No. 891059 and US DOE Grant No. DE-FG05-85ER45208 (JGS). We also thank our colleagues Professors G. Benedek and L. Miglio of Milan, Professors V. Celli, S. Safron and J.R. Manson from the USA, the many students and associates at the Max-Planck-Institut für Strömungsforschung, notably G. Brusdeylins, C. Heimlich, F. Hofmann, P. Ruggerone, D. Schmicker and R. Vollmer for many stimulating discussions. Finally we thank R.E. Palmer for sending us parts of J.L. Wilkes thesis and his correspondence on graphite phonons.

10. References

1. Pool, R., Science 247, 26 (1990).
2. Barbee, Jr., T. W., in Physics, Fabrication and Applications of Multilayered Structures, Ed. by Dhez, P. and Weisbuch, C., NATO ASI series, Vol. 182, pp. 17-32, Plenum Press, N.Y. (1988).
3. Benedek, G., Brusdeylins, G., Heimlich, C., Toennies, J. P. and Valbusa, U., Surf. Sci. 178, 545 (1986).
4. Boato, G., Cantini, P., Salvo, C., Tatarek, R. and Terreni, S., Surf. Sci. 114, 485 (1982).
5. Cantini, P. and Tatarek, R., Surf. Sci. 114, 471 (1982).
6. Benedek, G., Miglio, L., Skofronick, J. G., Brusdeylins, G., Heimlich, C. and Toennies, J. P., J. Vac. Sci. Technol., A5, 1093 (1987).
7. Benedek, G., Miglio, L., Brusdeylins, G., Heimlich, C., Skofronick, J. G. and Toennies, J. P., Europhys. Lett. 5, 253 (1988).
8. Benedek, G., Brusdeylins, G., Heimlich, C., Miglio, L., Skofronick, J. G., Toennies, J. P. and Vollmer, R., Phys. Rev. Lett. 60, 1037 (1988).
9. Brusdeylins, G., Heimlich, C., Skofronick, J. G., Toennies, J. P., Vollmer R. and Benedek, G., Europhys. Lett. 9, 563 (1989).
10. Brusdeylins, G., Heimlich, C., Skofronick, J. G., Toennies, J. P., Vollmer, R., Benedek, G. and Miglio, L., Phys. Rev. B41, 5707 (1990).
11. Cantini, P. in Dynamics of Gas-Surface Interactions, Ed by Benedek, G. and Valbusa, U., Springer Verlag, Heidelberg, p. 84 (1982).
12. Cantini, P. and Colella, R., Solid State Comm. 53, 77 (1985).
13. Cantini, P., Boato, G. and Colella, R., Physica 99B, 59 (1980).
14. Boato, G., Cantini P. and Colella, R., Phys. Rev. Lett. 42, 1635 (1979).
15. Brusdeylins, G., Hofmann, F., Ruggerone, P., Toennies, J. P., Vollmer, R., Benedek, G. and Skofronick, J. G., In Phonons 89, Ed. by Hunklinger, S., Ludwig, W. and Weiss, G., World Scientific, Singapore, 892 (1990).
16. Brusdeylins, G., Rechsteiner, R., Skofronick, J. G., Toennies, J. P., Benedek G. and Miglio, L., Phys. Rev. B34, 902 (1986).
17. Iannotta, S., Surf. Sci. (Submitted).
18. Schmicker, D., MPI, Private communication and to be published.
19. Brusdeylins, G., Doak R. B. and Toennies, J. P., Phys. Rev. B27, 3662 (1983).
20. Estermann, I., Frisch, R., and Stern, O., Z. Physik 73, 348 (1932).
21. Toennies, J. P., in Surface Physics, edited by Kress, W., Springer, Berlin, Chap. 5.
22. Engle, T. and Rieder, K. H., Structural Studies of Surfaces, Springer, Berlin, Vol. 91, pp. 55-90 (1982).
23. Hoinkes, H., Rev. Mod. Phys. 52, 933 (1980).
24. Toennies, J. P., in Dynamics of Gas-Surface Interactions, edited by Benedek, G. and Valbusa, U., Springer, Berlin, pp. 208-226 (1982).
25. Rieder, K. H., in Dynamic Phenomena at Surfaces, Interfaces and Superlattices, edited by Nizzoli, F., Rieder K. H. and Willis, R. F., Springer, Berlin, pp. 2-12 (1985).
26. Lovesey, S. W. and Springer, T., eds. Dynamics of Solids and Liquids by Neutron Scattering, Springer-Verlag, New York, 1977.
27. Allen, R. E., Alldredge, G. P. and de Wette, F. W., Phys. Rev. B4, 1648 (1971); B4, 1661 (1971); B4, 1682 (1971).
28. Vollmer, R., MPI, Private communication (on graphite).

29. Hofmann, F., MPI, Private communication (on 1T-TaS_2).
30. Toennies J. P. and Winkelmann, K., J. Chem. Phys. 66, 3965 (1977).
31. Ablebond 84-1, LMI-T, Ablestick Laboratories, 833 West 182nd Street, Gardinia, CA 90248.
32. Garibaldi, U., Levi, A. C., Spadacini, R., and Tommei, G. E., Surf. Sci. 48, 649 (1975).
33. Rieder, K. H., in Dynamics of Gas-Surface Interactions, edited by Benedek, G. and Valbusa, U., Springer, Berlin, pp. 61-78 (1982).
34. Strutt, J. W., Baron Rayleigh, The Theory of Sound, Vol II, Chap. XIII, Dover Publishing, New York, (1945).
35. Weare, J.H., J. Chem. Phys. 61, 2900 (1974).
36. Celli, V., Himes, D., Tran, P., Toennies, J.P., Wöll, Ch. and Zhang, G., Phys. Rev. Lett. 66 3160 (1991).
37. Manson, J.R., Phys. Rev. B43, 6924 (1991).
38. Klein, C. A., Chemistry and Physics of Carbon, Ed. by Walker, P. L., Jr., Vol 2, Marcel Dekker, New York, p.225, (1966).
39. Cole, M. W., Frankl, D. R. and Goodstein, D. L., Rev. Mod. Phys. 53, 199 (1981).
40. de Rouffignac, E., Alldredge, G. P., and de Wette, F. W., Phys. Rev. B23, 4208 (1981).
41. Nicklow, R., Wakabayashi, N., and Smith, H. G., Phys. Rev. B5, 4951 (1972).
42. Dolling, G., and Brockhouse, B. N., Phys. Rev. 128, 1120 (1962).
43. Al-Jishi, R. and Dresselhaus, G., Phys. Rev. B26, 4514 (1982).
44. Nemanich, R. J., Lucovsky, G. and Solin, S. A., Mater. Sci. Eng. 31, 157 (1977).
45. Friedel, R. A. and Carlson, G. C., J. Phys. C75, 1149 (1971).
46. Brillson, L. J., Burstein, E., Maradudin, A. A. and Stark, T., in The Physics of Semimetals and Narrow Gap Semiconductors, Ed. by Carter, D. L. and Bate, R. T., Pergamon Press, Oxford, p. 187 (1971).
47. Tuinstra, F. and Koenig, J. L., J. Chem Phys. 53, 1126 (1970).
48. Alldredge, G. P., de Rouffignac, E., Firey, B. and de Wette, F. W., Phys. Rev. B29, 3712 (1984).
49. Firey, B., de Wette, F. W., de Rouffignac, E. and Alldredge, G. P., Phys. Rev. B28, 7210 (1983).
50. Wilkes, J. L., Palmer, R. E. and Willis, R. F., J. Electr. Spect. Rel. Phenom. 44, 355 (1987). See also Wilkes, J. L., Chapter 4, Ph.D. Thesis, University of Cambridge, Cambridge, England (1990).
51. Oshima, C., Aizawa, T., Souda, R. and Ishizawa, Y., Sol. State Comm. 65, 1601 (1988).
52. Schneir, J., Sonnenfeld, R., Hansma, P. K. and Tersoff, J., Phys. Rev. B34, 4979 (1986).
53. Hennig, G. R., Chemistry and Physics of Carbon, Ed. by Walker, P. L., Jr., Vol 2, Marcel Dekker, New York, p.2, (1966).
54. Noda, T., Sumiyoshi, Y. and Ito, N., Carbon 6, 813 (1968).
55. Kish crystal obtained from H. Suematzu. Made by Toshiba Ceramics Co., Ltd. See Morita, S., Tsukada, S. and Mikoshiba, N., J. Vac. Sci.Technol. A6, 354 (1988).
56. Boato, G., Cantini, P., Guidi, C., Tatarek, R. and Felcher, G. P., Phys. Rev. B20, 3957 (1979).
57. Chung, S., Kara, N. and Frankl, D. R., Surf. Sci. 171, 45 (1986).
58. Ruiz, J. C., Scoles, G. and Jonsson, H., Chem. Phys. Lett. 129, 139 (1986).
59. Vidali, G., Ihm, G., Kim, H-Y. and Cole, M. W., Surf. Sci. Reports 12, 133 (1991).
60. Boato, G., Cantini, P. and Tatarek, R., Phys. Rev. Lett. 40, 887 (1978).
61. Taub, A. H., Passel, L., Kjeiws, J. K., Carueiro, K., McTague, J. P. and Dash, J. G.,

Phys. Rev. Lett. 34, 654 (1975).
62. Lauter, H. L., Godfrin, H., Tiby, C., Wichert, H. and Obermayer, P. E., Surf. Sci. 125, 265 (1983).
63. Frank, V. L. P., Lauter, H. J. and Leidener, P., Phys. Rev. Lett. 61, 436 (1988).
64. Toennies, J. P. and Vollmer, R., Phys. Rev. B40, 2495 (1989).
65. Faisal, A. Q. D., Hamichi, M., Raynert, G. and Venables, J. A., Phys. Rev. B34, 7440 (1986).
66. de Rouffignac, E., Alldredge, G. P. and de Wette, F. W., Phys. Rev. B24, 6050 (1981).
67. Hall, B., Mills, D. L. and Black, J. E., Phys. Rev. B32, 4932 (1985).
68. Kern, K., Zeppenfeld, P., David, R. and Comsa, G., Phys. Rev. B35, 886 (1987).
69. Motizuki, K., Editor for Structural Phase Transitions in Layered Transition Metal Compounds in Physics and Chemistry of Materials with Low-Dimensional Structure. Series A, Layered Structures, Published by D. Reidel Publishing Co., Dordrecht, Holland, (1986).
70. Levy, F. (Ed), Crystallography and Crystal Chemistry of Materials with Layered Structures, Reidel Publishing Co., Dordrecht, Holland, (1976).
71. Wilson, J. A., DiSalvo, F. J. and Mahajan, S., Adv. Phys. 24, 117 (1975).
72. Wilson, J. A., DiSalvo, F. J. and Mahajan, S., Phys. Rev. Lett. 32, 882 (1974).
73. Moncton, D. E., Axe, J. D. and DiSalvo, F. J., Phys. Rev. Lett. 34, 734 (1975); Phys. Rev. B16, 801 (1977).
74. Motizuki, K., Kimura, K., Ando, E. and Suzuki, N., J. Phys. Soc. Jpn. 53, 1078 (1984).
75. Coleman, R. V., Drake, B., Hansma, P. K. and Slough, G., Phys. Rev. Lett. 55, 394 (1985).
76. Giambittista, G., Slough, C. G., McNairy, W. W. and Coleman, R. V., Phys. Rev. B (1990).
77. Wilson, J. A., Phys. Rev. B17, 3880 (1978).
78. Fleming, R. M., Moncton, D. E., Axe, J. D. and Brown, G. S., Phys. Rev. B30, 1877 (1984).
79. Fleming, R. M., Moncton, D. E., McWhan, D. B. and DiSalvo, F. J., Phys. Rev. Lett. 45, 576 (1980).
80. Chen, C. H., Gibson, J. M. and Fleming, R. M., Phys. Rev. B26, 184 (1982).
81. Sugai, S., Physica 117B/118B, 587 (1983).
82. Friend, R. H. and Jerome, D., J. Phys. C12, 1441 (1979).
83. Steinitz, M. O. and Grunzweig-Genossar, J., Sol. State Comm. 29, 519 (1979).
84. Koyama, Y., Zhang, Z. P. and Sato, H., Phys. Rev. B36, 3701 (1987).
85. Fung, K. K., McKernan, S., Steeds, J. W. and Wilson, J. A., J. Phys. C14, 5417 (1981).
86. Pfeiffer, L., Kovacs, T. and DiSalvo, F. J., Phys. Rev. Lett. 52, 687 (1984).
87. Wilson, J. A. and Vincent, R., J. Phys. F14, 123 (1984).
88. Slough, C. G., McNairy, W. W., Coleman, R. V., Garuges, J., Prater, C. B. and Hansma, P. K., Phys. Rev. B42, 9255 (1990).
89. Coleman, R. V., McNairy, W. W., Slough, C. G., Hansma, P. K. and Drake, B., Surf. Sci. 181, 112 (1987).
90. Feldman, J. L., Phys. Rev. B25, 7132 (1982).
91. Holy, J. A., Klein, M. V., McMillian, W. L. and Meyer, S. F., Phys. Rev. Lett. 37, 1145 (1976).
92. Smith, Jr., J. E., Tsang, J. C. and Shafer, M. W., Solid State Commun. 19, 283 (1976).
93. Tsang, J. E., Smith, Jr., J. E. and Shafer, M. W., Solid State Commun. 27, 145 (1978).
94. Miglio, L. and Colombo, L., Phys. Rev. B37, 3025 (1988).

95. Withers, R. L., and Wilson, J. A., J. Phys. C (Solid State Physics) 19, 4809 (1986).
96. Wu, X. L., and Lieber, C. M., Phys. Rev. Lett. 64, 1150 (1990).
97. Brusdeylins, G., Heimlich, C. and Toennies, J. P., Surf. Sci. 211/212, 98 (1989).
98. Scruby, C. B., Williams, P. M., and Parry, G. S., Phil. Mag. 31, 255 (1975).
99. Sugai, S., Murase, K., Uchida, S. and Tanaka, S., Physica 105B, 405 (1981).
100. Ziebeck, K. R. A., Dorner, B., Sterling, W. G. and Schoellhorn, R., J. Phys. F7, 1139 (1977).
101. Uchida, S. and Sugai, S., Physica 105B, 393 (1981).
102. Duffey, J. R. and Kirby, R. D., Phys. Rev. B23, 1534 (1981).
103. Karecki, D. R. and Clayman, B. P., Phys. Rev. B19, 6367 (1979).
104. 351 K is the average of the hysteresis values reported by Thomson, R. E., Walter, U., Ganz, E., Clarke, J. and Zettl, A., Phys. Rev. B38, 10734 (1988).
105. Brouwer, R. and Jellinek, F., Physica 99B, 51 (1980).
106. Provided by R. Claessen, Institut fuer Experimentalphysik, D-2300 Keil 1, Olshausenstr. 40, FDR.
107. Brusdeylins, G., Heimlich, C., Skofronick, J., Toennies, J. P. and Vollmer, R.,Vuoto 16, 202 (1986).
108. Robbins, M. O. and Marseglia, E. A., Phil. Mag. B24, 705 (1980).
109. Claessen, R., Burandt, B., Carstensen, H. and Skibowski, M., Phys. Rev. B41, 8270 (1990 - II).
110. Thomson, R. E., Welter, U., Ganz, E., Clarke, J., Zettl, A., Rauch, P. and DiSalvo, F. J., Phys. Rev. B38, 10734 (1988 -II).
111. Barrett, R. C., Nogani, J. and Quate, C. F., Appl. Phys. Lett. 57, 992 (1990).
112. Ernst, J., Hulpke, E. and Toennies, J. P., Europhys. Lett. 10, 747 (1989).
113. Ernst, J., Hulpke, E. and Toennies, J. P., Phys. Rev B, Submitted.
114. Jericho, M. H., Simpson, A. M. and Frindt, R. F., Phys. Rev. B22, 4907 (1980).
115. Haga, T., Abe, Y. and Okwamoto, Y., Phys. Rev. Lett. 51, 678 (1983).
116. Suzuki, A., Yamamoto, R., Mizubayshi, M., Okuda, S., Endo, K. and Gonda, S., Solid. State Comm. 49, 1173 (1984).
117. Bak, P., Physica B99, 325 (1980); Rep. Prog. Phys., 45, 587 (1982).
118. Pettenkofer, C., Jaegermann, W. and Parkinson, B. A., Surf. Sci. 251/252, 583 (1991).
119. Jandl, S., Brebner, J.L., and Powell, B. M., Phys. Rev. B13, 686 (1976).
120. Brebner, J. L., Jandl, S. and Powell, B. M., Solid State Commun. 13, 1555 (1973).
121. Altshul, V. Ya., Bashenov, V. K., Marakov, D.I., and Petukhov, A. G., Phys. Stat. Sol. B98, 715 (1980).
122. Depeursinge, Y. and Baldereschi, A., Physica 105B, 324 (1981).
123. Weiting, J. J. and Schlüter, M., Electrons and Phonons in Layered Crystal Structures, Reidel, Dordrecht (1979).
124. Crystal obtained from Dr. F. Levy, Ecole Polytechnique Fédérale de Lausanne, Switzerland.
125. Brusdeylins, G., private communication
126. Lahee, A. M., Manson, J. R., Toennies, J. P. and Wöll, Ch., Phys. Rev. Lett. 57, 471 (1986).
127. Kress, W., de Wette, F.W., Kulkarni, A.D. and Schröder, U., Phys. Rev. B35, 5783 (1987).
128. Mooser, E., Schlüter, I .Ch. and Schlüter, M., J. Phys. Chem., Solids 35, 1269 (1978).
129. Schlüter, M., Camassel, J., Kohn, S., Voitchovsky, J. P., Shen, Y.R. and Cohen M.L.,

Phys. Rev. B13, 3534 (1976).
130. Pasternak, M., Baldereschi, A., Freeman, A. J. and Winner, E., Phys. Rev. Lett. 52, 863 (1984).
131. Bortolani, V., Franchini, A., Santoro, G., Toennies, J. P., Wöll, Ch. and Zhang G., Phys. Rev. B40, 3524 (1989-II).
132. Miglio, L. and Benedek, G., Europhys. Lett. 3, 619 (1987).
133. Bednorz, J. G. and Müller, K. A., Z. Phys. B64, 189 (1986).
134. Chu, C. W., Bechtold, J., Gao, L., Hor, P. H., Huang, Z. L., Meng, R. L., Sun, Y. Y., Wang, Y. Q. and Xue, Y. Y., Phys. Rev. Lett. 60, 941 (1988).
135. Terascon, J. M., McKinnon, W. R., Barboux, P., Hwang, D. M., Bagley, B. G., Greene, L. H., Hull, G. W., LePage, Y., Stoffel, N. and Giroud, M., Phys. Rev. B38, 8885 (1988).
136. Lindberg, P. A. P., Shen, Z.-X., Spicer, W. E. and Lindau, I., Surf. Sci, Rept. 11, pp 1-130 (1990).
137. Chen, C. H., Werder, D. J., Liou, S. H., Chen, S. H. and Hong, M., Phys. Rev. B37, 9834 (1988).
138. Claessen, R., Manzke, R., Carstensen, H., Burandt, B., Buslap, T., Skibowski, M. and Fink, J., Phys. Rev. B39, 7316 (1989).
139. Kirk, M. D., Eom, C. B., Spielman, S. R., Beaslez, M. R., Kapitulnik, A., Geballe, T. H. and Quate, C. F., Appl. Phys. Lett. 52, 620 (1988).
140. Schmicker, D., Toennies, J. P. and Vollmer, R., (In Preparation).
141. Mitzi, D. B., Lombardo, L. W., Kapitulnik, A., Laderman, S. S. and Jacowitz, R. D., Phys. Rev. B41, 6564 (1990).
142. Crystals obtained from C. Politis, Kernforschung Zentrum, Karlruhe.
143. Popovic, Z. V., Thomsen, C., Cardona, M., Liu, R., Stanisic, G., Kremer, R. and König, W., Solid State Commun. 66, 965 (1988).
144. Cardona, M., Thomsen, C., Liu, R., von Schnering, H. G., Harweg, M., Yan, Y. F. and Zhao, Z. X., Solid State Commun., 66, 1225 (1988).
145. Prade, J., Kulkarni, A. D., de Wette, F. W., Schröder, U. and Kress, W., Phys. Rev. B39, 2771 (1989).
146. Schmicker, D., Toennies, J. P. and Vollmer, R., 7th Workshop on Interactions of Molecular Beams and Surfaces, Hindas, Sweden, Sept 24, (1990).
147. Fauster, Th., Himpsel, F. J., Fischer, J. E. and Plummer, E. W., Phys. Rev. Lett. 51, 430 (1983).
148. Finnis, M. W. and Heine, V., J. Phys. F. Metal Phys. 4, L37 (1974).
149. Kaden, C., Ruggerone, P., Toennies, J. P., Zhang, G. and Benedek, G., Submitted to Phys. Rev. B.
150. Bortolani, V., Ercolessi, F., Tosatti, E., Franchini, A. and Santoro, G., To be published.
151. Ercolessi, F., Parrinello, M. and Tosatti, E., Philos. Mag. A58, 213 (1988).
152. Müller, J. E., Wuttig, M. and Ibach, H., Phys. Rev. Lett. 56, 1583 (1986).
153. Needs, R. J., Phys. Rev. Lett. 58, 53 (1987).
154. Benedek, G., Hofmann, F., Ruggerone, P. and Toennies, J. P., To be published.
155. Ellis, J., Reichmuth, A. and Toennies, J. P., To be published.
156. Frenken, J. W. M., Hinch, B. J., Toennies, J. P. and Wöll, Ch., Phys. Rev. B41, 938 (1990).
157. Kunkel, R., Poelsema, B., Verheij, L. K. and Comsa, G., Phys. Rev. Lett. 65, 733 (1990).
158. Duan, J., Bishop, G. G., Gillman, E. S., Chern, G., Safron, S. A. and Skofronick, J. G., Accepted by Surf. Sci.
159. Duan, J., Bishop, G. G., Gillman, E. S., Chern, G., Safron, S. A. and Skofronick, J. G., Accepted by J. Vac. Sci. Technol.

RARE GASES ON GRAPHITE

Neil D. Shrimpton
Milton W. Cole
William A. Steele
and
Moses H. W. Chan
Departments of Chemistry and Physics
The Pennsylvania State University
University Park, PA 16802

ABSTRACT. Rare gas films on graphite exhibit a wide variety of behavior, depending on coverage and temperature. Experimental data for monolayers have been interpreted variously as 2D gases, liquids, and solids. Solid phases are characterized as commensurate, incommensurate or showing orientational epitaxy or domain walls. Theoretical methods have been developed to treat these, beginning with empirical or _ab initio_ potentials. Great interest has been drawn to the nature of phase transitions, which show a greater diversity than the analogous 3D phenomena. While some behavior is understood quantitatively, many questions of interpretation remain.

1. Introduction

The adsorption of rare gases on graphite has been one of the most exciting areas of surface science research for the past two decades. This review aims to explain why there is so much interest in this field, describe the most fundamental results to date, and mention some of the outstanding problems to be resolved in future study. We may summarize the field succinctly by stating that the state of matter in two dimensions (2D) and the approach to 3D are becoming gradually understood.

While the graphite substrate has received the most attention in physical adsorption research, an extensive literature exists for a variety of other surfaces as well.[1-14] Reference to those studies will be made only occasionally here. The popularity of graphite can be attributed mainly to two (related) experimental advantages: the marked mechanical stability of the basal planes and their chemical inertness. As a consequence, _relatively_ little attention needs to be paid to the job of surface preparation. This situation is particularly important

G. Benedek (ed.), *Surface Properties of Layered Structures*, 219–260.

in the context of the need for high surface area materials for thermodynamic and scattering measurements. In the case of graphite, the rather brutal exfoliation process leaves large basal plane areas essentially intact. A typical exfoliated form is Grafoil, a Union Carbide product.[14] Its specific area is nearly 30 m^2/g. Since the 2D surface density of the basal plane is 0.38 carbon atoms per $Å^2$, Grafoil has ~10^{21} surface atoms/g. Taking into account the porosity (~50% empty), we conclude that more than 1 per cent of the C atoms reside at free surfaces of the material. In spite of this high value, the basal planes are measured to be uniform on a length scale ~100Å.[15] Moreover, fewer than 5 per cent of the adsorption sites are found to be heterogeneous (i.e., have an adsorption binding energy measurably distinct from that of the uniform plane).[16] Finally, there is a marked tendency for the planes to align, which is convenient for scattering experiments.

Because of this high specific area, when the surface is exposed to a gas of pressure P, the film provides a satisfactory "signal" (film) to "noise" (substrate) ratio. An additional convenient factor in the thermodynamic case is that the overlayer's excitations are generally of lower frequency than those of the substrate. Most significant of all, the homogeneity of the graphite surface means that data interpretation in terms of uniform surface models is feasible.

Graphite presents other characteristics rendering it particularly convenient or interesting for adsorption. Its thermal conductivity is higher than most insulators, expediting equilibration. It is relatively transparent to X-rays and neutrons. As shown in Table 1, the heat of adsorption of inert gases is higher on graphite than on nearly <u>any</u> other substrate,[17] meaning that ultra-low temperature is not needed to either produce a film or avoid desorption (except in the He case). Finally, the wavelength (~lattice constant) and amplitude of the periodic potential energy $V(\vec{r})$ experienced by an adatom are such that dramatic consequences - commensurate phases - occur in <u>some</u> cases, while in others this "corrugation" is sufficiently small as to legitimize an (ideally) smooth 2D description of the film. Indeed, a wealth of phenomena is seen for the various noble gas adsorbates; predicting these represents a challenge for the theories of both $V(\vec{r})$ and statistical mechanics.

Rare gas adsorption on graphite was first studied extensively by Polley, et al., and by Halsey and coworkers in the 1950's.[18,19] These data provided evidence of quasi-2D behavior (including condensation) of low coverage films and multilayer formation; these results stimulated the development of plausible, but simple, theoretical models. A great leap forward was made in the late 1960's by Thomy and Duval,[20,21] who observed unambiguous manifestations of 2D phase transitions in adsorption isotherms. Their low temperature (T) data exhibit "vertical risers", i.e., near discontinuities in film coverage N as a function of vapor pressure P. This behavior - variable N at fixed chemical potential μ - corresponds to two-phase coexistence in the film. One infers thereby the presence of ordering transitions, but explicit identification of the phases requires other techniques. Indeed, several groups showed the feasibility of heat capacity C(N,T), which has become a major tool in elucidating phase diagrams.[22,23] These techniques are partly overlapping due to a Maxwell relation

$$\left(\frac{\partial S}{\partial N}\right)_T = -\left(\frac{\partial \mu}{\partial T}\right)_N \tag{1}$$

where S is the entropy and

$$\mu = \beta^{-1} \ell n \left[n_v \lambda^3 / g\right] \tag{2}$$

$$\lambda^2 = 2\pi\beta\hbar^2/m \quad . \tag{3}$$

Here $\beta^{-1} = k_B T$ and $n_v = P\beta$ is the density of the coexisting vapor, assumed to be ideal. λ is the DeBroglie thermal wavelength and g is the spin degeneracy. In practice, the techniques are complementary because the C data are generally more reliable (or available) at low T, while P(N,T) is more straightforward generally at high T. With plausible interpolation, a merged, consistent data set can be obtained,[16] leading to a complete thermodynamic description in the (N,T) plane.

This reporting of more or less exotic data about 2D phase transitions was particularly timely because contemporaneous theoretical work (~1970) was then climaxing in the establishment of the renormalization group theory of critical phenomena.[24] Particularly useful for surface physics was a theory, due to Kosterlitz and Thouless[25,26] (KT), which reconciled an apparent contradiction between the experimental data and well-established (indeed, rigorous) theory. To wit, the work of Landau and Peierls had shown the impossibility of finite temperature crystalline order in 2D.[27,28] It was thus a little disconcerting that the experimental data provided evidence of 2D solid-like thermodynamic behavior (including a melting transition) in submonolayer films. A similar difficulty was the impossibility of Bose-Einstein condensation,[29,30] in apparent disagreement with ^{4}He films' superfluidity observed in the laboratory.[31] The KT idea (which built on earlier work of Mermin,[32] Stanley and Kaplan,[33] and Berezinskii[34]) was that a new topological long range order (TLRO) can exist in a 2D system; the argument is that experimental manifestations of TLRO had been misinterpreted as the "old-fashioned", conventional long range order, i.e., in nominal violation of the theorems cited above. The KT paper launched a revolution in concepts, including the possibility of a new hexatic[35] solid phase, and a wealth of new experiments. Theory and experiment have not yet converged on many of these issues,[36] in spite of inspired theoretical and experimental investigation.

The remainder of this chapter discusses these many topics at some length. For the most part, details are omitted. The focus is on general trends and the simplest guiding principles. We try to explain alternative hypotheses in currently ambiguous situations, and make critical assessments to the extent possible.

The intriguing subjects of molecular and multilayer adsorption will not be addressed, except in a few cases where deemed illuminating.

Gas	D(meV)	C_3(meV–Å³)	ΔV(meV)
He	16.5±1	175±15	2.5±15%
Ne	33±2	350±30	3
Ar	95±2	1210±100	4.1
Kr	125±5	1730±200	4.7
Xe	162±4	2500±200	5
H_2	52±1	520±50	6
CH_4	130±10	1750±200	

Table 1. Well depths D and asymptotic dispersion coefficients C_3 derived from various literature sources for gases on graphite (taken from Ref. 17). ΔV is the corrugation strength, i.e., the min-to-max variation across the substrate of the adsorption potential, estimated from Ref. 37. To convert meV to K, multiply by 11.6.

2. Interaction Potentials and Low Coverage Adsorption

2.1 OVERVIEW

The general theory of film structure and dynamics requires a specification of the Hamiltonian of the system, followed by a calculation of various properties by statistical mechanics, lattice dynamics, or other appropriate technique. The subject of physical adsorption on graphite is particularly convenient to treat theoretically and interpret data because[38]

a) the small binding energy (Table 1) means that the substrate is not strongly perturbed by the overlayer, so it may be assumed to be rigid,[39]

b) the adlayer interatomic interactions $U(\vec{r}_1, \vec{r}_2)$ are similar to their gas phase forms, and

c) the adatom's holding potential $V(\vec{r})$ is sufficiently strong and deep that it confines the monolayer within a narrow domain in the direction z perpendicular to the surface (typically[40] $\Delta z_{rms} \sim 0.1$Å).

Point (a) means that the adlayer and substrate motions are decoupled, to good approximation. This serves to justify the often assumed additivity of substrate and film thermodynamic properties. Point (b) omits any substrate mediation[41] of the potential, a ~10 per cent effect which we discuss below. Point (c) means that the 2D approximation, once carefully defined,[42,43] is relatively accurate. By way of contrast, none of these assumptions is at all correct in the case of chemisorption, for which it is believed necessary to include substrate rearrangement and high order many-body interactions,[44,45] which are not well known. Much closer to graphite is the case of physisorption on Ag, which differs in that the well depths and corrugation are both smaller.[5] The latter means that the smooth surface approximation is particularly reliable in that case.

In this section we describe and assess various potential energy models and their use in evaluating single atom properties and the adsorbate virial expansion, applicable to low coverage adsorption. The next section will discuss the use of such potentials to treat higher coverages.

2.2 GAS-SURFACE INTERACTION

The well depths in Table 1 are known quite reliably from scattering and/or thermodynamic experiments. This provides a fairly serious constraint on theoretical models of the gas-surface interaction. The potential energy of a single adatom at position $\vec{r}$ may be written as a 2D Fourier expansion[1,46]

$$V(\vec{r}) = V_o(z) + \sum_{\vec{G}\neq 0} V_{\vec{G}}(z)\, e^{i\vec{G}\cdot\vec{R}} \tag{4}$$

where z is the direction normal to the surface, $\vec{R}=(x,y)$ and $\vec{G}$ is a reciprocal lattice vector. For graphite, the Fourier series converges rapidly, so that usually one needs to include only the 6 smallest equivalent nonzero vectors; then Eq. (4) may be written approximately[46]

$$V(r) \simeq V_o(z) + 2V_1(z)\left[\cos(2\pi S_1) + \cos(2\pi S_2) + \cos[2\pi(S_1+S_2)]\right] \tag{5}$$

where $aS_1 = 2x/\sqrt{3}$ and $aS_2 = y-x/\sqrt{3}$, a is the spacing between basal plane hexagons (2.46Å), and the x axis points from the origin (hexagon center) toward a neighboring C atom. The quasi-2D approximation consists of omitting the corrugation term $V_1(z)$ entirely. Bruch has recently reviewed the effect of this term on monolayer phase diagrams.[37]

Potential energy functions for individual rare gas atoms on graphite have been calculated with both semiempirical and ab initio methods. The most commonly used model is based on a pair-wise sum of two body potentials

$$V(\vec{r}) = \sum_i v(\vec{r} - \vec{R}_i) \tag{6}$$

where $\vec{r}$ is the adatom position and $\vec{R}_i$ is the lattice site of the i-th carbon atom. Frequently, a Lennard-Jones functional form is adopted:

$$v(x) = 4\varepsilon \left[(\sigma/x)^{12} - (\sigma/x)^{6}\right] \tag{7}$$

where ε and σ are fit to low coverage data[47]. If Eqs. (6) and (7) are adopted, $V_o(z)$ and $V_1(z)$ may be evaluated explicitly. For example, the laterally averaged potential becomes[46]

$$V_o(z) = \left(4\pi\varepsilon\sigma^6/\left(a_s d^4\right)\right) \left[\frac{2}{5} (\sigma/d)^6 \zeta(10,z/d) - \zeta(4,z/d)\right]$$

$$\zeta(n,x) = \sum_{j=0}^{\infty} (j+x)^{-n} \tag{8}$$

where $a_s = 5.24Å^2$ is the area of the unit cell of a basal plane (containing 2 atoms) and $d=3.37Å$ is the interplanar separation. For $x \gg 1$ one may replace the sums by integrals and we get $\zeta(n,x) \to 1/[(n-1)x^{n-1}]$, so that

$$V_o \longrightarrow - \frac{4\pi\ \varepsilon\sigma^6}{3\ a_s\ d\left(z - \frac{d}{2}\right)^3}\ ,\quad z \gg d \tag{9}$$

with the help of the Euler-MacLaurin sum formula.[48] We will return to discuss this asymptotic dependence below.

This parameterized potential is a reasonable guess in the absence of detailed information about $V(\vec{r})$. From the theoretical perspective, however, one might question some implicit assumptions. The use of Eq. (6), for example, implies that many-body interactions are negligible; these have been found to be of order 15 per cent for the attractive part of the potential.[49] Eq. (8) is an isotropic interaction; calculations show, in contrast, that the considerable anisotropy of graphite ought to necessitate modification of this formulation.[50,51] A detailed investigation in the case of He/graphite revealed theoretical evidence for observable effects of this anisotropy, consistent with what was deduced from He scattering data.[50,52] A semiempirical potential explicitly incorporating this information was constructed; it replaces Eq. (7) with

$$v(\vec{x}) = 4\varepsilon \left\{ \left(\frac{\sigma}{x}\right)^{12}\left[1 + \gamma_R\left(1 - \frac{6\cos^2\theta}{5}\right)\right] - \left(\frac{\sigma}{x}\right)^{6}\left[1 + \gamma_A\left(1 - \frac{3\cos^2\theta}{2}\right)\right]\right\} \tag{10}$$

where θ is the angle of the $\vec{x}$ vector relative to the surface normal. The parameters, $\gamma_R \simeq -0.54$ and $\gamma_A = 0.4$, were derived from experiment and theory, respectively, for He. These functional forms were designed so that Eq. (8) is unaffected. The resulting He potential was found to be consistent also with thermodynamic data, as discussed below.[53–56] Semiempirical studies of other rare gases have provided evidence for anisotropic pair potentials as well.[37,57,59]

Several ab initio calculations of $V(\vec{r})$ also exist.[60–64] Freeman[60] determined $V(\vec{r})$ from the Gordon-Kim (GK) model.[65] This evaluates the energy by assuming that (a) the charge density for the atom-surface system is the sum of the independent densities, and (b) the energy is an integral of a (local) energy density which is that of a uniform electron gas. This model's deficiencies are well known, including an underestimation of the kinetic energy (due to the density gradient)[66,67] and an inadequate treatment of nonlocal correlation, which is responsible for the known asymptotic behavior[68,69]

$$V \to -C_3(z-z_0)^{-3} \tag{11}$$

where z_0 is the image plane location and

$$C_3 = \frac{\hbar}{4\pi}\int d\omega\ \alpha(i\omega)\left[\varepsilon(i\omega)-1\right]/\left[\varepsilon(i\omega)+1\right] \quad . \tag{12}$$

Here α is the adatom's polarizability and ε is the substrate dielectric function, both evaluated at imaginary frequency.

For accidental reasons, the GK model yields a potential of reasonable accuracy in most applications.[66,70] Indeed, this is the case of Freeman's results for adsorption on graphite; the exception is that the He binding energy was about a factor of 2 smaller than the experimental value.[52,54]

More recent studies of this problem have relied on an assumption that the total potential is separable into repulsive and attractive parts:[38,71]

$$V = V_R + V_A \quad , \tag{13}$$

Here V_R is derived from either a self-consistent or approximate theory involving overlap of the wave functions (or charge densities) of the solid and adatom. V_A is derived from models which reduce to Eq. (11) asymptotically. Liebsch, et al[61] calculated V_R for He from the Harris-Liebsch[72,73] pseudopotential theory of physical adsorption. This involves the perturbation of the substrate's electrons by the adatom's potential. Another scheme, the effective medium method,[74,75] was used by Toigo and Cole to determine V_R for He and Ne on graphite.[62] Both formalisms yield a proportionality between V_R and the substrate charge density $n(\vec{r})$ in the hypothetical limit when $n(\vec{r})$ is small and slowly varying:

$$V_R(\vec{r}) \simeq \alpha_- n(\vec{r}) \tag{14}$$

"Small" means perhaps $n \lesssim 10^{-3} \text{Å}^{-3}$, easily satisfied at physisorption distances ($\langle z \rangle \sim 3\text{Å}$ above the surface plane). The slow variation approximation means slow on an atomic length scale. This is not well satisfied, however, so more general procedures are applied. In any case, there is ambiguity in the value of α_-; estimates for He range from 200 to 600 $\text{eV}a_o^3$, where $a_o = 0.529$ Å.[72,76–78] Finally, there is not available particularly precise data for $n(\vec{r})$. The lowest Fourier coefficient (lateral average) is approximately fit[79,80] for $z \simeq 3\text{Å}$ by

$$n_o(z) \simeq A \exp(-Kz) \tag{15}$$

where $A \simeq 1.05$ and $K \simeq 1.94$ atomic units.

The various approaches differ also in their treatment of the attraction. One technique is to assume that the asymptotic form, Eq. 11, applies at all adatom distances. Another is based on an anisotropic active potential, i.e., the second term in Eq. (10). In the Toigo-Cole calculation, this dipole term was supplemented by a quadrupole term; it is a ~10 per cent well depth correction. The asymptotic behavior is still that of Eq. (9).

The many calculated potentials tend to be rather similar because of a "judicious" choice of method and/or parameters. In the most-studied case, He, the ab initio potentials have tended to err systematically in that the calculated equilibrium distance usually exceeds the experimental value[81] by ~0.2Å. If the data are correct, one must assume that the success of the models' other predictions is partly fortuitous.

The following sections will allude to various tests of potential models. A general deficiency in assessing their reliability is that theorists have needed to use quite simplified model potentials (e.g., Eq. 7) for the high coverage dynamical calculations, so that definitive conclusions are usually premature. It is desirable evidently to establish and assess the predictions at higher coverage of the more "sophisticated" potentials.

2.3 ULTRA-LOW COVERAGE FILMS

Experimental data in different regimes of coverage provide varying kinds of information. Data at low coverage allow one to assess interaction models most easily. Thermodynamic information includes the heat capacity, the pressure vs. coverage isotherm P(N,T), and the isosteric heat of adsorption;[1,10,13] per atom; the last is

$$q_{st} = -\left(\frac{\partial \ln P}{\partial \beta}\right)_N \quad . \tag{16a}$$

This is equal to the enthalpy change due to the evaporation of one atom. In the <u>ultralow</u> coverage limit of independent atoms, we may neglect all effects of interactions[82]; the partition function for N atoms is then

$$Q_N = Q_1^N/N! = \frac{1}{N!}\left[\sum_i e^{-\beta\varepsilon_i}\right]^N$$

where the sum is over single particle states of atoms bound to the surface. Then

$$\left[\frac{\partial}{\partial\beta}(\beta\mu)\right]_N = -\left[\frac{\partial}{\partial\beta}\left(\frac{\partial}{\partial N}\ln Q_N\right)_\beta\right]_N = \langle\varepsilon\rangle \quad , \tag{16b}$$

the average adatom energy. From Eqs. (2), (3), (16) and the ideal gas law, we obtain[84]

$$q_{st} = \frac{5}{2\beta} - \langle\varepsilon\rangle \qquad (N \to 0) \tag{17}$$

In the classical case

$$Q_1 = \lambda^{-3}\int d^3r \exp\left[-\beta V(\vec{r})\right] \quad ,$$

which simplifies considerably if we adopt a harmonic approximation to the surface potential in the z direction perpendicular to the surface; letting $k=m\omega^2$ be the force constant, the model is then

$$V \simeq -D + kz^2/2 \quad . \tag{18}$$

The Helmholtz free energy F and other functions are then

$$F/N \simeq \beta^{-1} \ell n(\theta\lambda^2\beta\hbar\omega/e) - D \tag{19}$$

$$\theta = N/A \tag{20}$$

$$q_{st} \simeq (2\beta)^{-1} + D \tag{21}$$

$$\mu = F/N + \beta^{-1} \quad . \tag{22}$$

Equating this to Eq. (2) yields the classical smooth surface isotherm

$$P = \theta\omega \left(\frac{m}{2\pi\beta}\right)^{\frac{1}{2}} \exp(-\beta D) \qquad (N\to 0) \tag{23}$$

The prefactor in this Arrhenius-type expression is typically of order 100 atmospheres (setting $\theta \sim 0.01 \text{Å}^{-2}$, $\hbar\omega \sim 3$ meV, m=20 amu, T=100K).

The preceding and subsequent low coverage expressions have an interpretation in terms of first order desorption rate models.[85] Equating the sticking and desorption rates in equilibrium, one finds a rate from Eq. (23)

$$\tau^{-1} = \frac{\omega s}{2\pi} \exp(-\beta D) \tag{24}$$

where s is the sticking coefficient. If s=1, the "attempt" frequency (prefactor) is the adatom's vibrational frequency.[86] To our knowledge, there are no extant measurements of these rates for graphite adsorption.

In the analogous quantum case, the one particle partition function is the product of 1D harmonic oscillator and 2D gas functions

$$Q_1 = A\lambda^{-2} n(\omega) \exp[\beta(D+\hbar\omega/2)]$$

$$n(\omega) = (e^{\beta\hbar\omega}-1)^{-1} \tag{25}$$

$$F/N = -\beta^{-1} - D - \hbar\omega/2 + \beta^{-1} \ell n \left[\theta\lambda^2/n(\omega)\right]$$

$$q_{st} = D + \frac{3}{2\beta} - \hbar\omega\,[n(\omega) + 1/2]$$

$$\mu = F/N + \beta^{-1} \quad .$$

These yield the previous classical expressions in the limit $\beta\hbar\omega \ll 1$. The quantum isotherm is

$$P = \frac{\theta \exp[-\beta(D+\hbar\omega/2)]}{\lambda\beta n(\omega)} \qquad (N\to 0) \qquad (26)$$

In the extreme quantum limit, $\beta\hbar\omega \gg 1$, only the ground vibrational state (of energy $E_o = -D + \hbar\omega/2$) is populated; the result is then

$$P \longrightarrow \frac{\theta \exp(\beta E_o)}{\lambda\beta} \quad . \qquad (27)$$

The activation energy is thus smaller than the well depth by the zero point energy $\hbar\omega/2$; compare Eq. 23. The desorption rate prefactor is skT/h, which differs from that of Eq. (24) by a factor $k_BT/\hbar\omega$, typically of order one.

These expressions for the pressure allow an easy determination of the binding energy from the measured T dependence. The prefactor values are more uncertain, in general, because the surface area is not known precisely for typical high specific area materials. In any event, the preceding model is only semiquantitatively reliable for adsorption on graphite because of the corrugation of the potential.[37] Corrections have been evaluated as expansions in the Fourier coefficients $V_G(z)$ in the classical case[42,43] and as band structure effects in the quantum case.[50,53,87,88] To illustrate the effects of corrugation, we consider a realistic, analytically tractable model. We assume that only the ν (=6 for graphite) equivalent smallest reciprocal lattic vectors (G_1) contribute to Eq. (4). Let them be described by

$$V_{G_1}(z) = V_1 \exp(-\gamma z) \quad ,$$

which is accurate in the vicinity of the equilibrium (z=0) plane. Thermodynamic perturbation theory gives the associated shift (relative to $V_1=0$):

$$\begin{aligned}\Delta\mu &\simeq -(\beta/2) < [V - V_o(z)]^2 > \\ &= -(\beta\nu\, V_1^2/2)\exp(\gamma^2 z_{rms}^2/2)\end{aligned}$$

where $z_{rms} = (\beta k)^{-1/2}$ in this harmonic model. For typical classical gas on graphite [$\gamma \sim G_1 = 2.95Å^{-1}$, $z_{rms} \sim 0.2Å$, $V_1 \lesssim 10K$, and $T \sim 200K$], $|\Delta\mu| \sim 3K$, which is quite small relative to both terms in the unperturbed result, Eq. 22. At lower T, however, the calculated $|\Delta\mu|$ increases since the adatoms settle into particularly attractive adsorption sites.

In the T=0 extreme quantum case of He, we may estimate the binding energy shift ΔE_o by second order perturbation theory.[88] The dominant contribution couples unperturbed wave vector $\vec{k}$ to $\vec{k} + \vec{G}$, so that (for k=0)

$$\Delta E_o \simeq \frac{-\nu|<0|V_1 e^{-\nu z}|0>|^2}{(\hbar^2 G_1^{\,2}/2m)}$$

$$= -\left[\nu\, V_1^{\,2}\; 2m/\left(\hbar^2 G_1^{\,2}\right)\right] \exp\left(\gamma^2 z_{rms}^{\,2}/2\right)$$

where $z_{rms}^2 = \hbar/(2m\omega)$ describes the surface-normal oscillation at T=0. Inserting typical numbers[52] for ^{4}He ($\hbar\omega \sim 6$ meV, $V_1 \sim 0.3$ meV), we find $\Delta E_o \sim -2K$. This is small compared to the smooth surface binding energy (~140K), but this shift has been measured.[55]

2.4 LOW COVERAGE

We turn briefly to the case of low coverage, with interactions. Here the theoretical approach of choice is a virial expansion.[1,23] This treatment is usually more complicated than in the familiar 3D case, however. One reason is that the second virial coefficient C_{AAS} depends on both atom-surface and atom-atom interactions. This is obvious from the fact that all effects of adatom interactions depend on the atoms' proximity to one another; this is related to their spread in the z direction. This dependence is manifest in the classical expressions[1]

$$N_s = N - N_o = B_{AS}\,(\beta p) + C_{AAS}\,(\beta p)^2 \ldots \tag{28}$$

$$B_{AS} = \int d\vec{r}\,\left[e^{-\beta V} - 1\right] \tag{29}$$

$$C_{AAS} = \int d\vec{r}_1\, d\vec{r}_2\, \left[e^{-\beta(V_1+V_2)}\right]\left[e^{-\beta U_{12}} - 1\right]$$

where the integration volume Ω is that defined as the film region and $N_o = n_v\Omega$ is the number of vapor atoms which would occupy the region in the absence of an attractive potential; thus, N_s is the surface excess coverage. For adsorption on graphite at relatively low T, one can invoke the 2D approximation to show that[89]

$$C_{AAS} \simeq 2B_{AS}^{\,2}\, B_{2D}/A$$

$$B_{2D} \simeq \frac{1}{2} \int d^2R \left[1 - e^{-\beta U(R)}\right]$$

The equation of state may be written in this approximation as

$$\ell n \left(N_s/pB_{AS}\right) = 2B_{2D} \left(N_s/A\right) + \frac{3}{2} C_{2D} \left(N_s/A\right)^2 \ldots$$

a form useful at low T and moderately low density.

Corrections due to the periodic potential and the z motion are known and values of B_{2D} and C_{2D} have been tabulated for Lennard-Jones potentials.[89] The first two terms of the virial expansion in the quantum case have been evaluated for both 2D and more general models, yielding successful explanations of low coverage He adsorption.[23,37,87–90]

The second complication in treating interactions between adatoms is a many body effect: they are modified by the substrate.[5,39,41] The substrate's elastic contribution to this is believed to be small for graphite,[39] while the electronic effect is non-negligible. The latter was derived by McLachlan;[41] in the case when the adatoms lie in a common plane z=L above the substrate, the substrate mediated energy is

$$\Delta U = \frac{4C_{S1}}{(rR)^3} \left(\frac{1}{3} - \frac{L^2}{R^2}\right) - \frac{C_{S2}}{R^6} \quad , \tag{30}$$

$$L = z - z_o$$

Here r is the adatoms' separation,

$$R = (r^2 + 4L^2)^{1/2} \tag{31}$$

is the distance between one atom and the other's image, and the C_{Si} are coefficients similar in form to C_3 described above.[91] This interaction can be interpreted in terms of one atom's dipole fluctuation interacting with the image of the other's. It is analogous to the three body triple dipole dispersion interaction needed to characterize accurately the equation of state of inert gases.[92,93] Indeed, one may compare Eq. (30) and the sum of the triple dipole dispersion energy $v_3(\vec{r}_1, \vec{r}_2, \vec{R}_i)$ over the carbon atoms at $\vec{R}_i$,

$$\Delta U_{sum} = \sum_i v_3(r_1, r_2, R_i) \tag{32}$$

This comparison allows one to identify the location of the image plane z_0 as lying about 1Å above the outermost basal plane,[94] for typical adatom distances z~3Å. The overall effect of this substrate-mediated energy is to reduce the adatom-adatom well depth by ~10 percent. This prediction is roughly in accord with a large body of experimental data.[5,41] Such a statement requires that we possess reliable information about gas phase interatomic potentials. This is indeed the case as a product of very extensive experimental and semiempirical investigation.[95,96] Also important is that intra-adsorbate many body interactions be known reasonably accurately. This is also true here; calculations[5,97] have relied on the same Axilrod-Teller-Muto triple dipole interaction found suitable for bulk phases.[93]

3. The Phases of Rare Gases on Graphite

The phases of rare gas monolayers on graphite have provided a fascinating focus of study for more than the last 20 years. They have theoretical interest partly because the critical properties of many of the possible phase transitions in 2D can be calculated explicitly. This, together with the principle of 2D conformal invariance, allows the observed phase transitions to be categorized in terms of a few general universality classes.[98] Because the interactions associated with physisorbed systems are relatively well understood, the location of the phase boundaries can be predicted and compared with experiment.

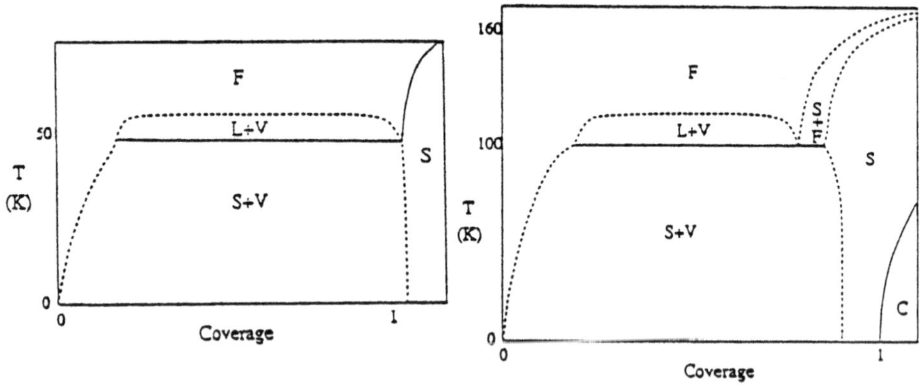

Fig. 1. The phase diagrams for a) Ar, and b) Xe. S is a solid incommensurate phase, L is a liquid phase, V is a gas phase, and F is the supercritical fluid phase. C is the $\sqrt{3} \times \sqrt{3}$ commensurate phase, defined here as coverage =1.

gas	T_c (K)	T_t (K)
Ne	15.8	13.57
Ar	55	48
Xe	115.5	100

Table 2: Measured temperatures for the critical point and triple point for the indicated rare gases on graphite.

In this section we outline the theoretical and experimental descriptions of these systems. The phase diagrams of He, Ne, Ar, Kr, and Xe display both similarities and differences. Ne (not shown), Ar, and Xe (figure 1) have phase diagrams that greatly resemble those of the 3D bulk. Solid, liquid and vapor phases are observed. The 2D triple point and gas-liquid critical point temperatures have been measured (table 2). The general topology of the phase diagrams can be matched by simple *floating monolayer* models that neglect the variation of the substrate potential energy. The melting transitions of Ne and Xe have been observed to be first order, as is always the case for 3D bulk. Ar, on the other hand, has a more complex melting transition. The critical point of the fluid provides a physical example of the critical behavior of the 2D Ising universality class.

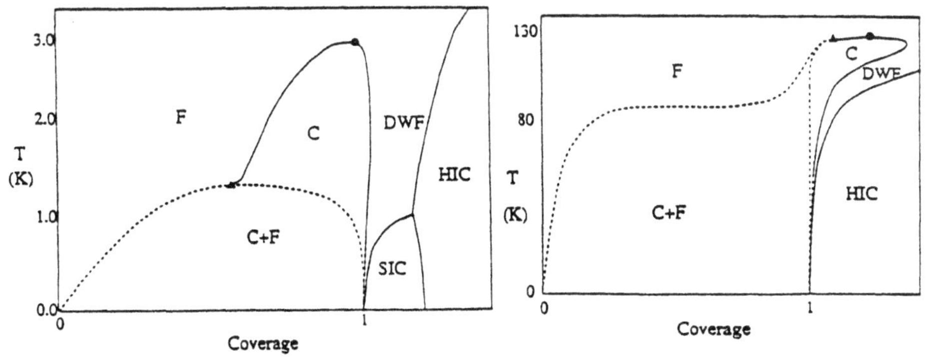

Fig. 2: The phase diagrams for a) ^{4}He, and b) Kr. C is the $\sqrt{3} \times \sqrt{3}$ commensurate phase, (corresponding to coverage one). F is a fluid phase. SIC is a uniaxial incommensurate phase with stripes of domain walls. HIC is a hexagonally symmetric incommensurate phase. DWF is a *domain wall fluid* incommensurate phase where the domain wall lattice is *melted*. The Potts critical point is indicated by the circle, while the tricritical point is indicated by the triangle.

The gases He and Kr are greatly influenced by the structure of the graphite substrate. An ordered phase is formed which is commensurate with the substrate. For He, this phase is so stabilized by the substrate that it preempts the liquid phase and hence precludes any possible superfluidity below monolayer coverage. At low temperatures it coexists with a gas phase. This coexistence terminates with increased temperature at a 3 state Potts tricritical point. Beyond this point the commensurate phase has a continuous order-disorder transition, with critical properties given by the 3 state Potts universality class, discussed below.

He, Kr, and Xe have a variety of possible solid configurations. They have commensurate phases, incommensurate phases with domain walls, and incommensurate phases that are either rotated or aligned with respect to the substrate. The melting of the commensurate phase may be to a domain wall fluid, discussed below.

The substrate interaction can be described by the potential form (5) which reflects the symmetry of the substrate shown in figure 3. The dominant effect is the averaged holding potential of the substrate, which strongly confines the adatoms to a minimal potential energy surface. The variation in the holding potential across this minimal energy surface defines a triangular lattice of minimum energy adsorption sites. The spacing of these adsorption sites is 2.46Å. The lateral variation of the substrate potential is of comparable magnitude as, and is in competition with, the strength of the adatom interactions. The values of the substrate corrugation that are effective at the averaged height $\langle z \rangle$ of the monolayer above the substrate are given in table 3.

Insight into the significance of the substrate corrugation can be obtained by comparing the gas-gas interaction information shown in table 3 with the substrate corrugation information provided in table 1. It is noted that the substrate corrugation should have the most effect for He, diminishing as one goes through the rare gases to Xe. Also relevant is the preferred spacing r_{min} of the gas-gas interaction. As this approaches a spacing commensurate with the substrate (see figure 3) the strain energy, associated with forming a commensurate solid, is reduced. For the lighter gases, the confinement to lattice sites reduces the quantum overlap between gas atoms. This decreases the effective interaction between gas atoms and thus enhances the significance of the substrate corrugation.[99,116,117]

Helium is so strongly influenced by the corrugation of the substrate that it forms a commensurate phase at a high temperature (~3K). The hard core repulsion prevents it from forming a 1×1 commensurate phase (that fills every adsorption site), but it does form a commensurate phase (shown in figure 3) based on a $\sqrt{3} \times \sqrt{3}$ R30° sublattice of adsorption sites. The effect of the substrate corrugation is not sufficient to induce Ne and Ar to form $\sqrt{3} \times \sqrt{3}$ commensurate phases. Kr, on the other hand, does form such a commensurate phase. While the preferred spacing of Kr atoms is 4.0Å, the strain energy associated with dilating the Kr monolayer to a lattice spacing of 4.26Å is relatively small compared to the energy gained from placing every Kr atom at an adsorption site. Xe also forms a commensurate phase. The preferred lattice spacing of Xe, 4.4Å, is only slightly larger than the commensurate spacing, and with compression a commensurate solid phase is formed.

As a basis from which the rare gas phase diagrams can be discussed and contrasted, the properties of the 2D floating Lennard-Jones (LJ) system are discussed in the next section. This is followed by a section which considers the effects of substrate corrugation on the properties of the incommensurate monolayer. Section 3.3 considers the commensurate phase and the transitions associated with it. Section 3.4 then describes each of the rare gases in detail, with the emphasis on the experimental understanding of the properties of the monolayers.

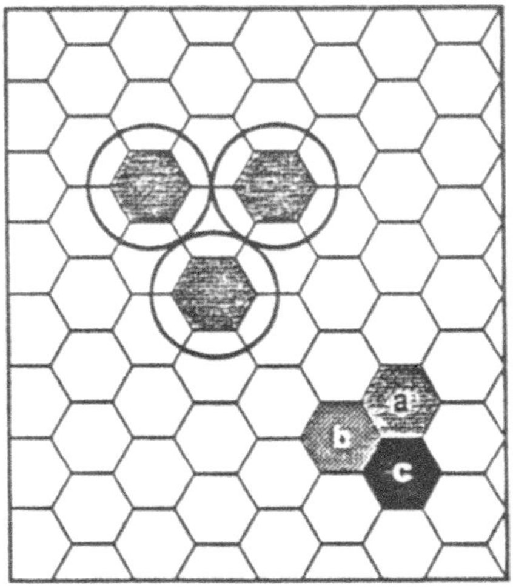

Fig. 3: Physisorbed gases on graphite. The honeycomb net shows the carbon bonds, which have length 1.42Å; the adsorption sites are at the centers of the hexagons and are spaced 2.46Å apart. The physisorbed gas shown by the circles forms a triangular sublattice with spacing of 4.26 Å, based on one of the three possible centers, a,b or c.

3.1 THE FLOATING MONOLAYER

In the LJ system, the interactions between atoms are described by simple 6-12 pair potentials,

$$V(r) = 4\varepsilon[(\sigma/r)^{12} - (\sigma/r)^6] \tag{33}$$

This relatively simple potential form, while providing a reasonable description of the interactions between adsorbed atoms, has the advantage of allowing all energies of the system to be expressed in reduced units of ε and all length scales to be expressed in reduced units of σ. Consequently a generic phase diagram can be constructed.

The details of the phase diagram are shown in figure 4. The general features of the floating 2D monolayer include liquid, vapor and solid phases. The critical temperature for the LJ system has been calculated to be around 0.54ε[100]. The triple point temperature T_t has been calculated to be roughly 0.46ε[101]. In the actual phase diagrams of the rare gases, it must also be remembered that rare gas atoms can condense on top of the solid monolayer, and the apparent increased density of the solid phase represents both a compressed monolayer solid and an increased number of second layer atoms.

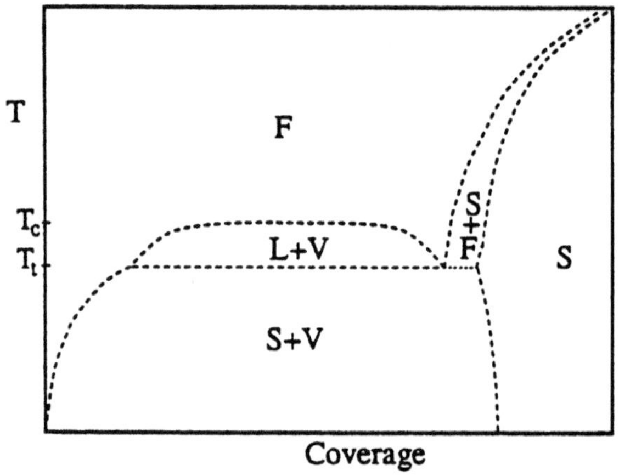

Fig. 4. This is a generic phase diagram for a rare gas adsorbed on a corrugationless substrate. S is a solid phase, L is a liquid phase, V is a gas phase, and F is the hypercritical fluid phase.

LJ parameters extracted from the bulk properties of rare gases are given in table 3. The critical point and triple point temperature of the LJ floating monolayer are also given in table 3. As a comparison, the actual values for the critical temperature, and triple point temperature measured for the Ne, Ar, and Xe on graphite are shown in table 2. Given the ~10% reduction in well depth of the bulk interaction potentials due to a substrate mediated dipole interaction (see section 2.4), the LJ floating monolayer provides a surprisingly accurate picture of the phase diagrams of Ne, Ar and Xe on graphite. In the case of He and Kr on graphite, the phase diagram of figure 4 for the floating monolayer is not appropriate.

The width in coverage of the L + V coexistence region is the difference in density between the liquid and vapor. This difference goes to zero at the fluid critical point. As in 3D, the transition belongs to the Ising universality class, with density being the order parameter. Near the critical point, the difference in density ρ of the liquid and vapor will vary with temperature as

$$\rho_\ell - \rho_V \sim (T_C - T)^\beta \qquad (34)$$

The value of β thus determines the shape of the curve which bounds the liquid–gas coexistence region. For a 2D system calculations[102,103] and measurements[104] show that $\beta = 1/8$ and the density range of the coexistence region is large even close to the critical point. For 3D systems $\beta \sim 0.325$ and the density range of the coexistence region near the critical point is much narrower.

gas	ε/k_B (K)	σ (Å)	T_c (K) $0.54\varepsilon/k_B$	T_t (K) $0.46\varepsilon/k_B$	r_{min} (Å) $\sigma 2^{1/6}$
He	10	2.56	5.4	4.6	2.8
Ne	35	2.78	19	16	3.1
Ar	120	3.38	65	55	3.8
Kr	171	3.60	92	79	4.0
Xe	236	3.92	127	108	4.4

Table 3: LJ parameters ε and σ for the interactions of rare gases.[41] The critical temperatures, critical densities, and triple point temperatures based on the 2D LJ description are shown, as is the radius at which the 6–12 potential is minimum.

Computer simulations of the LJ system indicate that the triple point melting transition is strongly first order with a large discontinuity, of order $0.11/\sigma^2$, between the liquid and solid phase densities.[101] In the solid region of increased density, the melting transition shifts to higher temperature because, under pressure, the solid phase is stabilized. Molecular dynamics calculations indicate the presence of a coexistence region S + F, and hence a first order melting transition. It must be pointed out, however, that the molecular dynamics conclusions are not without controversy[36]. In 1973, Kosterlitz and Thouless[25] proposed a mechanism for melting based on an incorporation of dislocations into the solid phase. This type of melting is proposed to be continuous and non-universal. An excellent review of the theory of dislocation-mediated melting is provided by Strandburg[36].

3.2 THE INCOMMENSURATE MONOLAYER

While the general features of the 2D floating monolayer are evident in the phase diagrams of Ne, Ar, and Xe, the substrate does add considerable complexity to the properties of these systems. A study of the impact of weak corrugations on the phase diagrams of the rare gas monolayers has been provided by Niskanen.[106] As shown by Monson, et al.[43] the temperature of the critical point can be increased by the substrate corrugation. This effect depends upon the relative size of the gas atom compared to the substrate lattice spacing. It is difficult to extract this information from experimental studies because a variation in the critical temperature due to corrugation is masked by a mediation of the gas interaction by the substrate.[5,41,97] Of course, the liquid and vapor near the critical point retain the same critical exponents even in the lattice gas limit[107].

The corrugation of the substrate can, on the other hand, have a radical effect on the solid phase of the monolayer. As is evident for He, Kr, and Xe, there can be at least two solid phases; commensurate, and incommensurate with the substrate. The degree of incommensurability is commonly identified by the misfit

$$\varepsilon = \frac{|d - d_c|}{d_c} \qquad (35)$$

of the averaged spacing d from that of the commensurate spacing d_c.

One of the intriguing facts is that even a vanishingly small corrugation has an orienting influence on the solid monolayer. All of the rare gases on graphite have incommensurate solid phases that are oriented with respect to the substrate. Novaco and McTague[108] showed that this orientation, surprisingly, need not match the orientation of the commensurate phase. They found that because the monolayer shears more easily than it compresses, more adatoms on average are placed closer to adsorption sites if the monolayer deforms with a shear modulation and rotates with respect to the substrate: This lowers the energy of the monolayer. The degree of rotation depends on the misfit of the monolayer. This theory was extended by Shiba[109] to consider a more realistic description of the substrate interaction and the monolayer was predicted to undergo an aligned to rotated transition as the misfit is increased from zero. From the theory of Shiba, one might expect that the value of the substrate corrugation could be determined from the misfit value at which the monolayer started to rotate.

Subsequent work by Shrimpton, et al.[110], however, showed that density modulations within the monolayer can influence the orientation behavior; the energy variation with orientation is small and thermal effects will influence the onset of rotation.

Aside from the shear modulations, which lead to an orientation of the monolayer, the substrate can also induce density modulations in the monolayer. These density modulations occur on length scales much larger than the interatomic distances. The regularity of the density modulation creates a superlattice periodicity within the monolayer: This is reflected by the splitting of the diffraction peaks into groups of satellite peaks which are clustered where the commensurate diffraction peak would occur.[111,112] The configuration of the satellite peaks about the commensurate position is determined by the superlattice periodicity. How rapidly the peak intensity diminishes away from the location of the commensurate peak indicates how sharply defined the density modulations are. At densities where the monolayer is close to being commensurate, large patches of adatoms are in fact registered in $\sqrt{3} \times \sqrt{3}$ sublattices of adsorption sites. These patches are separated from each other by *domain walls*, at which the density of the monolayer varies significantly from commensurate. Excellent reviews of the properties of modulated incommensurate monolayers are provided by Bak[113] and Pokrovsky and Talapov[114].

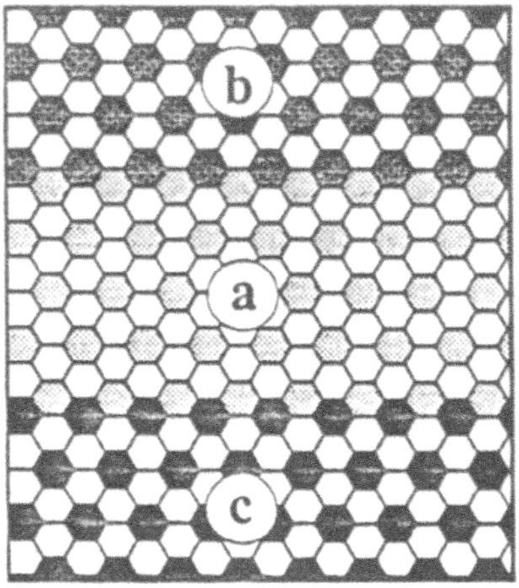

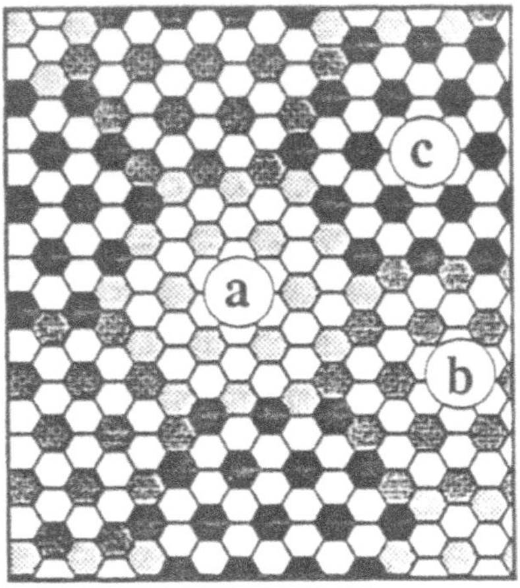

Fig. 5: Domain wall lattices in the incommensurate phase. a) shows the uniaxial arrangement of stripes of domain walls. b) shows the hexagonally symmetric configuration of a honeycomb domain wall network.

Several types of domain walls have been postulated.[115] These are illustrated in figure 6. Kr is observed to have a honeycomb lattice of super-heavy domain walls,[111] while Xe has a honeycomb lattice of super-light domain walls.[112] Theoretical calculations have predicted that He will have a uniaxially symmetric incommensurate phase with stripes of domain walls[116,117], and scattering studies have identified uniaxial stripes or superheavy domain walls in ^{3}He.[118]. The phase diagrams of He and H_2 show great similarities[119,150].

At low temperatures, and at coverages close to commensurate, D_2 also has a uniaxially symmetric phase with super-heavy domain walls aligned in stripes across the monolayer. However, D_2's phase diagram is more complex than that of He or H_2 because of the existence of a hexagonally symmetric phase with heavy domain walls.[120]

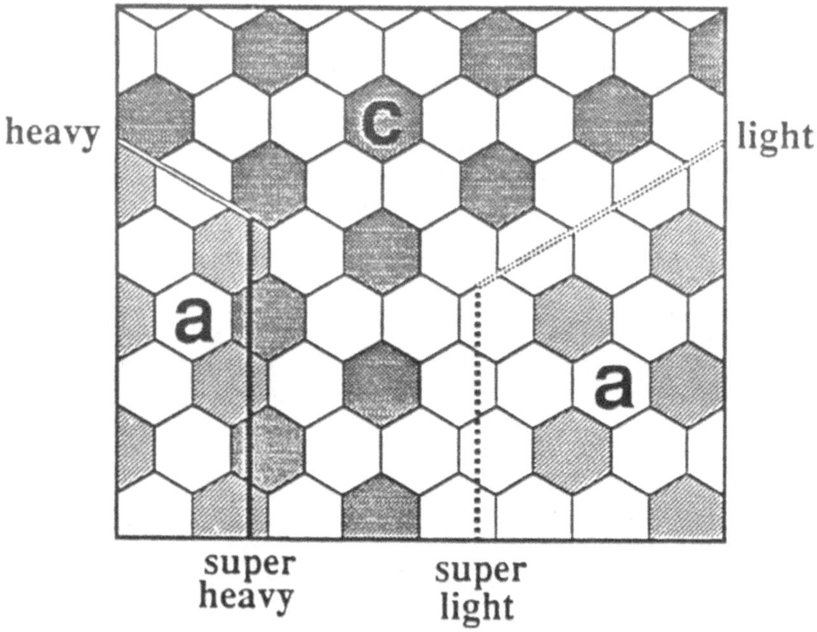

Fig. 6. Possible types of domain walls. This is a schematic representation, based on infinitely narrow domain walls. The thick solid line corresponds to a super-heavy domain wall, the double thin lines correspond to a heavy domain wall. The thick dashed line is a super-light domain wall, while the double thin dashed lines represent a light wall.

As shown in figure 5, the honeycomb lattice of domain walls involves intersections of the domain walls. At the point of intersection, the compressional fields of the domain walls become focussed, leading to density extremes at the vertices.[110] These density extremes are compensated by shifting the adatoms along the domain walls: A strain field interaction extends from each vertex along the domain walls. When the strain fields of two vertices overlap, there is a repulsive interaction that can effectively be described by a vertex–vertex pair potential.[121] This vertex interaction is surprisingly long ranged; away from the commensurate limit it produces a variation in the chemical potential μ of the Krypton monolayer with misfit ε of the form

$$\varepsilon = A|\mu - \mu_c(T)|^b \qquad (36)$$

where $b \sim 0.33$ and $A = 8\times10^{-4}K^{-b}$.[110] This is in agreement with the experimentally observed relationship found at much higher temperatures[122], and consequently indicates the dominant effect of the vertex interaction energy on the free energy of the monolayer. The fact that the experimental relationship is found to be invariant over a broad temperature range reinforces the fact that the monolayer energy variation stems from a strain field energy effect. For Xe, a similar relationship $\varepsilon \sim B(T - T_o)^b$ is observed.[123,124] Given a relatively linear relationship between μ and T over the fitted temperature range, a chemical potential relationship similar to Eq. (36) is implied. However, the value of b for Xe is far from certain, with values 0.80 ± 0.03[124] and 0.33 ± 0.08[123] being measured.

The question then remains as to when a honeycomb lattice of domain walls is energetically preferred over a uniaxial system of domain walls. Because of the focussed compression of the domain walls, the vertices are energetically unfavorable. Furthermore, the interaction between vertices is highly repulsive. However, a uniaxial lattice of domain walls affects the density in only one direction. To achieve a given monolayer density, the domain walls must be placed closer together in the uniaxial configuration than in the honeycomb configuration. Given the large repulsion between domain walls, at densities where the domain walls overlap, the honeycomb configuration will be favored. This does not, however, explain why Kr and Xe maintain a honeycomb configuration at densities approaching commensurate.

It was shown by Villain[125] that the honeycomb lattice of domain walls could alter the size of any given hexagonal domain without altering the total length of the domain walls. Even with a large vertex repulsion, this breathing motion of the domain walls remains very low in energy.[121] Consequently, there is considerable entropy associated with the honeycomb configuration, and at sufficiently high temperatures, the honeycomb configuration is energetically favored. The domain wall phase of He, on the other hand, exists only at very low temperatures. As a consequence, the entropic contribution of the domain wall breathing is not significant, and the honeycomb lattice is not favored.

As the misfit of the incommensurate monolayer is increased, however, the size of the commensurate patches decreases, until the domain walls merge into each other. At this point, the incommensurate monolayer is better described as being weakly modulated, rather than as having a lattice of domain walls. In the weakly modulated regime, shear modulations are more easily tolerated than density modulations, and the weakly modulated monolayers are rotated.[108–110] On the other hand, when the monolayer has a clearly defined lattice of domain walls, any rotation of the monolayer away from the commensurate orientation involves shearing the atoms along the domain walls. This shear increases the energy of the domain walls.[121] Thus, for the domain wall lattice, non-rotated configurations are energetically favored. The transition from a non-rotated to a rotated incommensurate monolayer is therefore related to the separation and distinctiveness of the domain walls.[110] Because the rotated phase is hexagonally symmetric and the domain wall lattice of He is uniaxially symmetric this transition must be discontinuous. For Kr and Xe, which have hexagonally symmetric domain wall lattices, this is not the case.

Considerable attention has been focussed on the nature of the commensurate to incommensurate transition. Theories based on Fermi statistics have been particularly useful in understanding this transition.[126] The configurational meanderings of the domain walls are analogous to the motions of a one dimensional lattice of fermions. Instead of a domain wall energy, one has a fermion mass. Instead of a domain wall line tension, one has a fermion hopping probability. Pauli exclusion of the fermions merely represents a hard core repulsion between domain walls. Using fermion theory, Pokrovsky and Talapov's[127] showed that the commensurate to uniaxial incommensurate transition should be continuous. Simple arguments based on entropy associated with the configurational meanderings of the domain walls by Fisher and Fisher[128], by Coppersmith et al.[129], and by Huse and Fisher[130] reproduced Pokrovsky and Talapov's results. These arguments indicate that the transition is continuous because collisions between the domain walls reduce the total entropy of the system. Consequently, there is a relationship between the averaged domain wall spacing L and the monolayer chemical potential μ

$$L \sim (\mu - \mu_c)^{\frac{1}{2}} \qquad (37)$$

as μ increases beyond a critical value μ_c. The relationship between misfit $\varepsilon \sim 1/L$ and the chemical potential implied by (37) has not in fact been measured for He on graphite, but has been observed for the comparable uniaxial system of Br-intercalated graphite.[131,132]

The honeycomb domain wall lattice, on the other hand, has been predicted to have a first order commensurate–incommensurate transition.[125,129] The entropy associated with breathing domain walls is such that the total free energy of the system is reduced as the domain wall spacings are reduced. A heat capacity study of the system of CO on graphite tends to confirm this prediction.[133]

Both the uniaxial and honeycomb lattices can be treated as renormalizable objects and, as such, they have elastic properties. They are prone to defects and can in fact melt.[126,129,130,134] Examples of dislocation defects in the uniaxial domain wall lattice and the honeycomb domain wall lattice are shown in figure 7. Domain wall fluid phases have been identified for both Kr[122,139,159] and He.[140] It appears that the commensurate to incommensurate solid transition proceeds via a domain wall fluid. This is in accord with theory[126,129,130,139], although a Lifshitz point has been proposed for temperatures below which the commensurate–incommensurate transition is to a solid domain wall lattice.[126,130]

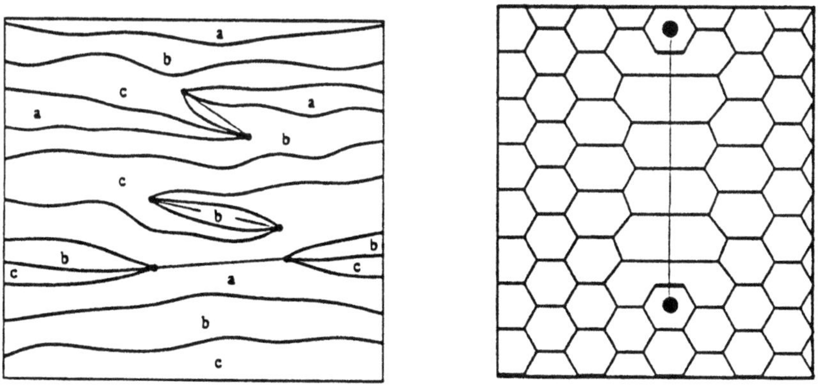

Fig. 7. Domain wall dislocations a) in a striped phase, and b) in a honeycomb domain wall lattice. When the dislocation pairs unbind, this is defined to be a domain wall fluid.

3.3 THE MONOLAYER AS A LATTICE GAS

The commensurate phases of He, Kr, and Xe place adatoms on a $\sqrt{3} \times \sqrt{3}$ R30° sublattice of adsorption sites, illustrated in figure 3. The spacing of this sublattice is 4.26Å. As shown in the figure, there are three possible origins for this sublattice on the substrate. A given commensurate phase can be identified as being based on an a, b, or c center. If the adsorption sites are split up into triplets of sites on a $\sqrt{3} \times \sqrt{3}$ sublattice, an order parameter can be defined which specifies which type of site is occupied by an adsorbed atom. While Xe merely expands with increased temperature into an incommensurate monolayer with a honeycomb lattice of reduced density domain walls, the commensurate phases of He and Kr melt. The properties of this latter transition fall in the universality class of the three state Potts model.[135] The three state Potts model is known to have a continuous order–disorder transition with a specific heat exponent of 1/3.

It is not necessary, of course, that every site of the sublattice be occupied by an adatom. This possibility is taken into account by defining an order parameter which specifies whether or not a triplet is occupied or not. This extended model is known as a dilute Potts model. The combination of the occupancy order parameter with the 3 state Potts order parameter allows the possibility of tricritical behavior. At the tricritical point, the transition belongs to the the tricritical 3-state Potts universality class. The phase diagram expected for this system is shown in figure 8.

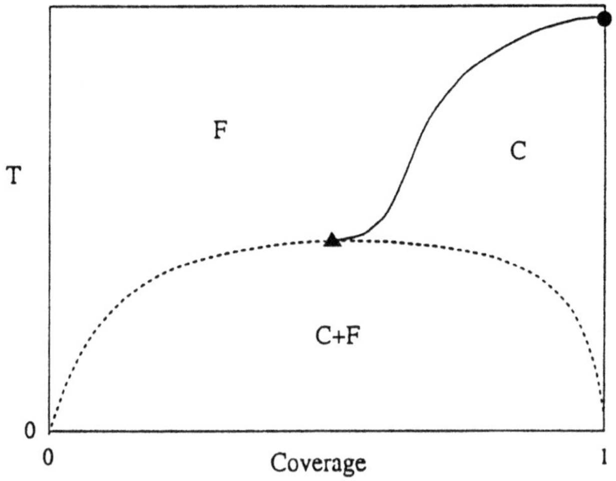

Fig. 8. This is a generic phase diagram for the 3–state Potts model with vacancies. C is the commensurate phase, F is the disordered fluid phase. The circle indicates the Potts critical point, while the triangle indicates the tricritical point.

At low temperatures, the commensurate phase coexists with the disordered phase. With increasing temperature, the coexisting disordered phase will have a steadily increasing density, and the commensurate phase, due to an increasing number of vacancies, will have a steadily decreasing density. At the tricritical point, the two densities coincide, and the coexistence region terminates. Since this behavior is concerned with whether sites are occupied, it involves the triplet occupancy order parameter. This parameter is analogous to the density parameter of the liquid-gas transition. The coexistence region is a first order transition of this order parameter, which becomes continuous at the tricritical point in the same way that the liquid-gas transition becomes continuous at the critical point. The difference in density $\rho_c - \rho_d$ between the commensurate and disordered phase varies with temperature near the tricritical temperature T_{TCP} as

$$\rho_c - \rho_d \sim (T_{TCP} - T)^{\beta_T}$$

However, the 2D 3-Potts tricritical exponent is [136] $\beta_T = 1/2$, in contrast to the 2D Ising exponent $\beta = 1/8$.

Unlike the case of the liquid-vapor system, there is a structural difference between the commensurate phase and the disordered phase above the tricritical temperature. The transition between them involves the order parameter which specifies which of the triplet sites is occupied. On the commensurate side of the transition, the order parameter identifies one of the three possible commensurate configurations. On the disordered side, the order parameter vanishes, showing that on average all sites become equally probable. It should be pointed out that the disordered state need not have the adatoms confined to adsorption sites. The distinction of whether or not the adatoms in the disordered phase are confined to adsorption sites will not alter the critical properties of the commensurate-disordered transition.[137]

The phase diagram shown in figure 8 may, however, be distorted greatly from that shown. An indication of possible variations to the phase diagram is provided by Ostlund and Berker[107] and by Niskanen.[106] The disordered lattice gas phase can have an Ising-type critical temperature T_c, below which two possible disordered phases are possible: vapor and liquid. The commensurate phase can also be thought of as melting at a *triple point* temperature T_t. If T_t is less than T_c a phase diagram similar to figure 4 is possible, the only difference being that there is no fluid-solid coexistence region. This transition, from ordered commensurate to disordered, is continuous, with critical behavior in the 3-state Potts universality class.

While the liquid and gas phases are not greatly influenced by the substrate corrugation, the commensurate solid phase is. This phase may be stabilized by the the substrate to the extent that its melting temperature, T_t, can exceed the critical temperature T_c of the liquid–gas coexistence region. The phase diagram then is described as having an "incipient triple point".[138] There is no liquid–gas coexistence, and the phase diagram is modified from the form shown in figure 4 to that shown in figure 9.

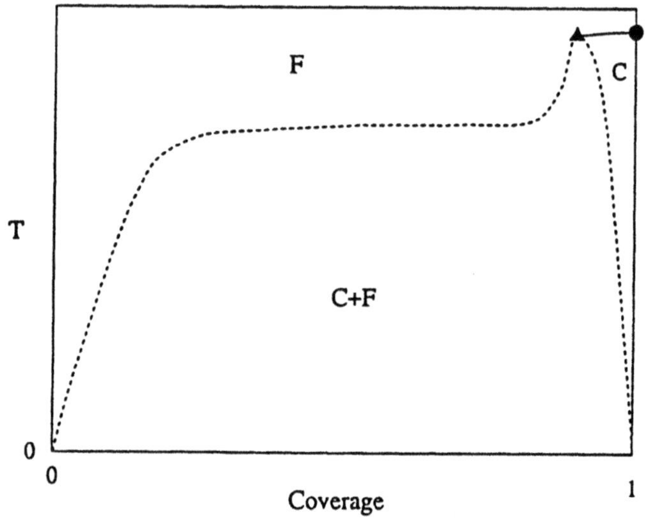

Fig. 9. This is a generic phase diagram for a system with an Incipient triple point. C is the commensurate phase, F is the disordered fluid phase. The circle indicates the Potts critical point, while the triangle indicates the tricritical point.

At high temperatures, where the commensurate phase does melt, the transition will be in the 3–state Potts universality class. Ostlund and Berker[107] showed how the proximity of the nearby triple point to the commensurate–disordered phase boundary distorts the 3–Potts phase diagram from that shown in figure 8 to that shown in figure 9.

The He case is distinguished by the relatively weak adatom–adatom interaction. This, along with quantum effects[193], would place the critical point of the fluid phase at a very low temperature, if it had not been preempted by the commensurate phase. The "triple point temperature" greatly exceeds the "critical temperature" and the commensurate–fluid phase boundary remains well away from the fluid critical point. Consequently, the envelope of the coexistence region does not display rapid variations of density with temperature except near the tricritical point, and the phase diagram of figure 8 is appropriate to He for coverages below commensurate.

Kr atoms, on the other hand, have interaction energies which are comparable to the corrugation of the substrate attraction. The "triple point temperature" thus does not greatly exceed the nominal "critical temperature" and near this "critical point", the fluid will display rapid variations in density with temperature along the commensurate–fluid phase boundary. Consequently, the phase diagram of figure 9 is appropriate to Kr.

The phase diagrams shown in figures 8 and 9 do not consider densities greater than commensurate. With increased density, the monolayer will undergo a commensurate to incommensurate phase transition, where melting must be considered at the length scales of the domain wall lattice. As pointed out by Huse and Fisher[130], the introduction of domain wall droplets increases the monolayer entropy and leads to logarithmic singularities in the specific heat near "melting". The commensurate phase of Xe does not in fact "melt"; instead it undergoes a commensurate to incommensurate transition and forms a domain wall lattice.[126]

With this in mind, one may consider the "melting" of the commensurate phase. Given that the correlation length of the monolayer is ε, in the commensurate limit, a comparison can be made with the length L (which scales as $1/\varepsilon$) associated with the domain walls of the incommensurate phase. A useful classification is provided by the critical relationship[126]

$$\lim_{\varepsilon \to 0} L \sim \varepsilon^{x_Q} \qquad (39)$$

The commensurate transition is predicted to be to a well ordered domain wall lattice if $x_Q < 1$; to a domain wall fluid (see Fig. 7) if $x_Q = 1$, or to a conventional fluid with correlations specified on interatomic length scales if $x_Q > 1$. While a well ordered domain wall lattice implies a solid phase, the scaling relationship (39) with $x_Q < 1$ is not appropriate for a solid phase. Thus the classification afforded by Eq. (39) is to distinguish between domain wall fluids and fluids disordered on atomic length scales. Heat capacity studies do not identify any feature associated with this transition.[133]

If $x_Q = 1$, as has been inferred from studies of the melting transition of Kr[139], the nature of the transition must be considered at length scales of the domain wall lattice. This lattice differs from the 3-state Potts lattice in one important respect. As shown in figure 6, there is more than one type of domain wall. The differing energies of the domain walls lower the symmetry of the domain wall lattice from that of the 3-state Potts model. For a uniaxial phase, this is described by a chirality order parameter, while the "honeycomb" domain wall system with three 120° domain wall orientations has a helicity parameter. The commensurate to uniaxial domain wall fluid transition has critical properties different from the 3-Potts universality class, while commensurate melting with helicity has the critical properties of this class.[126] As shown by Rabedeau[140], however, the melting of commensurate He may have critical properties altered by chirality from the 3-Potts prediction.

4 Experimental Review

Argon

Ar does not form a commensurate phase due the small size of the argon atom. The solid phase is always denser than the commensurate phase. From table 3 the near neighbor spacing can be inferred to be about 3.8Å. The phase diagram has a general similarity to that of the floating LJ system shown in figure 4. The solid phase of the monolayer is rotated with respect to the substrate, as shown by Shaw et al.[141] and D'Amico et al.[142] The rotational behavior of the monolayer agrees well with the theory of Shiba.[109] Scattering profiles of the solid show a power law decay which indicates that the solid is a 2D elastic medium.[144]

Ar is distinguished from the other "floating" rare gases by the nature of its melting transition. The 2D triple point temperature has been identified to be around 48K, with a liquid-gas critical point at 55K. Unlike Ne and Xe, Ar does not appear to have a region of solid-fluid coexistence. The triple point temperature is ~ 0.57 of the 3D value[145] and is difficult to locate precisely, because of the broad nature of the transition. An unexplained sharp heat capacity spike has been observed at 47.2K in the region of coexistence, but this feature is small.[146] Adsorption isotherms[147,148] do not show vertical risers (which would indicate coexistence). Scattering studies of the melting transition reveal that any density discontinuity, if present, is less than 0.1.[144] Scattering measurements have been taken over a temperature span which includes the triple point temperature.[141,142] The scattering peaks reveal that the the film above the melting transition is not isotropic; it has an orientation that evolves continuously from that of the solid towards that of the substrate over the melting temperature range.

That there should be an oriented fluid is one of the predictions of the theory of Kosterlitz, Thouless, Nelson, Halperin, and Young (KTHNY)[25,35,143] and the melting of Ar has been considered to be an example of such a continuous melting transition.[36] Through scattering studies, the variation in inverse correlation length κ of the *fluid* with temperature has been examined[144,142]; it is found to vary as

$$\kappa(T) = \kappa_0 \exp(-Bt^{-\nu}) \qquad (40)$$

where $t = (T - T_M)/T_M$ and $\nu = 0.369$, near the melting temperature T_M as predicted by the KTHNY theory. It must be pointed out however, that the numbers quoted by the two groups for the parameters κ_0 and B of (40) at comparable Ar densities are not in agreement.

Given the structural symmetry of the substrate, it is not surprising that the fluid phase should reflect some effect of orientation. What is surprising is that the fluid orientation should rotate to that of the solid phase. This behavior is consistent with the dislocation unbinding mechanism of KTHNY, where the unbound dislocations destroy positional correlation, while leaving the *fluid* with the same orientation of the solid. This is perhaps the only melting mechanism capable of displaying this feature. Other mechanisms, such as grain boundaries, dislocation–disclination, and cavitation, do not constrain the fluid to prefer the *orientation* of the solid phase.

That the transition is continuous was contested by Abraham[101] who showed MD calculations which mimic experimentally observed results, while revealing regions of coexistence. The melting of Argon on graphite is complicated by the fact that the fluid density is close to that of a commensurate monolayer, and the melting temperature of 48K is comparable to the effective substrate corrugation well depth (of order 50K). However, given the strength of the substrate corrugation and the near *commensurate* density of the fluid, one might expect the fluid to have a preferred orientation aligned with the substrate. This is not what is observed.

Helium

Both ^{3}He and ^{4}He have been studied. The phase diagrams of these two gases are similar. ^{3}He's tricritical point is much reduced in temperature from that of ^{4}He. Furthermore, H_2 and HD on graphite have phase diagrams that are greatly similar to He. Unlike Ar, He forms a commensurate solid phase. At low temperatures, this phase coexists with a dilute 2D gas phase, and because of vacancies, its density at T>0 can be less than that of a perfect $\sqrt{3} \times \sqrt{3}$ lattice. With increasing temperature, the density of the gas phase increases, and the density of the commensurate phase decreases, until at the tricritical point the two densities coincide. Past this temperature, the commensurate phase melts continuously to a *fluid* phase. This temperatuare (3K) is nearly identical for the two isotopes and quite high for He, indicative of the large magnitude of the substrate corrugation.

With increased density, there is a phase transition from the commensurate phase to phases that are incommensurate with the substrate. At low temperature, the incommensurate phase is solid and has uniaxial symmetry. This phase, illustrated in figure 6, is characterized by density modulations which form stripes of increased density domain walls cross the monolayer. With increased temperature, this phase melts into a hexagonally symmetric *domain wall fluid*. At lower temperatures and higher densities, the striped phase undergoes a transition to a solid incommensurate phase that has hexagonal symmetry but is rotated with respect to the substrate.

An excellent summary of the early work on adsorbed He is provided by Bretz et al.[22] Elgin and Goodstein[16] considered the incommensurate rotated solid phase, studied its elastic properties and considered the melting transition in the light of the then emerging theory of Kosterlitz and Thouless.[25] Work by Hering et al.[149], which considered the transitions of the incommensurate monolayer as it decreased in density toward that of the commensurate phase showed that the monolayer had a new phase, later determined to be the uniaxial striped phase.[118]

Bretz[151], followed by Tejwani et al.[152], Ecke and Dash[153], Carey and Fahey[154], and Ecke et al.[136], identified and mapped out the transitions associated with the commensurate monolayer. The commensurate phase was shown to melt continuously with a specific heat exponent of 1/3, in accordance with the order-disorder transition of the 3 state Potts model. At lower temperatures the coexistence of the commensurate phase with the disordered phase was shown, and the tricritical point identified.

Meanwhile, Feile et al.[155] using neutron scattering, and Widom et al. [136] using NMR, considered the melting of the incommensurate solid, and determined that the transition had the appearance of a KT transition. Specific heat results of Ecke and Dash[153] and later Hurlbut and Dash[157] showed that the transition could be first order rounded by finite size effects. The nature of this melting-transition remains undecided. Finally, through compressibility measurements, Rabedeau[140] considered the effect of chirality on the melting transition. While showing that the melting transition may be modified by the effect of domain wall chirality, the results were not conclusive.

Krypton

Kr has a commensurate phase. There is a great similarity between the Kr phase diagram and that of CO.[133] There is coexistence of the commensurate phase with a disordered gas that extends in temperature up to the termination of the commensurate phase. This places the tricritical point at the same temperature as the commensurate melting critical point ~ 130K.[139] Because of the relatively strong Kr interactions, the fluid phase has near critical behavior and its phase diagram displays the features of an incipient triple point, as shown in figure 9. With increased density, the commensurate phase undergoes a transition to an incommensurate domain wall fluid phase. With a further increase in density, this domain wall fluid solidifies into a honeycomb domain wall lattice. With further compression, the domain walls merge, and the monolayer rotates with respect to the substrate.

It is interesting to note that the compressed solid phase has a continuous melting transition.[133,139] At lower coverages, the melting of the solid is accomplished by first a continuous melting of the domain wall lattice, which then with increased temperature evolves continuously to a conventional fluid. At increased coverage, the domain wall picture no longer applies, but the melting is still continuous.

Aside from the distortions in the coexistence phase boundary due to the incipient triple point, there is great similarity of the Kr system with the He system. Because the density of the monolayer shown in figure 2 includes second layer atoms the shape of the commensurate phase is greatly distorted. Because with increased temperature, the number of second layer adatoms increases, the phase boundaries are shifted to higher densities. This effect also distorts the shape of the commensurate-fluid coexistence region below the tricritical point. Without the effect of second layer and interstitial atoms, the commensurate monolayer should have a phase diagram as shown in figure 9, where the tricritical point has a density discontinuity given by Eq. (38).

Krypton was first studied by Thomy and Duval.[21] They found that different phases are possible within the monolayer. LEED scattering studies by Chinn and Fain[158], showed that the Kr forms a commensurate phase. X-ray diffraction studies showed the existence of a domain wall lattice[111], and that this lattice is disordered at the commensurate-incommensurate transition.[122,159] The presence of a domain wall fluid at the transition is also identified by the observed chiral relationship, Eq. (39). A relatively simple relationship, Eq. (36), was discovered between the misfit of the monolayer and its chemical potential.[111,122,139,160-162] The commensurate-incommensurate transition has at most a small density jump.[162] With increased density, the krypton monolayer was observed to change from being aligned to being rotated with respect to the substrate.[160,161,163] Adsorption isotherms[21,164,165], and specific heat measurements[166] were instrumental in locating phase boundaries. The location of the tricritical point at nearly 130K was shown from X-ray diffraction studies.[139] The critical nature of the phase transitions has only recently been confirmed through a specific heat study of the analagous system of CO on graphite.[133]

Neon

Ne does not have a $\sqrt{3} \times \sqrt{3}$ commensurate phase. Its phase diagram is similar to that of the floating monolayer. Solid, liquid and vapor phases have been identified. The triple point temperature is 13.6K.[167,168,170,171] The melting is first order.[167] The critical temperature for the gas-liquid coexistence is 16K.[168,169] These values are about 15% below the smooth surface prediction of table 3, presumably because of the substrate mediation of the repulsive interactions as discussed in section 2.4. The fluid phase does have a lorentzian scattering profile; however, the inverse correlation length does not vary with temperature as predicted by Eq. (40).[171] The orientation of the solid monolayer is in agreement with the predictions of Novaco and McTague.[170] At low temperatures (less than 4K) a commensurate $\sqrt{7} \times \sqrt{7}$ solid phase is inferred from neutron scattering data.[171]

Of all the rare gases, the phase diagram of Ne has received the least attention. The phase diagrams presented by Huff and Dash[167], Tiby et al.[171] and Rapp et al.[168] show qualitative differences. Particularly surprising is the broad nature of the solid-fluid coexistence region identified by Rapp et al.

Xenon

Xenon's phase diagram closely resembles that of the floating LJ system. The incommensurate solid melts by a strong first order transition.[172,173] The triple point of melting is 100K, while the liquid-gas critical point temperature is 116.[175] With increased density, the Xe monolayer undergoes a commensurate transition.[123,124,161,176]

At high temperature, and increased coverage, the transition from the compressed solid to fluid has been claimed to be an example of KTHNY melting[36], and much work on the nature of the melting transition has been done. The melting transition does have a continuous appearance.[173,177-181] The solid melts to an oriented fluid phase and the correlation lengths in the fluid region follow the form predicted by Eq. (40). The orientation of the fluid is in alignment with the substrate.[179] However, as was shown by Abraham[101], this type of behavior is also found in MD simulations even though there is coexistence between solid and fluid. This solid-fluid coexistence has been shown by Gangwar et al.[174], and Jin et al.[175], to extend to temperatures much higher than expected. In fact it appears from the results of Jin et al. that Xe always has a first order melting transition.

The incommensurate phase of Xe, like Kr, has a modulated structure.[112] At the commensurate transition, the incommensurate phase of Xe is aligned with the substrate.[123,142] With reduced coverage, the monolayer rotates, although not as much as could be expected from Shiba's[109] theory. With a further reduction in density and an increase in temperature, the monolayer becomes aligned again with the substrate. This has been taken to be indicative of the effect of domain wall breathing.[142] However, it must be remembered that the domain walls in Xe are regions of depleted density, and that large fluctuations can lead to an accumulation of domain walls. Such a depletion can be the precursor to cavitation, and the fluid phase is aligned with the substrate.[179] Thus the realignment of the monolayer may be due simply to its transition to a fluid. This type of behavior has been observed by molecular dynamics simulations through the melting transition.[182]

Acknowledgement

We are grateful to many collaborators for contributing to our understanding of these issues. Helpful comments on the manuscript by M. J. Bojan, J. G. Dash and M. Shrimpton were appreciated. This research was supported in part by NSF grant DMR-8718771.

References

1. Steele, W. A., The Interaction of Gases with Solid Surfaces (Pergamon, Oxford(1974).
2. Nicholson, D. and N. G. Parsonage, Computer Simulation and the Statistical Mechanics of Adsorption (Academic, London, 1982).
3. Statistical Mechanics of Adsorption, Surf. Sci. 125, 1 (1982).
4. Gas-Surface Interactions and Physisorption, Surf. Sci. 148, 1 (1984).
5. Bruch, L. W. and M. B. Webb in Interfacial Aspects of Phase Transformations, ed. B. Mutaftschiev (Reidel, Dordrecht, the Netherlands, 1982).
6. Ordering in Two Dimensions, S. K. Sinha, editor, N. Holland, Amsterdam, 1980.
7. Phase Transitions in Surface Films, ed. by J. G. Dash and J. Ruvalds (Plenum, NY, 1980).
8. Phase Transitions in Adsorbed Layers, J. Chem. Soc. Far. Trans. II 82, 1569 (1986).
9. Vilches, O. E., Ann. Rev. Phys. Chem. 31, 463 (1980).
10. Thomy, A., X. Duval and T. Regnier, Surface Sci. Repts. 1, 1 (1981).
11. Birgeneau, R. J. and P. Horn, Science 232, 329 (1986).
12. Faraday Disc. Chem. Soc. 80, 1-313 (1985).
13. Marx, R., Phys. Repts. 125, 1 (1985).
14. Dash, J. G., Films on Solid Surfaces (Academic Press 1975); Physics Repts. 38C, 177 (1978).
15. Birgeneau, R. J., P. A. Heiney, J. Pelz, Physica 109 and 110, B&C, p. 1785, 1982.

16. Elgin R. L. and D. L. Goodstein, Phys. Rev. A9, 2657 (1974).
17. A review of physisorption well depths is G. Vidali, G. Ihm, H. Y. Kim and M. W. Cole, Surface Sci. Repts, to be published.
18. Polley, M. H., W. D. Schaeffer and W. R. Smith, Jr., J. Phys. Chem. 57, 469 (1953); J. H. Singleton and G. D. Halsey, Jr., J. Phys. Chem. 58, 330 and 1011 (1954).
19. Avgul, N. M. and A. V. Kiselev, in Chemistry and Physics of Carbon, Vol. 6, ed. P. Walker, Jr., Dekker, NY (1970).
20. Thomy, A. and X. Duval, J. Chim. Phys. Physicochim. Biol. 66, 1966 (1969).
21. Thomy, A. and X. Duval, ibid. 67, 286 and 1101 (1970).
22. Bretz, M., J. G. Dash, D. C. Hickernell, E. O. McLean and O. E. Vilches, Phys. Rev. A8, 1589 (1973); A9, 2814 (197); W. A. Steele and R. Karl, J. Coll. Int. Sci. 28, 397 (1968).
23. Dash, J. G. and M. Schick, in the Physics of Liquid and Solid Helium, ed. K. H. Bennemann and J. B. Ketterson (Wiley, NY 1977), p. 497.
24. Wilson, K. G., Rev. Mod. Phys. 55, 383 (1983).
25. Kosterlitz, J. M. and D. J. Thouless, J. Phys. C6, 1181 (1973).
26. Kosterlitz, J. M. and D. J. Thouless, in Prog. Low Temp. Phys., ed. D. F. Brewer (N. Holland, Amsterdam), Vol. VIII B.
27. Landau, L. D., Sov. Phys. JETP 7, 627 (1937).
28. Peierls, R., Helv. Phys. Acta 7, Suppl. II, 81, 1934; Ann. Inst. H. Poincaré 5, 177 (1935).
29. Hohenberg, P. C., Phys. Rev. 158, 383 (1967).
30. Chester, G. V., M. E. Fisher, N. D. Mermin, Phys. Rev. 185, 760 (1969).
31. For reviews, see articles by G. V. Chester and A. Eggington in the Helium Liquids, ed. J. G. M. Armitage and I. E. Farquhar (Academic, London 1975); A. J. Leggett, Phys. Fennica 8, 125 (1973); D. F. Brewer in The Physics of Liquid and Solid Helium, ed. K. H. Bennemann and J. B. Ketterson (Wiley, London)Vol. 2, 1978.
32. Mermin, N. D., Phys. Rev. 176, 250 (1968); E B20, 4760 (1979).
33. Stanley H. E., and T. A. Kaplan, J. Appl. Phys. 38, 975 (1967).
34. Berezinskii, V. L., Sov. Phys. JETP 32, 493 (1970).
35. Nelson D. R. and B. I. Halperin, Phys. Rev. B19, 2457 (1979); D. R. Nelson in Fundamental Problems in Statistical Mechanics V, ed. E. G. D. Cohen (N. Holland Amsterdam 1980), p. 53.
36. Strandburg, K. J., Rev. Mod. Phys. 60, 161 (1988).
37. G. Vidali and M. W. Cole, Phys. Rev. B29, 6736 (1984); L. W. Bruch in Proceedings of the Erice Conference, 1990, ed. by H. Taub, et al.
38. A general review of physical adsorption potentials is presented by L. W. Bruch, Surf. Sci. 125, 194 (1983).
39. Deviations from this include adatom scattering by phonons and polaron-type self-energy or interaction effects; see M. Schick and C. E. Campbell, Phys. Rev. A2, 1591 (1970), and F. Ancillotto and F. Toigo, Sol. St. Comm. 48, 593-596 (1983).

40. Cole, M. W. and J. R. Klein, Surf. Sci. 124, 547 (1983). Note that the well depth values therein were revised in G. Vidali, M. W. Cole and J. R. Klein, Phys. Rev. B28, 3064 (1983).
41. The earliest approximate treatment is that of O. Sinanoglu and K. S. Pitzer, J. Chem. Phys. 32, 1279 (1960). Later work is by McLachlan, A. D., Surf. Sci. 7, 381 (1964); G. L. Price and J. A. Venables, Surf. Sci. 59, 509 (1976); S. Rauber, J. R. Klein and M. W. Cole, Phys. Rev. 27, 1314 (1983) and references therein; J. R. Klein and M. W. Cole, Faraday Disc. Chem. Soc. 80, 71 (1985).
42. Steele, W. A., in Chemistry and Physics of Solid Surfaces, ed. R. Vanselow and S. Y. Tong (CRC Press, 1977, Cleveland, OH), p. 139.
43. Monson, P. A., W. A. Steele and D. Henderson, J. Chem. Phys. 74, 6431 (1981).
44. Einstein, T. L., CRC Crit. Rev. Sol. State Mat. Sci. 7, 261 (1978).
45. Tsong, T. T., Repts. Prog. Phys. 51, 759–782 (1988); P. J. Feibelman, Phys. Rev. Lett. 58, 2766 (1987) and Ann. Rev. Phys. Chem. 40, 261 (1989).
46. Steele, W. A., Surf. Sci. 36, 317 (1973);
47. Steele, W. A., J. Phys. Chem. 82, 817 (1978).
48. A more complete expansion appears in Ref. 47.
49. Nicholson, D., Surf. Sci. 184, 255 (1987); H. Y.Kim and M. W. Cole, Phys. Rev. B35, 3990 (1987).
50. Bonino, G., C. Pisani, F. Ricca and C. Roetti, Surf. Sci. 50, 379 (1975).
51. Carlos, W. E. and M. W. Cole, Surf. Sci. 91, 339 (1980).
52. Boato, G. P., P. Cantini, C. Guidi, R. Tatarek and G. P. Felcher, Phys. Rev. B20, 3957 1979); G. D. Derry, D. Wesner, W. E. Carlos and D. R. Frankl, Surf. Sci. 87, 629 (1979).
53. Carlos, W. E. and M. W. Cole, Phys. Rev. B21, 3713 (1980).
54. Silva-Moreira, A. F., J. Codona, and D. L. Goodstein, Phys. Lett. A76, 324 (1980).
55. Cole, M. W., D. R. Frankl, and D. L. Goodstein, Rev. Mod. Phys. 53, 199 (1981).
56. Subsequent scattering studies show small shifts which do not affect the conclusions of the preceding reference; see Chung, S., A. Kara and D. R. Frankl, Surf. Sci. 171, 45 (1986); and J. C. Ruiz, G. Scoles and H. Jonsson, Chem. Phys. Lett. 129, 139 (1986).
57. See, e. g., Sander, L. M. and J. Hautman, Phys. Rev. B29, 2171 (1984); and Crowell, A. D. and J. S. Brown, Surf, Sci. 123, 296 (1982).
58. Joshi, Y. P. and D. J. Tildesley, Mol. Phys. 55, 999 (1985).
59. A recent comparison of such potentials is provided by Sharma, S. R., S. F. O'Shea and W. J. Meath, Phys. Rev. B40, 6356–6367 (1989).
60. D. L. Freeman, J. Chem. Phys. 62, 941 (1975).
61. Liebsch, A., J. Harris and M. Weinert, Surf. Sci. 145, 207 (1983).
62. Toigo, F. and M. W. Cole, Phys. Rev. B32, 6989 (1985).
63. Vidali, G. and M. Karimi, Surf. Sci. 208, L73 (1989).

64. Ihm, G. and M. W. Cole, Langmuir 5, 550 (1989).
65. Gordon, R. G. and Y. S. Kim, J. Chem. Phys. 56, 3122 (1972).
66. This model is reviewed by M. J. Clugston, Adv. Phys. 27, 893 (1978).
67. Waldman, M. and R. G. Gordon, J. Chem. Phys. 71, 1325 (1979) and 71, 1340 (1979).
68. For an anisotropic medium, ε is the geometric mean of values parallel and perpendicular to the c axis; see T. Kihara and N. Honda, J. Phys. Soc. Japan 20, 15 (1965).
69. C_3 values are tabulated in Refs. 17 and 38.
70. An exception, where the G–K model yields dramatically incorrect answers, is the problem of atomic immersion in jellium; see Ihm, G. and M. W. Cole, Phys. Rev. A40, 1153 (1989).
71. Zaremba, E. and W. Kohn, Phys. Rev. B15, 1769 (1977).
72. Harris, J. and A. Liebsch, J. Phys. C15, 2275 (1982).
73. A. Chizmeshya and E. Zaremba, Surf. Sci., in press.
74. Norskov, J. K. and N. D. Lang, B21, 2131 (1980); Norskov, J. K., Phys. Rev. B26, 2875 (1982); Lang, N. D. and J. K. Norskov, Phys. Rev. 27, 4612 (1983).
75. Stott, M. J. and E. Zaremba, Phys. Rev. B22, 1564 (1980).
76. Puska, M. J., R. M. Nieminen and M. Manninen, Phys. Rev. B24, 3037 (1981).
77. Manninen, M., J. K. Norskov, M. J. Puska and C. Umrigar, Phys. Rev. B29, 2314 (1984).
78. I. P. Batra, Surf. Sci. 148, 1 (1984).
79. Posternak, M., A. Baldererchi, A. J. Freeman, and E. Wimmer, Phys. Rev. Lett. 52, 863 (1984).
80. Baldereschi, A., private communication
81. Carneiro, K., L. Passell, W. Thomlinson and H. Taub, Phys. Rev. B24, 1170 (1981).
82. We include here statistical interactions in the quantum case.
83. Note that there are quasi-stationary states in which the perpendicular energy is negative but $\varepsilon_1>0$; these are seen in scattering as bound state resonances. See D. R. Frankl, Prog. Surf. Sci. 13, 285 (1983).
84. See e.g., Schwartz, C. and M. W. Cole, Phys. Rev. B34, 1250 (1986).
85. Kreuzer, H. J. and Z. W. Gortel, Physisorption Kinetics (Springer Verlag, Berlin, 1985).
86. Indeed the preceding simple behavior does not even coincide with numerical results for classical systems because s is not constant. See, e.g., Tully, J. C., Surf. Sci. 111, 461 (1981).
87. Hagen, D. E., A. D. Novaco and F. J. Milford, in Adsorption–Desorption Phenomena, ed. by F. Ricca (Academic, London, 1972); Chow, H., Surf. Sci. 79, 157 (1979).
88. A finite T expansion of the heat capacity has been derived by L. W. Bruch, Phys. Rev. B23, 6801 (1981); see also Z.–C Guo and L. W. Bruch, J. Chem. Phys. 77, 1417 (1982).
89. W. A. Steele, Surf. Sci. 39, 149 (1973).

90. R. L. Siddon and M. Schick, Phys. Rev. A9, 907 and 1753 (1974).
91. Simplied expressions for these coefficients are presented by Rauber, S., J. R. Klein, M. W. Cole and L. W. Bruch, Surf. Sci. 123, 173 (1982).
92. Meath, W. J. and R. A. Aziz, Mol. Phys. 52, 225 (1984).
93. Barker, J. A., Phys. Rev. Lett. 57, 230 (1986) and Barker, J. A. in Simple Molecular Systems at Very High Density, ed. by A. Polian, P. Loubeyre and N. Boccara (Plenum, 1989), p. 341-351.
94. Kim, H. Y. and M. W. Cole, Surf. Sci. 194, 257 (1988).
95. Maitland, G. C., M. Rigby, E. B. Smith, and W. A. Wakeham, Intermolecular Forces (Clarendon Press, Oxford, 1981).
96. Aziz, R. A. in Inert Gases, ed. M. L. Klein (Springer-Verlag, Berlin, 1984); Scoles, G., Ann. Rev. Phys. Chem. 31, 81 (1980).
97. See e. g., S. Rauber, J. R. Klein and M. W. Cole, Phys. Rev. B27, 1314 (1983).
98. A review of the symmetry classes found in adsorbed monolayers is M. Schick, Prog. Surf. Sci. 11, 425 (1981). For a review of conformal invariance, see J. L. Cardy in Phase Transitions and Critical Phenomena, Vol. 11, Editors C. Domb and J. Lebowitz (Academic Press, London, 1987).
99. Siddon, R. L. and M. Schick, Phys. Rev. A9, 907 (1974); L. K. Moleko, B. Joos, T. M. Hakim, H. R. Glyde and S. T. Chui, Phys. Rev. B34, 2815 (1986); T. M. Hakim, H. R. Glyde and S. T. Chui, Phys. Rev. B37, 974 (1988).
100. Reddy, M. R. and S. F. O'Shea, Can. J. Phys. 64, 677 (1986).
101. For a review see F. F. Abraham, Adv. Phys. 35, 1 (1986). Also see Phillips, J. M., L. W. Bruch and R. D. Murphy, J. Chem. Phys. 75(10), 5097 (1981).
102. Onsager, L., Phys. Rev. B65, 117 (1944).
103. Yang, C. N., Phys. Rev. B85, 809 (1952).
104. Zhang, Q. M., Y. P. Feng, H. K. Kim, and M. H. W. Chan, Phys. Rev. Lett. 57, 1456 (1986).
105. van Swol, F., L. V. Woodstock, and J. N. Cape, J. Chem. Phys. 73, 913 (1980).
106. Niskanen, K. J., Phys. Rev. B33, 1830 (1986).
107. Ostlund, S. and A. N. Berker, Phys. Rev. B21, 5410 (1980).
108. Novaco, A. D. and J. P. McTague, Phys. Rev. Lett. 38, 1286 (1977).
109. Shiba, H., J. Phys. Soc. Jpn. 46, 1852 (1979); J. Phys. Soc. Jpn. 48, 211 (1980).
110. Shrimpton, N. D., B. Joos, and B. Bergersen, Phys. Rev. B38, 2124 (1988).
111. Stephens, P. W., P. A. Heiney, R. J. Birgeneau, P. M. Horn, D. E. Moncton, and G. S. Brown, Phys. Rev. B29, 3512 (1984).
112. Mowforth, C. W., T. Rayment, and R. K. Thomas, J. Chem. Soc. Faraday Trans. 2, 82, 1621 (1986).
113. Bak, P., Rep. Prog. Phys. 45, 587 (1982).
114. Pokrovsky, V. L. and A. L. Talapov, Theory of Incommensurate Crystals, (Harwood Academic Publishers, New York, 1984).
115. Kardar, M. and A. N. Berker, Phys. Rev. Lett. 48, 1552 (1982).

116. Broughton, J. Q. and F. F. Abraham, J. Phys. Chem. 92, 3274 (1988).
117. Gottlieb, J. M. and L. W. Bruch, Phys. Rev. B40, 148 (1989).
118. Lauter, H. J., H. Godfrin, V. L. P. Frank and H. P. Schildberg, ^{3}He-Layers on Graphite, 19th Int. Conf. on Low Temp. Phys., Brighton (1990)
119. Freimuth, H. and H. Wiechert, Surf. Sci. 162, 432 (1985).
120. Cui, J. and S. C. Fain, Jr., Phys. Rev. B60, 1848 (1988).
121. Shrimpton, N. D. and B. Joos, Phys. Rev. B40, 10564 (1989).
122. Stephens, P. W., P. Heiney , R. J. Birgeneau, P. M. Horn, Phys. Rev. Lett. 43, 47 (1979).
123. Hong, H., C. J. Peters, A. Mak, R. J. Birgeneau, P. M. Horn, and H. Suematsu, Phys. Rev. B40, 4797 (1989).
124. Hamichi, M., A. Q. D. Faisal, J. A. Venables, and R. Kariotis, Phys. Rev. B39, 415 (1989).
125. Villain, J., Surf. Sci. 97, 219 (1980).
126. den Nijs, M., in Phase Transitions and Critical Phenomena, Vol. 12, editors C. Domb and J. L. Lebowitz (Academic Press, Londong, 1988).
127. V. L. Pokrovsky and A. L. Talapov, Phys. Rev. Lett. 42, 65 (1979).
128. Fisher, M. E. and D. S. Fisher, Phys. Rev. B25, 3192 (1982).
129. Coppersmith, S. N., D. S. Fisher, B. I. Halperin, P. A. Lee and W. F. Brinkman, Phys. Rev. B25, 349 (1982).
130. Huse, D. A. and M. E. Fisher, Phys. Rev. B29, 239 (1984).
131. Erbil, A., A. R. Kortan, R. J. Birgeneau, M. S. Dresselhaus, Phys. Rev. B28, 6329 (1983).
132. Mochrie, S. G. J., A. R. Kortan, R. J. Birgeneau, and P. M. Horn, Z. Phys. B62, 79 (1985).
133. Feng, Y. P. and M. H. W. Chan, Phys. Rev. Lett. 64, 2148 (1990).
134. Shrimpton, N. D. and B. Joos, Phys. Rev. B41, 4764 (1990).
135. Alexander, S., Phys. Lett. A54, 353 (1975).
136. Ecke, R. E., Q. S. Shu, T. S. Sullivan and O. E. Vilches, Phys. Rev. B31, 448 (1985).
137. Huse, D. A. , Phys. Rev. B29, 5031 (1984).
138. Butler, D. M., Litzinger, J. A., G. A. Stewart, and R. B. Griffiths, Phys. Rev. Lett. 42, 1289 (1979).
139. Specht, E. D., A. Mak, C. Peters, M. Sutton, R. J. Birgeneau, K. L. D'Amico, D. E. Moncton, S. E. Nagler, and P. M. Horn, A. Phys. B-Condensed Matter 69, 347 (1987).
140. Rabedeau, T. A., Phys. Rev. B39, 9643 (1989).
141. Shaw, C. G., S. C. Fain, M. D. Chinn, Phys. Rev. Lett. 41, 955 (1978).
142. D'Amico, K. L., J. Bohr, D. E. Moncton, and D. Gibbs, Phys. Rev. B41, 4368 (1990).
143. Young, A. P., Phys. Rev. B19, 1855 (1979).
144. Nielsen, M., J. Als-Nielsen, J. Bohr, J. P. McTague, D. E. Moncton, and P. W. Stephens, Phys. Rev. B35, 1419 (1987).
145. J. P. McTague, J. Als-Nielsen, J. Bohr, and M. Nielsen, Phys. Rev. B25, 7765 (1982).

146. Migone, A. D., Z. R. Li, and M. H. W. Chan, Phys. Rev. Lett. 53, 810 (1984).
147. Larher, Y., Surf. Sci. 134, 469 (1983).
148. de Souza, J. L. M. Demetrio and E. Lerner, J. Low Temp. Phys. 66, 367 (1987).
149. Hering, S. V., O. E. Vilches, J. Low Temp. Phys. 25, 793 (1976).
150. Motteler, F. C., 1985 Ph.D. thesis (Univ. of Washington).
151. Bretz, M., Phys. Rev. Lett. 38, 501 (1977).
152. Tejwani, M. J., O. Ferreira, and O. E. Vilches, Phys. Rev. Lett. 44, 152 (1980).
153. Ecke, R. E. and J. G. Dash, Phys. Rev. B28, 3738 (1983).
154. Crary, S. B. and D. A. Fahey, Phys. Rev. B35, 2101 (1987).
155. Feile, R., H. Wiechert, and H.J. Lauter, Phys. Rev. B25, 3410 (1982).
156. Widom, A., J. R. Owers-Bradley, and M. G. Richards, Phys. Rev. Lett. 43, 1340 (1979).
157. Hurlbut, S. B. and J. G. Dash, Phys. Rev. Lett. 53, 1931 (1984).
158. Chinn, M. D. and S. C. Fain, Phys. Rev. Lett. 39, 146 (1977).
159. Moncton, D. E., P. W. Stephens, R. J. Birgeneau, P. M. Horn, G. S. Brown, Phys. Rev. Lett. 46, 1533 (1981).
160. Fain, S. C., M. D. Chinn, R. D. Diehl, Phys. Rev. B21, 4170 (1980).
161. Schabes-Retchkiman, P. S. and J. A. Venables, Surf. Sci. 105, 536 (1981).
162. Chung, S., A. Kara, J. Z. Larese, W. Y. Leung and D. R. Frankl, Phys. Rev. B35, 4870 (1987).
163. D'Amico, K. L., D. E. Moncton, E. D. Specht, R. J. Birgeneau, S. E. Nagler, P. M. Horn, Phys. Rev. Lett. 53, 2250 (1984).
164. Larher, Y. and A. Terlain, J. Chem. Phys. 72, 1052 (1980).
165. Suter, R. M., N. J. Colella, and R. Gangwar, Phys. Rev. B31, 627 (1985).
166. Butler, D. M., J. A. Litzinger, G. A. Stewart, Phys. Rev. Lett. 44, 466 (1980).
167. Huff, G. B. and J. G. Dash, J. Low Temp. Phys. 24, 155 (1976).
168. Rapp, R. E., E. P. deSouza and E. Lerner, Phys. Rev. B24, 2196 (1981).
169. Hanono, F., C. E. N. Gatts, and E. Lerner, J. Low Temp. Phys. 60, 73 (1985).
170. Calisti, S., J. Suzanne and J. A. Venables, Surf. Sci. 115, 455 (1982).
171. Tiby, C., H. Wiechert, and H. J. Lauter, Surf. Sci. 119, 21 (1982).
172. Litzinger, J. A. and G. A. Stewart, in *Ordering in Two Dimensions*, edited by S. K. Sinha, 267 (North-Holland, New York, 1980).
173. Heiney, P. A., P. W. Stephens, R. J. Birgeneau, P. M. Horn, and D. E. Moncton, Phys. Rev. B28, 6414 (1983).
174. Gangwar, R., N. J. Colella, and R. M. Suter, Phys. Rev. B39, 2459 (1989).

175. Jin, A. J., M. R. Bjurstrom, and M. H. W. Chan, Phys. Rev. Lett. 62, 1372 (1989).
176. Faisal, A. Q. D., M. Hamichi, G. Raynard, and J. A. Venables, Phys. Rev. B34, 7440 (1986).
177. Rosenbaum, T. F., S. E. Nagler, P. M. Horn, and R. Clarke, Phys. Rev. Lett. 50, 1791 (1983).
178. Dimon, P., P. M. Horn, M. Sutton, R. J. Birgeneau, and D. E. Moncton, Phys. Rev. B31, 437 (1984).
179. Nagler, S. E., P. M. Horn, S. K. Sinha, and D. E. Moncton, Bull. Am. Phys. Soc. 30, 334 (1985).
180. Specht, E. D., R. J. Birgeneau, K. L. D'Amico, D. E. Moncton, S. E. Nagler, and P. M. Horn, J. Phys. (Paris) Lett. 45, L561 (1985).
181. Colella, N. J. and R. M. Suter, Phys. Rev. B34, 2052 (1986).
182. Suh, S.-H., N. Lermer, and S. F. O'Shea, Chem. Phys. 129, 273 (1989).

MONOLAYER ADSORPTION OF Ar, Kr, Xe AND CH_4 ON LAYERED HALIDES

YVES LARHER
SCM-DRECAM-DSM-CEA, Centre d'Etudes de Saclay,
91191 Gif-sur-Yvette, Cedex France

ABSTRACT. Volumetric measurements of adsorption isotherms over a certain range of temperatures, as well as a few structural studies, have been carried out for the monolayers of Ar, Kr, Xe and CH_4 adsorbed upon the basal face of numerous layered dihalides. 2D phase transitions are observed, which, in addition to being of intrinsic interest, are particularly helpful in the molecular interpretation of the thermodynamic results. The structure and the thermodynamic properties of the monolayers turn out to depend in a radical way upon the value of the so-called size incompatibility, i.e. the relative difference between the period of the surface corrugation and the diameter of the adspecies. This shows itself in the remarkable correlation between this parameter and a number of thermodynamic quantities, mostly critical temperature of 2D condensation and entropy. We claim that, within the range in which the size incompatibility can be varied, from about - 20 to + 20 percent, almost all the structures a physisorbed monolayer of atoms can assume are encountered, including the two simplest : corrugation-independent and lattice gas type monolayer. These two cases are discussed in some detail, and suggestions are made for further theoretical studies which should lead to a better understanding of monolayer adsorption.

1. Historical

Maybe, no event ever prompted so great a renewal in the field of physisorption as the discovery of phase transitions within adfilms did. This happened around the middle of the century. The early indication appeared in adsorption isotherms [1-3]. Figure 1a shows

G. Benedek (ed.), Surface Properties of Layered Structures, 261–315.

results obtained by Bonnetain, Duval and Letort [3] in 1952. Next to it, in Figure 1b [4], is another isotherm determined about 25 years later which demonstrates beyond doubt two essential characteristics, already noticed by Bonnetain *et al.*, for these so-called stepwise isotherms : i) the layer by layer growth of the film, strongly suggested by steps of almost equal height ; ii) the existence of first order layering transitions, manifested by the discontinuous variation of the adsorbed amount versus pressure. Admittedly, the smoothly shaped steps of Figure 1a did not demonstrate the existence of phase transitions, and yet Bonnetain *et al.* wrote : "Ces nouvelles données devraient être importantes pour l'analyse théorique du phénomène d'adsorption et, particulièrement, des changements de phase". In the following three years, adsorption isotherms were published [5-8] showing almost "vertical" steps, which, today anybody would unhesitatingly consider as evidence for first order layering transitions, the slight deviation from verticality resulting from imperfections of the substrate or size effects.

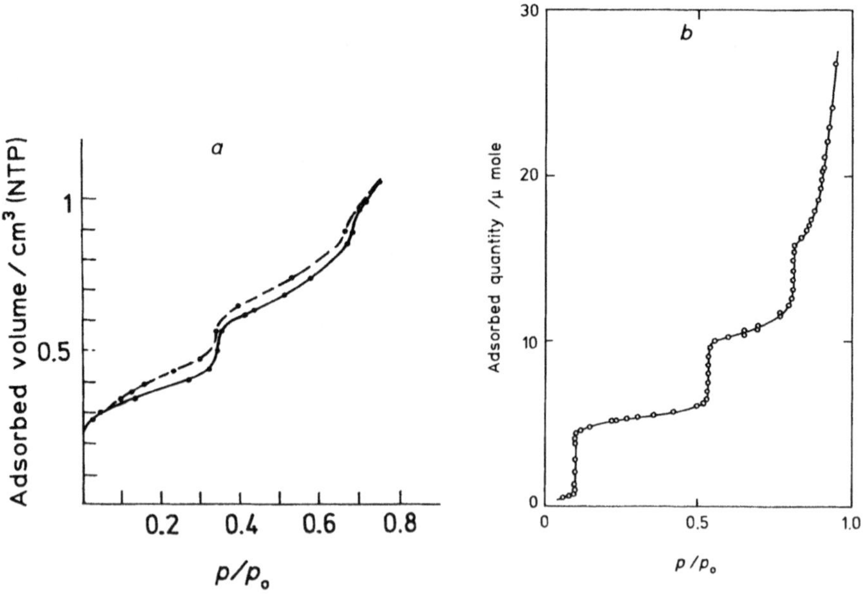

Fig. 1.(a) One of the first stepwise adsorption isotherms, showing the 2nd and 3rd layering transitions of CH_4 on $MoS_2(0001)$, published in 1952 [3]. (b) An isotherm of CF_4 on $PbI_2(0001)$ [4], typical of what could already be found in the literature a few years later.

In all the studies we have mentioned so far, graphite was used as a substrate. However, Bonnetain *et al.* [3] also observed stepwise isotherms by using other substrates, vz MoS_2, $CdCl_2$, CdI_2 and $HgBr_2$, which share with graphite the property of having a layered structure, with a marked trend to forming crystals with a platelet shape. From that moment it became obvious that the basal face of these platelets, which represent the prominent part of their external surface, was responsible for the appearance of phase transitions in the adfilm. In Section 3, in which we will discuss the structure of the layered dihalides, we will thoroughly argue this view. Incidentally, it could be pointed out that by choosing such substrates, a few physical chemists were initiating clean surface studies, in a clever way, with no use of ultra-high vacuum techniques, which were not yet common at that time.

An important step in the expansion of such studies occurred at the end of the fifties, when systematic thermodynamic investigations were carried out, i.e. when adsorption isotherms were measured at more than one or two temperatures, as it was usual earlier. The first significant contribution, due to Fisher and McMillan [9], was a set of seven adsorption isotherms, measured from 66.85 to 80.93 K and showing the condensation of a monolayer of Kr on NaBr(100). The cryostat used by these authors was particularly simple, and we could easily satisfy our inclination for thermodynamics by building a similar one, at Saclay, to investigate adsorption on lamellar halides in the early sixties. Shortly thereafter, Thomy and Duval, at Nancy, devised a cryostat of a different type to investigate adsorption on graphite. The working-out of uniform substrates during the sixties also favored progress in that field (see Section 3).

Later, adsorption studies either on graphite or on lamellar halides did not develop likewise. Following the heat capacity measurements of Dash and coworkers [10], in the early seventies, on the He-graphite system, there was a burst of investigations of films of various gases adsorbed onto graphite, using a variety of experimental techniques. On the other hand, few adsorption studies were done on layered halides in addition to our volumetric measurements of adsorption isotherms. This obviously contributes to reduce their interest. However, the layered halides have a unique advantadge over graphite : they form a family of substrates expected a priori to be not very different from each other from the point of view of physical adsorption, and yet it turns out that strikingly different behaviours are observed depending on the size incompatibility between the substrate and the adsorbate. We will give later a quantitative definition of this notion. For the moment being, we just point out the existence of remarkable correlations between a number of thermodynamic properties of the adfilm and the size incompatibility which permit to sketch a first qualitative picture of the structure of the adfilm.

We would like to show from a selection of specific results how the adsorption work on layered halides has contributed, jointly with similar studies on graphite (see previous Chapter), to the understanding of subjects as diverse as two-dimensional phase transitions, wetting, roughening and finally modified in a radical way the approach to the molecular theory of physisorption, mainly by bringing more realism into a field in which existing theories such as Langmuir, BET, etc... had strongly undervalued the complexity of the physical world.

The first part of our contribution to the subject is devoted to the monolayer adsorption of classical rare gases, Ar, Kr and Xe, as well as methane. We would like to complete it later by a second part dealing with multilayer adsorption of the same gases.

Ar, Kr and Xe are the simplest adspecies which can be envisaged. But, in the temperature range we have been working in, from about 30 to 55 percent of the critical temperature of the bulk adsorbates, the quasi-spherical methane molecule is rotating almost freely, so that the properties of its adlayers are expected not to differ much from those of Ar, Kr and Xe, and their discussion does not appear at all misplaced here. But in fact, the main reason for including methane in this chapter is that it is the single adspecies for which we carried out a structural investigation of the 2D melting transition.

2. Thermodynamic analysis of stepwise adsorption isotherms.

As stated in the introduction, most of the experimental results to be presented in this chapter are adsorption isotherms, determined at various temperatures. The object of this Section is mainly to sketch the methods used to analyse the information that can be derived from them, especially when phase transitions are observed in the adfilm. Our presentation will be in keeping with two comparatively recent papers [11,12] devoted to surface thermodynamics.

Let us briefly recall what an adsorption isotherm is. In the volumetric method, we are currently using for its determination, an initial amount of gas is put into contact with the substrate. The adsorbed mole number N is the difference between this amount and that remaining in the bulk gas phase at pressure p, once equilibrium has been attained. The representation of N as a function of p at a constant temperature T is called an adsorption isotherm. If one works at various temperatures, one gets a set of adsorption isotherms, or equivalently a tabulation of the function $N = N(T,p,A)$, in which A is the surface area of the substrate.

This function turns out to be an equation of state of the adfilm, i.e. a first partial derivative of one of its fundamental equations. If N is expressed as a function of the chemical potential, μ, instead of p, it can be shown indeed that :

$$N(T, \mu, A) = (\partial \Omega / \partial \mu)_{T,A} , \tag{1}$$

Ω being the grand potential of the adfilm [12], derived by the appropriate Legendre transformation from its primitive fundamental equation :

$$U = U(S,A,N) . \tag{2}$$

The derivation of the useful thermodynamic formulae is straightforward, once the formal similarity between the fundamental equation of an adfilm and of a bulk phase has been noticed. Indeed one goes from one to the other by simply interchanging the geometrical variables A (area) and V (volume). We will see later that, when the adfilm is a monolayer, the similarity goes even further. For the adfilm, the variable equivalent to the bulk pressure, p, is the so-called spreading pressure,

$$\phi = - (\partial U / \partial A)_{S,N} . \tag{3}$$

The first tool which will prove useful is the Gibbs phase rule. If the gaseous adsorbate is a mixture of c components, then we will have $c + 3$ intensive parameters, vz $\mu_1, ..., \mu_c$ T, p, ϕ. They will be related by $1 + \psi$ Gibbs-Duhem equations in case of one single gaseous phase and ψ coexisting 2D phases, so that the variance, i.e. the number of independent intensive parameters, will be :

$$\upsilon = c + 2 - \psi \tag{4}$$

Only systems involving pure adsorbates will be discussed here, in which case the variance will be $\upsilon = 3 - \psi$. If a single phase exists on the surface, i.e. $\psi = 1$, then $\upsilon = 2$, which means that once the temperature is fixed, the pressure can still be varied. This is what is done when measuring continuous parts of the adsorption isotherms. On the other hand, if two surface phases coexist, i.e. $\psi = 2$, then $\upsilon = 1$, and the pressure becomes a function of temperature. Coverage can be changed at constant T and p, resulting in discontinuous steps on the adsorption isotherms. Conversely these steps give evidence for 2D phase coexistence or equivalently first order phase transitions.

To each surface phase is associated a Gibbs-Duhem equation which writes, in the entropy scheme of thermostatics :

$$d(\phi/T) = -\Gamma u\, d(1/T) + \Gamma d(\mu/T). \tag{5}$$

u is the molar energy and Γ the surface density. For two coexisting phases, denoted as ' and ", a Clausius-Clapeyron equation results in :

$$d(\mu/T) / d(1/T) = (\Gamma''u'' - \Gamma'u') / (\Gamma'' - \Gamma'). \tag{6}$$

For all the systems we will be studying, the 3D gas phase could be considered as perfect, so that its Gibbs-Duhem equation writes :

$$\mathrm{d}(\mu/T) = h_g\, \mathrm{d}(1/T) + R\mathrm{d}(\ln p), \tag{7}$$

h_g being the molar enthalpy of the gas. Comparing Equations (6) and (7), one gets :

$$R\mathrm{d}(\ln p)/\mathrm{d}(1/T) = -\{h_g - (\Gamma''u'' - \Gamma'u') / (\Gamma'' - \Gamma')\}. \tag{8}$$

From relations such as

$$S = U/T + (\phi/T) A - (\mu/T)N \tag{9}$$

expressing the homogeneity of the fundamental equation of the phases, and from the particular expression μ/T for a perfect gas,

$$\mu/T = h_g/T - s_{g0} + R\ln p, \tag{10}$$

in which s_{g0} is the molar entropy at unit pressure, the integral counterpart to equation (8) is easily obtained :

$$R\ln p = -\{h_g - (\Gamma''u'' - \Gamma'u')/(\Gamma'' - \Gamma')\}/T + s_{g0} - (\Gamma''s'' - \Gamma's')/(\Gamma'' - \Gamma') \quad (11)$$

Note in passing that it would be more satisfying to use the variable μ/T rather than $\ln p$ in the presentation of our experimental results, because contrarily to the former the latter is not an intensive variable of the adfilm. However, we find more convenient to keep the second, because it is the experimentally measured quantity. Additionally, it can be noticed that the qualitative aspect of a phase diagram of the adfilm is independent upon the use of either variable, since, according to equation (10), the difference between both involves a T dependent expression of the bulk gas phase which varies smoothly.

In the practically important case where the 2D transition is a layering transition it proves convenient to define the two quantities u_n and s_n by :

$$u_n = (\Gamma''u'' - \Gamma'u')/(\Gamma'' - \Gamma'), \quad (12)$$

$$s_n = (\Gamma''s'' - \Gamma's')/(\Gamma'' - \Gamma'), \quad (13)$$

so that equation (8) and (11) write :

$$R\mathrm{d}(\ln p_n)/\mathrm{d}(1/T) = -(h_g - u_n), \quad (14)$$

$$R\ln p_n = -(h_g - u_n)/T + (s_{g0} - s_n). \quad (15)$$

The reasons for introducing this notation is that under certain conditions u_n and s_n can be considered as good estimates for the energy and entropy of the nth dense layer in equilibrium with its 2D vapor. For the first layer this is almost obvious : if the temperature is well below the layering critical temperature, Γ', $\Gamma'u'$ and $\Gamma's'$ will become negligible in comparison with Γ'', $\Gamma''u''$ and $\Gamma''s''$. For $n \neq 1$, it is necessary in addition that no change occurs in the (n-1) already adsorbed layers as the nth layering transition occurs. We will discuss these approximations in due time.

The vapor pressure of the condensed phase of the bulk absorbate is governed by equations analogous to (14) and (15), namely :

$$R\mathrm{d}(\ln p_0)/\mathrm{d}(1/T) = -(h_g - u_0), \quad (16)$$

$$R \ln p_0 = -(h_g - u_0)/T + s_{g0} - s_0 . \tag{17}$$

We will use them to determine the energy u_0 and the entropy s_0 of this 3D phase to which we will compare u_n and s_n

We have set out equations which are valid only for the monovariant regions of the phase diagram. No other is needed since, for reasons to be presented in the next section, we will not analyse our results in bivariant regions of the phase diagram.

3. Layered dihalides as substrates for adsorption

3.1 CRYSTALLINE STRUCTURE

The trend of layered halides to crystallise under the form of platelets originates from their strong structural anisotropy. Like graphite, these halides can be considered as the stacking-up of 2D crystals attracting themselves through van der Waals forces only. For graphite the 2D crystals are covalently bonded graphitic planes, while for the layered halides they are sandwiches of two planes of anions surrounding one plane of cations. Figure 2 shows a section of the crystal perpendicular to the direction of the planes of the sandwiches. The anions are in a quasi-close packing, either of hexagonal or of face-centered-cubic type, while the cations occupy one half of the octahedral sites. Such structures belong respectively either to the hexagonal or to the rhombohedral system. Both types can be represented with a hexagonal cell, which proves convenient for us since the parameter of the basal plane, a, is sufficient to describe the close-packed plane of anions, which is the surface of greatest interest in our adsorption studies.

The van der Waals bonding between the planar units of the structure is important in two respects. Firstly, the basal face of anions limiting a crystal will have a very low surface free-energy in comparison with other faces, so that the equilibrium shape should show a large extension of such faces. Arguments, involving growth kinetics would lead to the same conclusion [13]. Secondly, no vacant chemical bond exists on such a face, so that the risk for chemical contamination is considerably reduced. Consequently, the layered halides, as well as graphite, offer an exceptional opportunity for obtaining substrates for which a given crystal face is largely prevailing and could be easily prepared chemically clean and also free from defects. Such adsorbents are often said to be uniform.

Fig. 2. The layered dihalides can be considered as a packing of 2D crystals MX_2 (two planes of halide anions X^- surrounding a plane of cations M^{++}) attracting themselves through van der Waals forces. The microcrystals of our powder substrates will be mainly limited by planes of X^- in a triangular arrangment of parameter *a*. We will be interested exclusively in the adsorption occuring on such crystal faces.

Additionally, such surfaces are low index faces, so that they are atomically the smoothest. Consequently, the hindrance to lateral interaction between adatoms on such faces is minimized and the best chance exists for phase transitions to occur within an adlayer. For graphite this property was discovered accidentally by physical chemists [1-3] who were measuring adsorption isotherms in order to determine the surface area of their substrate. However, in their choice of the layered halides as substrates for adsorption studies, Bonnetain, Duval and Letort [3] were already guided by the structural analogy with graphite.

3.2 UNIFORM SUBLIMED SAMPLES.

Let us take a commercial sample of a layered halide, dehydrate it if necessary, then grind it and heat it under vacuum at a temperature somewhat below two third of the melting temperature so that no appreciable sintering occurs. If we now measure an adsorption isotherm on such a substrate a stepwise adsorption is obtained. An example is shown in Figure 3a representing the first layering transition for Kr on $CdBr_2$ at 87.74 K and 89.91

K. Next to it, in Figure 3b, is another isotherm at a neighbouring temperature, vz 86.99 K, of the same gas, again on $CdBr_2$ but on a sample prepared in a different and slightly more involved way, namely by sublimation in a rapid current of inert gas. Before discussing the striking difference between these two isotherms let us first shortly describe the sublimation technique [13].

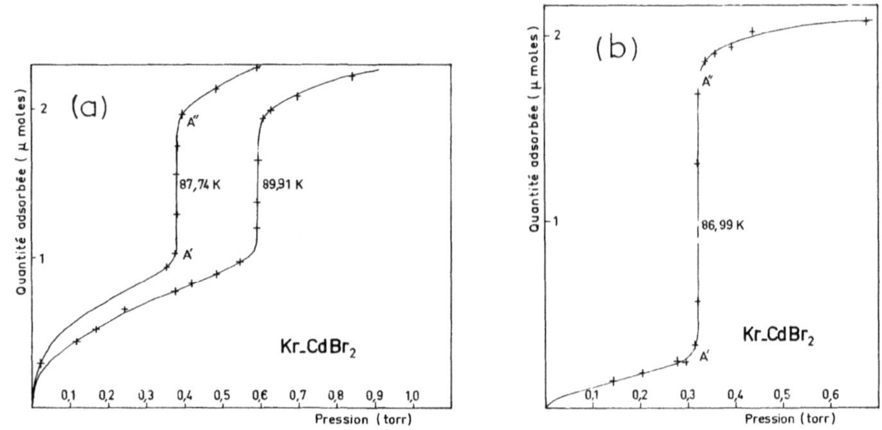

Fig. 3. Adsorption isotherms of Kr on $CdBr_2$ [13] : a) ground sample ; b) sublimed sample.

It was first used by Craig and Mc Intosh [14], then by Young and Morrison [15] to prepare powders of sodium chloride having large specific surface areas, but neither of these works gave any evidence that the adsorbents obtained were uniform. In both experiments the smokes of sodium chloride were recovered by electrostatic precipitation. We found it easier to proceed by filtration as van Dongen [16] and Ross *et al.*[17] also did. In our case the filter was a nylon cloth. The sublimation is carried out at a temperature at which the vapor pressure of the halide is about 0.1 to 1 Torr. The flow rate of the inert gas, which has always been nitrogen, is about 1 l min^{-1}.

Let us come back to Figure 3. The vertical part of the step, corresponding to adsorption on the basal face, is much more extended in Figure 3b, which means that the sublimed sample has a larger degree of uniformity, i.e. the microcrystals are bounded by a larger fraction of basal faces. The convexity towards the ordinate axis, occurring at the origin of the isotherms, corresponds to adsorption upon ill-defined surfaces. For the sample prepared without any particular care, they represent about one half of the total surface area. For the sublimed sample they are much less important but not necessarily negligible. If they were, the isotherm should obey the Henry law at the origin. We spent a long time trying to observe such a behavior by a careful preparation of $NiCl_2$ samples, but were unsuccessful. As a consequence we decided not to try to obtain quantities such as energies and entropies from the bivariant regions of our phase diagrams, since we did not know to what extent they could be spoiled by a contribution from amounts of vapor adsorbed upon the ill-defined parts of the surface. An additional reason prompted us to do so : the equations which should be used involve a derivation at constant coverage and we were not sure to have a sufficient accuracy in the measurement of adsorbed amounts. An excellent accuracy would be necessary since, outside the vertical parts of the steps, the slope of the isotherms is very small. Finally, we must also mention capillary condensation which can become important as saturation is approached, and which makes even more risky the analysis of the isotherms in the bivariant regions of the phase diagram.

We feel it worth emphasizing now a practically important point. The first order transitions we observe in our adsorption isotherms are by themselves a most interesting object of investigation, but they also constitute a remarkably simple and efficient mean for obtaining quantities concerning only the uniform part of the substrate surface . They will be the more interesting, the more perfect this uniform part is.

So far, we discussed the relative extension of the uniform part of the surface of a substrate. From the deviation to verticality of the steps of the isotherms we can also assess its degree of perfection. We will dwell on this question when discussing the problem of the determination of layering critical temperatures. For the moment suffice it to say that except in the vicinity of critical points it can be considered as satisfying in most cases.

From our discussion it appears that it is not absolutely necessary to obtain substrates in which the non-uniform parts of the surface have been reduced to a negligible amount. Nevertheless this is desirable, not only because the isotherms on such substrates are more beautiful, but also for a number of other more pragmatic reasons. We would like to mention capillary condensation which increases steeply close to saturation : clearly there is a better chance to detect a layering transition if it involves an important density change. So far we discussed only measurements of adsorption isotherms. Of course, when other experimental techniques are used, such as for instance X-ray or neutron diffraction, which are less efficient in isolating the first order transitions, it is even more important that the substrate be limited mostly by its basal face.

To end this subsection we wish to quote another interesting property of the sublimed samples : they have a specific surface area varying from about 10 to 50 m^2g^{-1} which is two orders of magnitude higher than those of adsorbents obtained by grinding, and sufficient to make possible X-ray or neutron diffraction investigations. In such studies it is important to have a maximum surface to volume ratio, since the scattering intensity contains contributions from both the substrate and the adsorbed film.

3.3 SURFACE CORRUGATION

The efficiency of statistical mechanics in calculating thermodynamic properties of condensed matter from interaction potentials between the molecules has considerably increased during the last two or three decades. The physical chemistry of physisorbed films has benefited from such progress (see, in particular, the previous chapter). Unfortunately, the knowledge of potential parameters is often still too sketchy to permit detailed realistic calculations to be carried out on specific experimental systems. This is particularly true for adsorption upon layered halides, since we know little about the interaction of even the rare gases atoms with such substrates. However, in view of a qualitative interpretation of our results, it still remains useful to try to sketch a rough picture of an important aspect of the interaction potential of the adatoms with the substrate, namely its modulation by the surface structure, generally called surface corrugation.

The potential energy of an adatom, $V(\mathbf{r})$, depends upon its three coordinates : x, y (parallel to the surface), and z (perpendicular to it). V achieves a minimum value V_m at a distance z_m from the surface. We will describe the corrugation by the values of z_m and V_m at the point A, S and SP defined by their (x,y) coordinates in the unit mesh of the surface whose origin is taken above the center of an anion (Figure 4) : A (0,0) ; S (2/3, 1/3) and (1/3, 2/3) ; SP (0, 1/2) and (1/2, 0). We use the same terminology as Steele in his treatise on "The Interaction of Gases with Solid Surfaces" [18] : A, S and SP are respectively the initials of atom, site and saddle point.

Clearly, above an anion A V_m has a maximum, while at a site S it has a minimum. Consequently the attraction by the substrate will tend to localize the adatoms at sites S. But this tendancy could be counterbalanced by the attraction between adatoms which will favor a structure of the adlayer which can be quite distinct from that of the lattice of sites. The efficiency of the substrates to influence the structure of the monolayer will depend upon the difference (V_m (A) - V_m (S)) relative to the adatom - adatom lateral cohesion energy.

As a rough estimate of the latter we will consider the quantity $[V_m(A) - V_m(S)]/3\varepsilon$, in which ε is the minimum energy of the (6,12) Lennard-Jones two-body potential we will use to describe the adatom-adatom interaction :

$$u(r) = 4\,\varepsilon\,[\,(\sigma/r)^{12} - (\sigma/r)^{6}\,] \qquad (18)$$

Since we will later compare the behavior of the various rare gases, it is important that a coherent set of values is taken for ε and σ. This is ensured by relating them to the critical constants as suggested by Rowlinson [19] : $kT_{3c}=1.25\varepsilon$, $\upsilon_{3c} = 3.14\,N_A\sigma^3$, $p_c = 0.116\,(\varepsilon/\sigma^3)$. The values of the critical constants will be taken in ref. [20] throughout this paper.

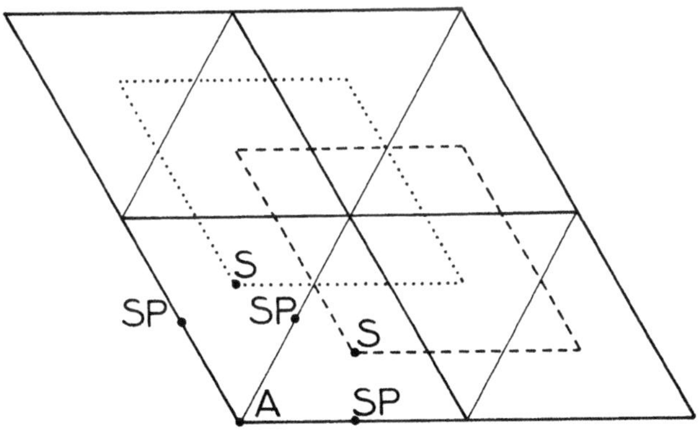

Fig. 4. The lattice of the halide ions is represented in full lines ; the two non-equivalent sublattices of sites are represented in dotted and dashed lines.

The saddle points SP are of interest in that the difference $(V_m(SP) - V_m(S))$ yields the minimum potential barrier an adatom must overcome to go from a site to a neighbouring one and consequently an information about the mobility of an adatom.

The quantities defined so far are related to the amplitude of the modulation of the potential energy. As will appear later, an even more important quantity is the period of this modulation. Indeed the surface has two triangular sublattices of sites, geometrically identical, except for a translation, to the lattice of anions of crystalline parameter *a*.. The densest plane of a rare gas bulk crystal, the (111) plane, has also the same structure, but a different lattice parameter, which we call *d*. Clearly the smaller the relative difference between a and d, the more probable the existence a simple 1x1 registered adlayer. We define the size incompatibility as their relative difference : $i = (a - d) / d$. The outstanding importance of this factor will appear later. With the family of layered halides, where a varies from 3.48 Å for $NiCl_2$ to 4.56 Å for PbI_2, and the three classical rare gases, Ar, Kr, and Xe with d going from about 3.80 Å to 4.40 Å, a range of incompatibilities from about - 0.20 to + 0.20 can be investigated. As in all our previous publications we take for d its value at the average temperature at which the adsorption experiments have been carried out. We derive it from Figure 1 to 3 in ref. [21]. Most often we have values of a at room temperatures only [22]. This is not a problem since this parameter varies little with temperature : for instance for $CoCl_2$ it varies by less than 0.2 percent from room temperature down to 4.2 K [23].

We have made semi-empirical estimates of the parameters V_m and z_m at A, S and SP for a number of systems chosen so as to grossly cover the range of incompatibilities of interest to us. The details of the calculations are given in an appendix and the results in Table I. Note that in two cases the calculations have been carried out for a rare gas atom

TABLE I

Surface corrugation of layered halides.

System	i	Height/Å			Energy/K			$\frac{(V_m(A)-V_m(S))}{3\varepsilon}$
		S	SP	A	S	SP-S	A-S	
Ar-Xe (111)	0.15	3.25	3.42	4.02	- 789	75	291	0.80
Ar-Ar (111)	0	3.04	3.17	3.64	- 607	49	202	0.55
Ar-$CdCl_2$	0	2.95	3.08	3.59	- 883	76	310	0.85
Xe-$CdCl_2$	- 0.13	3.30	3.41	3.84	- 1369	88	395	0.57
Xe-$FeCl_2$	- 0.18	3.38	3.45	3.81	- 1369	66	328	0.47
Xe-$NiCl_2$	- 0.21	3.41	3.47	3.79	- 1405	58	305	0.44
Kr-Graphite [a]	0.046				- 1373	38	41	0.08
" [b]	0.046				- 1450		54.5	0.11

[a] Réf. [18] ; [b] Previous chapter.

adsorbed upon the (111) face of a rare gas crystal. Since the van der Waals interaction parameters of the halide ions and of rare gas atoms are comparable, such models should mimic well our experimental systems. They present the advantage that the interaction parameters are better known. Good agreement of our results with those of Steele [18] for Ar-Xe (111) served as a test for our calculations.

We mentioned earlier the existence of two sublattices of sites S. Even for rare gases they are not equivalent. However Steele [24] noticed that lattice sums show they are "essentially identical". This view could also be deduced from the well-known fact that for rare gases the relative potential energy difference between hexagonal and cubic-face-centered close-packing is extremely small, of the order of 0.01 percent [25]. As a consequence the two sublattices of sites S of the basal face of the layered halides should be considered as distinct because the metal ions can be found only below one of them. For this sublattice the adsorption potential will be more favorable. Even if it is only by a few percent this is sufficient to destroy the honeycomb symmetry and consequently the critical behaviour associated with this type of symmetry. In an earlier article [26] we did not realize this in a clear way.

4. Corrugation-independent adsorption

As we already stated, it is mainly from sets of adsorption isotherms that we attempt to outline a qualitative molecular picture of the first monolayer. Since a variety of behaviours are observed, depending upon the size incompability i between adsorbate and substrate, it seems logical to describe in turn the various aspects of the monolayer, beginning from those appearing to us as the simplest. However, for certain physical quantities it will prove convenient to present at one time the experimental values over the whole range of i, although the detailed discussion will occur, in the following sections, over limited ranges of i. This is, for instance the case, for the critical temperature of the monolayer condensation which will be first discussed.

4.1. CRITICAL TEMPERATURE OF 2D CONDENSATION

The variation with size incompatibility of the critical temperature gives a first striking evidence for the influence of the surface corrugation upon the properties of monolayers. Its determination from adsorption isotherms raises difficulties which require an examination of the problem of the critical exponents associated with 2D condensation. This will be done in the next Subsection. In the following we will present the experimental values of the critical temperatures and start their discussion.

4.1.1. Critical Exponents Associated with 2D Condensations.

The first noteworthy attempt to determine a critical exponent from adsorption isotherms was carried out in 1979 [27] for the second layer of Ar adsorbed on cadmium chloride. The second layer rather than the first was chosen mainly because, around its critical temperature, the layering transition occurred in a range of pressures one order of magnitude higher, about 40 Torr, and consequently measurable with a much better accuracy than for the first. As far as the properties of universality were concerned it could be conjectured that no significant difference with respect to a first layer was to be expected.

The adsorption isotherms are shown in Figure 5 and the inverse slope of the steps versus temperature in Figure 6. Both Figures suggest a critical temperature around 72.5 K. The steps are not vertical below this temperature, but the change in pressure from bottom to top remains low, about 1.2 %, corresponding to a variation in chemical potential $\Delta\mu$ of only 7 J mol^{-1} for a process involving an energy change of 8 kJ mol^{-1}, which means a very good energetic homonegenity of the substrate as well as unimportant size effects.Actually $\Delta\mu$ could be accounted for by a range of sizes of the surface elements from 600 Å to infinity [27].

However, as we already noticed in 1973 [28], close to the critical temperature it might prove worrying, particularly if the critical exponents take the values of the 2D Ising model as it could be conjectured. This is indeed supported by the estimate of β obtained from our study of the layering transition of Ar on $CdCl_2$, namely 0.16 ± 0.05, to be compared to the Ising value 0.125. We remind the reader that β is the exponent associated with the curve limiting the coexistence region: $\Delta\Gamma \propto (T_{2c}-T)^{\beta}$.

3D condensations are well-known to belong to the universality class of the 3D Ising model, so that our result was by no means unexpected. A subsequent study of the methane monolayer [29] adsorbed on graphite, claiming to establish this result for the first time, was actually raising at least one difficulty,as it will be shown in Section 4.1.2. The exponent δ, governing the variation with temperature of the inverse slope of the steps, should then be equal to 1.75, the 2D Ising value, so that the curve in Figure 6 should be

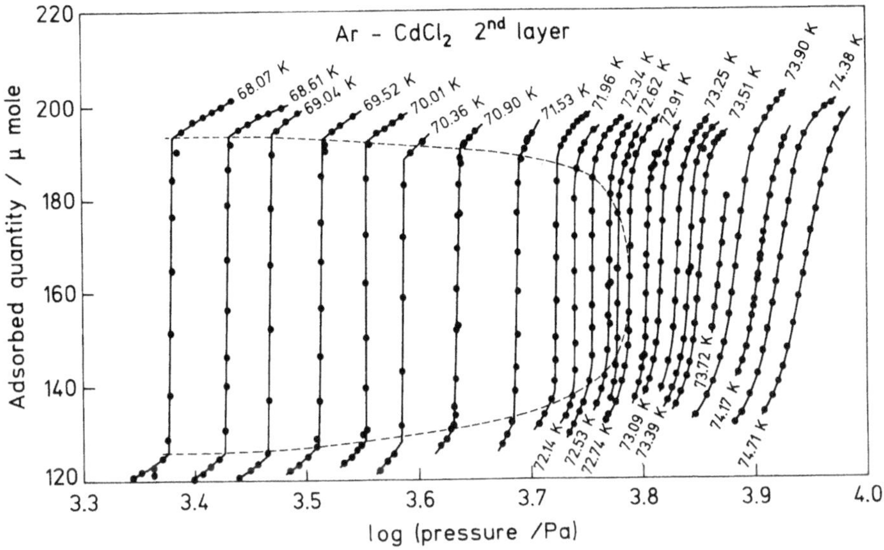

Fig. 5. Adsorption isotherms of Ar on the basal face of cadmium chloride at coverages corresponding to the formation of the second layer [27].

tangent to the T axis at T_{2c}. Since this is not the case, a rigorous treatment of the results implies a due account of the imperfections of the substrate. This appeared to us very difficult to carry out realistically, so that we proposed in 1974 [28] to take as an approximation for T_{2c} the abscissa of the intersection of the two straight lines drawn as in Figure 6. This yields a value T_{2c} = 72.8 K for the second layering transition of Ar on $CdCl_2$. A less crude method, consisting to minimize the mean-square deviation of the linear regression used to determine β, yielded T_{2c} = 72.6 K [27], very close to 72.8 K, which suggested that our crude procedure did not work too badly.

Attemps to take into account the imperfections of the substrate have been made recently by Nham and Hess [30] for the second layering transition of tetrafluoromethane on graphite. We recently repeated measurements on the same system to try to assess the error which could result from our way of treating the data [31] : in that case our estimate would be too high by at most 1.5 %. From this comparison we feel that our proposed values should not be significantly revised, particularly because quite large margins of errors had been already taken.

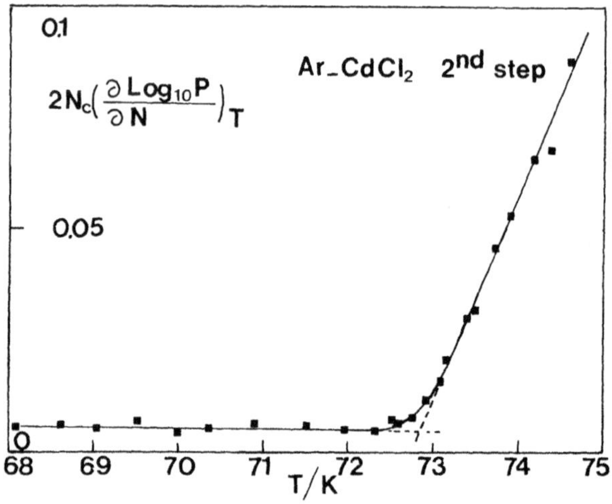

Fig. 6. Inverse slope of the adsorption isotherms at their inflexion point [27].

We must add that in our measurements there is a second cause of error. In our samples there certainly exist very small grains for which the critical temperature is decreased with respect to what it is for an infinite system. Such an effect, which has been theoretically studied by Au-Yang and Fisher [32], can induce a non-negligible error. This however is opposed in sign to the former one leading to some compensation.

4.1.2. Variation with Size Incompatibility of the Critical Temperatures of 2D Condensation.

Our experimental results [26] are given in Table II. The values of $\Delta\mu$ in the last column represent the variation, below T_{2c}, of the chemical potential from the bottom to the top of the steps. We notice that for almost one third of the systems this quantity is three times higher than for the others. This is not an indication that the substrate is less homogeneous. In fact, these systems were studied with a different cryostat, and the $\Delta\mu$ is due to a temperature gradient of about 0.2 K throughout the sample. Clearly it affects the uncertainty on the estimate of T_{2c} which is about twice higher for these systems, i.e. about 4 percent instead of 2 percent for the others.

TABLE II

Critical temperatures of 2D condensation for the first monolayers of Ar, Kr or Xe adsorbed on layered halides.

System	Size Incompatibility	T_{2c}/K	T_{2c}/T_{3c}[a]	$\Delta\mu$ /(cal mol^{-1})
Xe-$NiCl_2$	- 0.210	112	0.386	4
Xe-$CoCl_2$	- 0.196	114	0.393	6
Xe-$FeCl_2$	- 0.185	112	0.386	4
Xe-$CoBr_2$	- 0.166	115	0.397	5
Kr-$NiCl_2$	- 0.144	81	0.387	5
Xe-$CdCl_2$	- 0.128	113	0.390	5
Ar-$NiCl_2$	- 0.088	58	0.385	10
Ar-$FeCl_2$	- 0.058	63	0.42	10
Kr-$CdCl_2$	- 0.057	90	0.43	5
Ar-$MgBr_2$	- 0.008	77	0.51	4
Ar-$MnBr_2$	- 0.009	> 79	> 0.52	5
Ar-$CdCl_2$	- 0.003	79	0.523	3
Ar-CoI_2	0.027	81.5	0.54	5
Ar-$CdBr_2$	0.033	83	0.55	4
Ar-FeI_2	0.048	80.5	0.53	4
Ar-MnI_2	0.078	> 75	> 0.50	12
Ar-CdI_2	0.105	71	0.47	11
Ar-CaI_2	0.174	60	0.40	9
Ar-PbI_2	0.194	60.5	0.40	9

[a] Ref. [20]

In Figure 7 we have represented the values of T_{2c} scaled by the critical temperatures T_{3c} of the vapor-liquid transition of the bulk adsorbate as a function of the size incompatibility i. The choice of the scaling factor T_{3c}, which is a measure of the strength of the attractive energy between adatoms, aims at taking into account the obvious idea that the lateral attraction between adatoms is the main cause responsible for the 2D condensation. Considering that the investigated ranges of i for Xe, Kr and Ar slightly overlap, the fact that the experimental points stay on a single curve, within the experimental uncertainty, supports this idea. However the important variations of the ratio T_{2c}/T_{3c} – there is about 40 % difference between the extreme values – and their correlation with the size incompatibility i is a clear indication that the surface corrugation strongly influences the properties of the monolayers. There is a range however in which it does not. Indeed, below $i < -0.09$, T_{2c}/T_{3c} is constant to better than the experimental uncertainty : for the seven investigated systems in that range the average value is 0.388 with a maximum deviation of 0.008. This suggests a behavior simpler than in other ranges of i. This is the reason why it seemed to us wise to first discuss this particular range, in order to try to deepen our understanding of the ratio T_{2c}/T_{3c}. A thorough analysis of the variations of this ratio over the whole range of i necessitates a priori knowledge of the

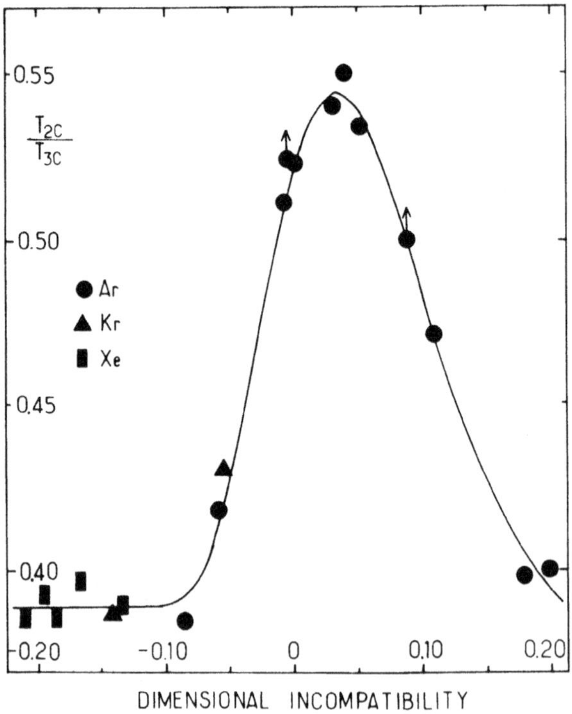

Fig. 7. Variation with size incompatibility of the critical temperature of 2D condensation of the monolayers of Ar, Kr and Xe adsorbed on various layered dihalides [26].

behavior of other thermodynamic quantities and consequently will be done in successive steps in the following Sections.

The constancy of T_{2c}/T_{3c} for $i <- 0.09$ suggests that in certain conditions, we will specify later, the law of corresponding states can be extended to 2D systems [33]. This law is very well obeyed by the 3D phases of the classical rare gases [34,35]. A simple statistical mechanical justification can be given for an assembly of particles interacting through a pair potential of the form : $u(r) = \varepsilon f(r/\sigma)$, in which f is a reference function from which the potential for distinct systems can be obtained by choosing appropriate scaling factors ε and σ [35]. It is well-known today that three-body forces are by no means negligible in condensed phases of rare gases [36]. However a reasonably good rationalization of their thermodynamic properties, ignoring such effects, can still be carried out on condition that an effective pair potential, not the real one, is used [37]. Most calculations utilize the (6-12) Lennard-Jones potential, as defined earlier (Equation (18) in Subsection 3.3).

If the interaction potential between adatoms within a monolayer is not perturbed by the substrate, then of course it is expected that the law of corresponding states should also be obeyed by the 2D phases, which involves that T_{2c}/ε should be a constant in the same way as T_{3c}/ε, and consequently the ratio T_{2c}/T_{3c} is also a constant. Thus our results suggest that below $i \approx -0.09$, where T_{2c}/T_{3c} is constant, the surface corrugation of layered halides does not significantly affect monolayers of classical rare gases, at least at sufficiently high temperatures. And therefore, in the ratio $T_{2c}/T_{3c} = 0.39$ derived from our results, T_{2c} should be considered as characteristic of a planar assembly of rare gas atoms, completely independent on any substrate. This view is supported by a number of additional findings.

First of all, on the basal face of graphite, a surface less corrugated than that of the layered halides (see Table I), T_{2c}/T_{3c} takes for Ar and Xe values close to 0.39 : for Ar two adsorption isotherms investigations lead respectively to $T_{2c}/T_{3c} = 58/150.8 = 0.385$ [38] or $59/150.8=0.391$ [39] ; for Xe the old value of Thomy and Duval [40], $117/289.73 = 0.404$ deviates a bit more from the average, but is probably in agreement with it, within experimental uncertainty.

Additionally, the constancy of T_{2c}/T_{3c} is not limited to the classical rare gases. This is not too surprising, because in 3D the law of corresponding states is also obeyed for simple molecules at sufficiently high temperatures, as long as the thermal motion blurs out orientational effects. There are indications that it is true at least down to T_{3c} for O_2, N_2,

CH_4 etc... For instance their value of the ratio $p_{3c}v_{3c}/RT_{3c}$ is the same as for the classical rare gases [34]. If it remains down to T_{2c}, then the constancy of T_{2c}/T_{3c} is again expected. A number of results support this view. We give here three values of T_{2c} for CH_4 or CD_4, which we think, are determined with good accuracy from adsorption isotherms : on the basal face of BN Tessier and Larher [41] find T_{2c} (CH_4) = 74 K and T_{2c} (CD_4) = 73 K, while on $NiCl_2$ T_{2c} (CH_4) = 73.5 K [42]. The corresponding values of the ratio T_{2c}/T_{3c} are respectively : 0.388, 0.386 and 0.389, showing remarkable agreement with the average value 0.388, mentioned above.

From all these data it seems safe to conclude that in the vicinity of their critical temperature of 2D condensation a number of monolayers either of atoms or of simple molecules behave ideally as if they were adsorbed on perfectly smooth surfaces. The list of systems we have mentionned here is not exhaustive, but we feel that, as far results derived from adsorption isotherms are concerned, there seems to be no exception to the rule, once due account is taken of the uncertainty on the estimate of T_{2c}.

It is on purpose that we mentioned above the value $T_{2c}/T_{3c} = 0.388 \pm 0.002$ for three monolayers of either CH_4 or CD_4, in order to compare it with the heat capacity estimate of Kim and Chan [29], for a monolayer of CH_4 on graphite, yielding T_{2c}/T_{3c} = 68.8/190.6 = 0.361, about 7 percent below our average value. Since for obvious reasons, a better accuracy is expected from this second method of measurement [29,43], the beautiful constancy we observe appears then questionable. However, we feel that this view is weakened by the recent measurements by Nham and Hess [30], mentioned above, of the critical temperature of the second layering transition of CF_4 on graphite, which seems very reliable.Indeed their value T_{2c} = 99.9 K is 7 percent higher than a heat capacity estimate of Zhang, Kim and Chan [43], vz 93 K. We wonder wether the dynamical method used by Chan and coworkers to measure the heat capacity [44] is not open to criticism because of critical slowing down. It is well known that this phenomenon makes the study of the heat capacity at constant volume, for bulk systems, particularly difficult in the vicinity of critical points [45].

We stated earlier in the introduction that a bad knowledge of the interaction parameters was often the major obstacle for the calculation of the thermodynamic properties of a condensed phase. However, this is not true for the classical rare gases, for which the two-body as well as the three-body forces are very well-known [36], so that very many thermodynamic functions have been calculated with good accurancy. However no attempt to calculate critical constants, such as temperature T_c, density n_c, pressure p_c, in 2D or 3D from these potentials has yet been carried out. This is the more surprising that critical phenomena have raised such tremendous an interest for more than two decades. But the efforts in this field seem to have been directed mostly towards lattice models which prove simpler to investigate than continuous models (see for instance ref.[46]). In fact, such constants have been calculated only for the (6-12) Lennard-Jones potential, which constitutes a reasonably good effective potential. Unfortunately, it appears that even for such a crude potential no precise estimate exists neither in 3D (see Table 6.4 in ref. [47]), nor in 2D (see Table 2 in ref. [48]). These tables show differences of about 10 percent between the extreme calculated values of kT_c/ε in 3D as well as in 2D, and for the other critical constants the situation is much worse.

In 1979 [39], we thought we would minimize the uncertainty on the theoretical estimate of T^*_{2c}/T^*_{3c} ($T^* = kT/\varepsilon$) by taking similarly determined values for T^*_{2c} and T^*_{3c}, namely estimates from virial expansions truncated after the fourth virial coefficient. Two years later [26], we proceeded similarly, using now virial series up to the fifth term. This gave $T^*_{2c}/T^*_{3c} = 0.475/1.291 = 0.368$ which is 5.6 percent below the experimental value. Unfortunately the reliability of critical temperatures derived from virial expansions might be poor. It is particularly worrying to see that the virial expansion estimate of the critical density would be, if one refers to Barker [48], about one third of the best estimate of this parameter (see Table 2 in this reference) !

Therefore it might be more judicious to take for T^*_{2c}, the value 0.533, derived by Barker by using jointly virial coefficients and Monte Carlo results, which he considers as being the best. The difficulty is that we do not know which value of T^*_{3c} we should choose to compare to it. Additionally, we know that, close to critical points, computer simulations are hampered by small size-effects because of the divergence of the correlation length. According to a recent article by Rovere *et al.* [49] devoted to a Monte Carlo investigation of a 2D Lennard-Jones systems a finite-size scaling analysis should give a better assessment of the critical region of the phase diagram. In the next section we will come back to this question, because we think the study of 2D melting gives the opportunity to obtain semi-empirically the best estimate of T^*_{2c} for (6-12) Lennard-Jones particles.

We have ignored the possible influence of three-body forces upon the values of T^*_{2c}. It seems to us that attempts such as that of Klein *et al.* [50] to take into account such forces are premature, so long as the statistical problem is not yet mastered for assemblies of particles interacting according to a very simple two-body potential.

4.2. 2D MELTING (OR FREEZING) AND REFERENCE PHASE DIAGRAM.

From the extensive review paper of Strandburg [51] the reader can get an idea of the considerable interest raised by 2D melting and realize that the derivation of an elegant theory for this transition, the so-called KTHNY (Kosterlitz, Thouless, Halperin, Nelson, Young) theory, is responsible for very many studies whose major goal was precisely to try to verify the predictions of this theory. In her introduction Strandburg writes : "..., the question as to the nature of 2D melting remains unresolved", suggesting that there would not exist any final piece of evidence for the KTHNY theory.

We feel that this statement, although probably correct, does not concern but a fraction, very likely the smallest, of known experimental results. Indeed, numerous melting transitions appear very clearly first order, while a few others might be continuous or weakly first order. Figure 1 in ref. [52], in which one adsorption isotherm of N_2 on exfoliated graphite is compared with adsorption isotherms of Ar on the same substrate, gives clear evidence for these two distinct types of melting. For the N_2 monolayer a discontinuity of about 7 percent in the density at melting is observed, unequivocally showing that the transition is first order. For Ar, if a discontinuity exists, it is smaller by an order of magnitude, and melting could be at best weakly first order. Precise structural studies favor a continuous melting transition for the Ar monolayer [53]. In this article, we will be mainly concerned with first order melting since it has been observed for a number of rare gases as well as methane monolayers on layered halides. We will also mention a system for which continuous melting is suspected, but we will not try to discuss it in detail, mainly because our results lack too much the degree of excellence necessary to obtain evidence for the very subtle effects associated with the KTHNY melting.

Figure 8 [42,54] represents a set of adsorption isotherms for Xe on $FeCl_2$, which is typical of the systems showing first order monolayer melting. The other systems behaving similarly are Xe on $NiCl_2$ and on $CoCl_2$, as well as methane on $NiCl_2$ [42,54], all these systems having a size incompatibility below -0.17. The phase diagram disclosed by this figure closely resembles that of a bulk one-component system, having three coexistence domains, a triple point and a critical point. This resemblance suggests that these three distinct 2D phases of the monolayer are gas, liquid and solid. This conjecture is strongly supported both by the thermodynamic analysis of the experimental results and by a few structural investigations.

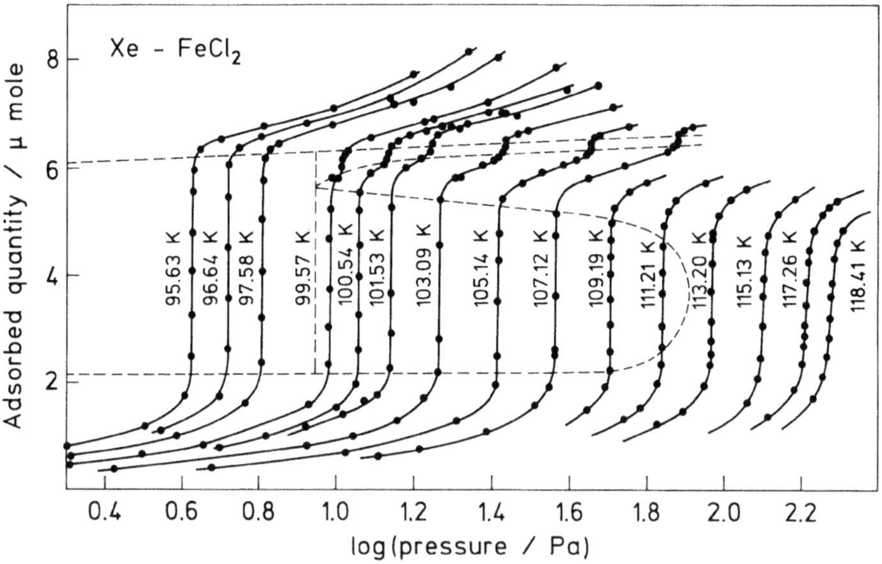

Fig. 8. Monolayer adsorption isotherms of Xe on $FeCl_2$ [54].

A structural study of a CD_4 monolayer adsorbed upon $NiCl_2$ has been carried out using neutron diffraction [42,55]. Deuterated methane was used, mainly to get rid of the very large incoherent signal arising from the scattering of neutrons by protons. A schematic phase diagram indicating the coverages and temperatures at which the experiments were carried out, together with three spectra are shown in Figure 9. At 30 K the sharp asymmetric peak at a value of the scattering vector $Q \approx 1.7$ Å^{-1} is characteristic of a powder of 2D crystallites randomly oriented in space. At 58.3 K this peak has flattened, indicating that the 2D solid has melted. Under compression, up to a state indicated by number 8 in the phase diagram, a solid-like spectrum, not shown here, appears again. The triple point temperature derived from a set of adsorption isotherms [42] is 56.6 K. At 54.8 K, i.e. 1.8 K below this temperature, the diffraction spectrum indicates that a small fraction of the crystal monolayer has already melted. Either a size effect or a temperature gradient within the cell containing the sample might explain this observation, so that the structural study can be considered as supporting our conjecture concerning the nature of the 2D phases suggested by the phase diagram. This point of view can be considered as definitively founded by a more recent X-ray investigation by Morishige *et al.* [56] of the monolayer of Xe adsorbed on $NiCl_2$, for which we observed

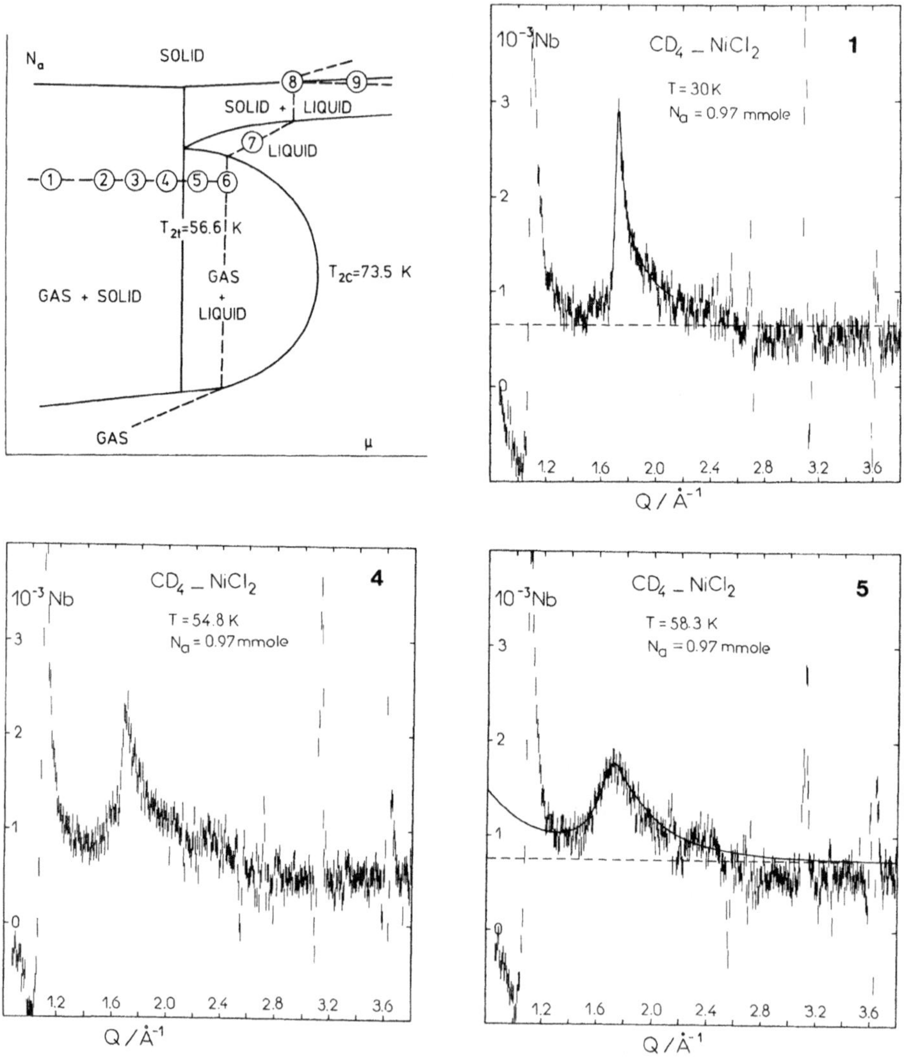

Fig. 9. Neutron diffraction spectra of a monolayer of CD_4 adsorbed on $NiCl_2$(0001)(measured at ILL, Grenoble, France). The numbered points in the phase diagram indicate the temperature and coverage conditions in which the spectra have been measured. Only three spectra, 1, 4 and 5 are shown (see text). The scattering vector is defined as a function of the wavelength λ and of the Bragg angle θ by $Q = 4\,\pi \sin \theta / \lambda$.

earlier a similar phase diagram.

Let us come back to the adsorption isotherms. We claimed above that 2D melting was first order for a number of systems. In Figure 8 the substeps of the adsorption isotherms corresponding to the liquid-solid transition look less vertical than those of the steps representing the gas-liquid transition. However the chemical potential variation from bottom to top of the step have the same order of magnitude in both cases, and, in particular, Figure 10, showing at a larger scale the melting transition of the Xe monolayer adsorbed upon $NiCl_2$ [57], clearly supports that melting is first order. Rounding of the upper part of the step is similar below and above the triple point temperature T_{2t}, and consequently cannot be invoked in favor of a continuous melting, since the gas-solid transition below T_{2t} is obviously first order. We think it is ascribable to a small size effect, probably the presence within the powder of a few very small grains.

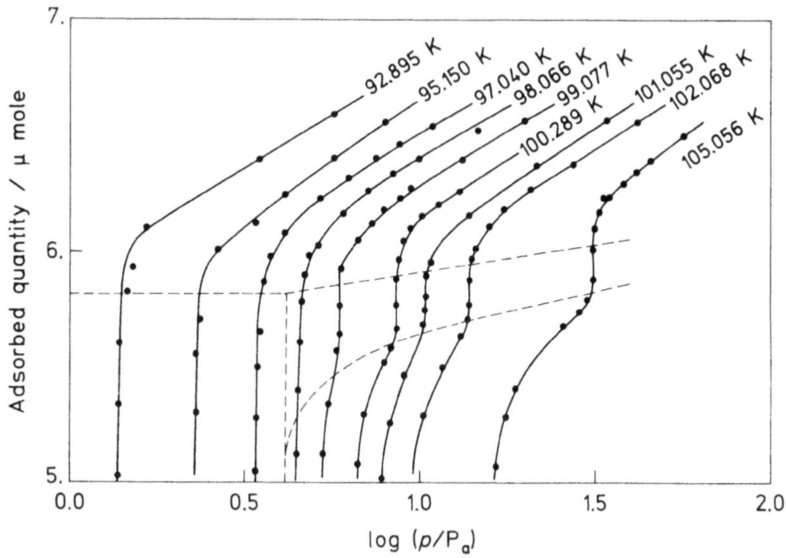

Fig. 10. Adsorption isotherms of Xe ou $NiCl_2$ in the range of coverages in which melting occurs. The isotherms for the full monolayer range are published in ref. [57].

The existence of a first order melting - or freezing - transition for a monolayer is by no means a surprise. Indeed, as soon as 1962, Alder and Wainwright [58] had evidence for such a transition from a computer simulation investigation of a 2D system consisting of 870 hard disk particles. Subsequent similar studies [59,60], using particles interacting according to more realistic pair potentials, in particular (6-12) Lennard-Jones potentials, also suggested a first order melting or freezing. Toxvaerd [61] carried out the first thorough investigation of the 2D fluid-solid transition of (6-12) Lennard-Jones particles, using molecular dynamics. He suggested a value of the 2D triple point temperature, $kT_{2t}/\varepsilon = 0.41$, which was confirmed by a further more detailed study due to Barker, Henderson and Abraham [62]. The analysis of our experimental results will be based mainly upon the latter. These authors have carried out both calculations and computer simulations. The calculations involve perturbation theory for the liquid phase and self-consistent cell theory for the solid phase, while the computer simulations utilize the Monte-Carlo method.

Table III summarizes our results concerning monolayer melting on layered halides : it mentions all the systems for which such a transition has been observed and yields the values of the triple point temperature and melting entropy, when available. We also give for comparison the theoretical estimates of these quantities [62] as well as the values they take for the Xe-graphite system [40].

TABLE III

Thermodynamic quantities associated with 2D melting [42].

System	i	T_{2t}/K	T_{2t} / T_{3t}		$(s_l - s_c)$ /(J mol^{-1}K^{-1})		
			exp.	th.	exp.[a]	exp.[b]	th.
Xe-$NiCl_2$	- 0.210	97.9	0.607		6.9	6.5	
Xe-$CoCl_2$	- 0.196	98.3	0.609		6.0	6.2	
Xe-$FeCl_2$	- 0.185	98.8	0.612	0.61	6.3	6.3	6.8
CH_4-$NiCl_2$	- 0.171	56.5	0.623				
Ar-$NiCl_2$	-0.088	50	0.60				
Xe-graphite[c]	- 0.035	99	0.613			5.8	

[a] From 2D vaporization and melting curves. [b] From 2D sublimation and melting curves.
[c] From ref. [40]

Let us discuss first the triple point temperature. For Xe its value is constant within experimental accuracy : for adsorption on layered halides the maximum deviation from the average value 98.3 K is only 0.5 K, suggesting a negligible influence of the surface corrugation. The good agreement with the value determined by Thomy and Duval [40] for Xe on graphite, namely 99 K, reinforce this view, since the corrugation of the basal face of this substrate is less pronounced than that of the basal face of layered halides (see Section 3.3). The size incompatibility indicated in the Table III for Xe on graphite, -0.035, is calculated with respect to the $\sqrt{3}\times\sqrt{3}$ sublattice of sites of the (0001) face of graphite. We can notice that, at such a small incompatibility, no gas-liquid coexistence of rare gases monolayers is observed on layered halides, because of their stronger corrugation.

Three estimates exist for the 2D triple point temperature of (6-12) Lennard-Jones particles : from computer simulations it is found that $kT_{2t}/\varepsilon = 0.41$ [61] or 0.415 [62], while a calculation gives 0.41 [62]. It is gratifying to see there is much less scatter than in theoretical estimates of critical point temperatures. This seems also to be true in 3D, since two estimates, the first from theoretical calculation [63], the second from a Monte-Carlo simulation [64] both give $kT_{3t}/\varepsilon = 0.68$. This situation clearly prompts us to use the quantity T_{2t}/T_{3t} rather than T_{2t}/T_{3c} for the comparison between the experimental results and the theoretical predictions. This is done in Table III . The agreement between these quantities is remarkable. The fact that it is also good for the CH_4-$NiCl_2$ system might be somewhat fortuitous, since the 3D triple point temperature of CH_4 does not obey the law of corresponding states. The good agreement between theory and experiment reinforces the view that the (6-12) Lennard-Jones potential is a good effective potential to account for the interaction between rare gas atoms and additionally that small-size effects are unimportant in computer simulations far from the critical temperature.

From these conclusions we can propose a semi-empirical way to obtain T^*_{2c} for (6-12) Lennard-Jones particles, by using the experimental value of T_{2t}/T_{2c}. From the data contained in Table II and III, we conclude that for the three monolayers of rare gases adsorbed upon layered halides, showing a first order melting, $T_{2t}/T_{2c} = 0.865 \pm 0.01$. The value of this ratio for the Xe-graphite system, 0.85 [40], agrees with it within experimental accuracy. Since we expect the effect of three-body interaction to be largely cancelled in a ratio such as T_{2t}/T_{2c}, it seems probable that 0.865 is a good estimate for (6-12) Lennard-Jones particles. Since the 2D triple point temperature seems reliably determined for that potential, $T_{2t}^* = 0.415$, we propose the critical temperature derived from these two values, $T_{2c}^* = 0.48$, as the best estimate of this quantity. At any rate it will be interesting to compare to it, the results of future computer simulations, in which the difficulty of finite-size effect will have been overcome as suggested by Rovere *et al* [49].

From the ln$p$ versus $1/T$ curves corresponding to 2D sublimation and vaporization, we can obtain by using equations (14) and (15), estimates of $s_{go} - s_1 \approx s_{go} - s_c$ for $T < T_{2t}$, $s_{go} - s_1 \approx s_{go} - s_l$ for $T > T_{2t}$, and consequently of the melting entropy $s_l - s_c$. In these expressions the lower indexes c and l refer to the 2D crystal and liquid phases. Unfortunately, the accuracy achieved in this procedure is poor because it involves differences between quantities differing only by a few percent, about 3 percent. It can be significantly increased by making use of the ln$p$ versus $1/T$ 2D melting curve, from which we can get according to equations (8) and (11) : $s_{go} - (\Gamma_c s_c - \Gamma_l s_l) / (\Gamma_c - \Gamma_l)$. The differences between this quantity and either $s_{go}-s_c$ or $s_{go} - s_l$ are respectively : $(s_l - s_c) \times \Gamma_l/(\Gamma_c - \Gamma_l)$ and $(s_l - s_c) \times \Gamma_c/(\Gamma_c - \Gamma_l)$, i.e. the 2D melting entropy multiplied by a factor which is about 10. As a consequence, by using the melting curve jointly with either the vaporization or the sublimation curve, we could in principle increase the accuracy on the melting entropy by an order of magnitude. Unfortunately the substrates are not sufficiently uniform to obtain a reasonably good estimate of Γ_l and Γ_c, which prevents us from getting a purely experimental value of $s_l - s_c$. In fact, we will take for Γ_c and Γ_l the values proposed by Barker *et al.*, namely $\Gamma_l/(\Gamma_c - \Gamma_l) = 11.4$ and $\Gamma_c/(\Gamma_c - \Gamma_l) = 12.4$ at $kT/\varepsilon = 0.45$. In this way we obtain for each system two estimates of $s_l - s_c$ (see Table III). The maximum relative difference between both sets, about 6 percent, gives an order of magnitude of the accuracy, so that our results suggest that the melting entropy can be considered as independent on the substrate, as already was the triple point temperature. In addition, the difference with the values derived from Thomy and Duval results for Xe on graphite [40] is also probably beyond experimental uncertainty. So is the difference with the theoretical estimates of Barker *et al.* [62], although it can attain about 10 percent, but the accuracy on the latter is probably not significantly better than that, as could be guessed from a representation of their values $s_l - s_c$ versus kT/ε. Such a plot also shows that having an estimate of $s_l - s_c$ at a temperature $kT/\varepsilon = 0.45$, somewhat above the triple point temperature $kT/\varepsilon = 0.41$, does not entail our comparison.

We have not included the values of the melting enthalpy in Table III. Since $h_l - h_c = T_{2t}(s_l - s_c)$ the agreement between the experimental and theoretical values follows trivially from the agreement between those of T_{2t} and $(s_l - s_c)$.

Our estimates of T_{2c}/T_{3c}, discussed in the previous Subsection, remain constant up to a size incompatibility of about - 0.09, suggesting a gas-liquid equilibrium, unaffected by the surface structure. But, no indication of the freezing of the liquid monolayer is detected on the adsorption isotherms for size incompatibilities ranging from - 0.17 to -0.9. However, at the extreme right of this range, for the Ar-$NiCl_2$ system, we have structural evidence for the existence of melting [42]. This is obtained from a neutron diffraction study of a monolayer of ^{36}Ar on $NiCl_2$. ^{36}Ar was used because it has an unusually large coherent cross-section for neutron scattering. The results are similar to those observed for CD_4 on the same substrate, indicating in particular coexistence. Of course this could be due to a temperature gradient, so that we cannot claim that the melting of the Ar monolayer is first order. In fact, we rather suspect either a continuous or a weakly first order melting, which is the only way of explaining that no indication of the transition would appear in the adsorption isotherms. Table III shows that the triple point temperature for this transition has the same reduced value as those for clearcut first order melting. This was already the case for the "continuous melting" of the Ar monolayer on graphite.

Probably the more striking outcome of subsection 4-1 and 4-2 is the evidence for the existence of a standard phase diagram for monolayers of rare gases. The excellent constancy of the quantities T_{2c}/T_{3c} and T_{2t}/T_{3t}, and a number of other facts discussed above, leave little doubts that this phase diagram is that of a 2D assembly of rare gas atoms unaffected by the surface corrugation. For layered halides this could difficultly be predicted a priori. We will come back to this problem later, once we will have discussed the strong influence of corrugation upon the properties of the first monolayer in other ranges of size incompatibilities.

This standard phase diagram has a limited predictive value, since only the classical rare gases obey the law of corresponding states. On the other hand it can prove useful as a reference phase diagram, since the analysis of deviations from it for more complicated systems can serve as a guide in the search for a molecular interpretation.

Although qualitatively similar to the phase diagram of bulk rare gases, the 2D phase diagram shows with respect to it a significant quantitative difference in the value of the ratio T_t/T_c, which is 0.555 for 3D systems and 0.865 for 2D systems. If for some reason the relative solid-liquid stability is increased with respect to what it is for a rare gas system, then the ratio T_t/T_c can achieve much larger values. For instance, for bulk CO_2, an orientational order of the molecules in the crystal phase favorable for quadrupole interaction, results in an increase of T_{3t}/T_{3c} up to 0.71 which is 29 percent above the value for rare gases. For a similar increase in 2D, T_{2t}/T_{2c} would be larger than 1 because the value for rare gases is already close to 1. What really results in is a phase diagram with no gas-liquid coexistence domain, of a type never been observed in 3D for a one-component system. But it is by no means exceptional in 2D, since it might account for the behaviour of at least of four distinct monolayers adsorbed upon graphite, vz Kr [65,66], N_2 [67], CO [68] and C_2N_2 [69,70].

5. Sitewise adsorption

In the range of size incompatibility i examined in detail in the previous section, the dependence of the properties of the adsorbed monolayer upon surface corrugation appeared negligible. Here, on the contrary, we will focus our attention upon a range in which the influence of corrugation can be considered maximum in the sense that the surface imposes its own structure to the monolayer in certain conditions of temperature and pressure. As we argued in Section 3.3, this is expected to occur for favorable size compatibility, i.e. for i close to zero.

The thermodynamic quantity which best suggests what happens is the entropy s_1 defined by Equation (13) in Section 2. From the behaviour of s_1 versus i for Kr, we obtained, as soon as 1968 [71-73], the first evidence for the influence of the surface corrugation upon the structure of the monolayer. Subsequent investigations of CH_4 and Ar adsorption [74,76] displayed a very similar behaviour. Here we will discuss the results obtained for Ar, mainly by Millot [74,76], because for that gas, a larger number of experimental points are available in the range of size incompatibilities of interest in this Section.

5.1 ADSORPTION ENTROPY

The values of s_1-s_0 as a function of size incompatibility i are given in Table IV. They are plotted in Figure 11. A sharp minimum is observed at $i = 0$, and $s_1 - s_0$ increases on both sides of this minimum, so steeply as to suggest the existence of an angular point.

Let us consider first the increase of s_1-s_0 from $i = 0$ to about 0.12, approximately accounted for by the linear relation :

$$(s_1 - s_0) / (\text{J mol K}^{-1}) = 110\, i - 4.2. \tag{19}$$

Larher and coworkers [72-76] suggested to explain this increase by the existence of a registered monolayer. Such a monolayer has the structure of a (111) plane of the bulk crystal of adsorbate isotropically expanded by i. Clearly, to such an expansion should correspond an entropy increase, somewhat lower, but of the same order of magnitude as that associated to the isotropic expansion by i of a bulk crystal of adsorbate. The latter writes at first order :

$$\Delta s = (3\alpha \upsilon/\chi) \times i, \tag{20}$$

υ, α and χ being respectively, the molar volume, the thermal expansion coefficient and the isothermal compressibility of this phase. At 70 K, which is close to the average temperature at which the adsorption isotherms have been measured, at least for $0 < i < 0.12$, the coefficient of this equation, $3\alpha\upsilon/\chi$, is equal to 195 J mol^{-1}K^{-1} [21,77]. The coefficient of equation (19) accounting for the entropy increase of the registered monolayer with size incompatibilities is 56 percent of this value, which seems a good

TABLE IV

Variation of the entropy s_1-s_0 with the size incompatibility i for Ar.

Adsorbent	Size incompatibility	(s_1-s_0)/(J mol^{-1}K^{-1})
$NiCl_2$	- 0.068	10.7
$FeCl_2$	- 0.058	8.4
$MgBr_2$	- 0.007	- 2.2
$MnBr_2$	- 0.005	- 2.0
$CdCl_2$	0	- 4.2
CoI_2	0.031	- 0.9
$CdBr_2$	0.039	- 0.1
FeI_2	0.052	1.8
MnI_2	0.089	6.7
CdI_2	0.110	7.8
CaI_2	0.179	11.6
PbI_2	0.199	13.1

order of magnitude since only two degrees of freedom are involved in the expansion of the monolayer instead of three for the bulk crystal.

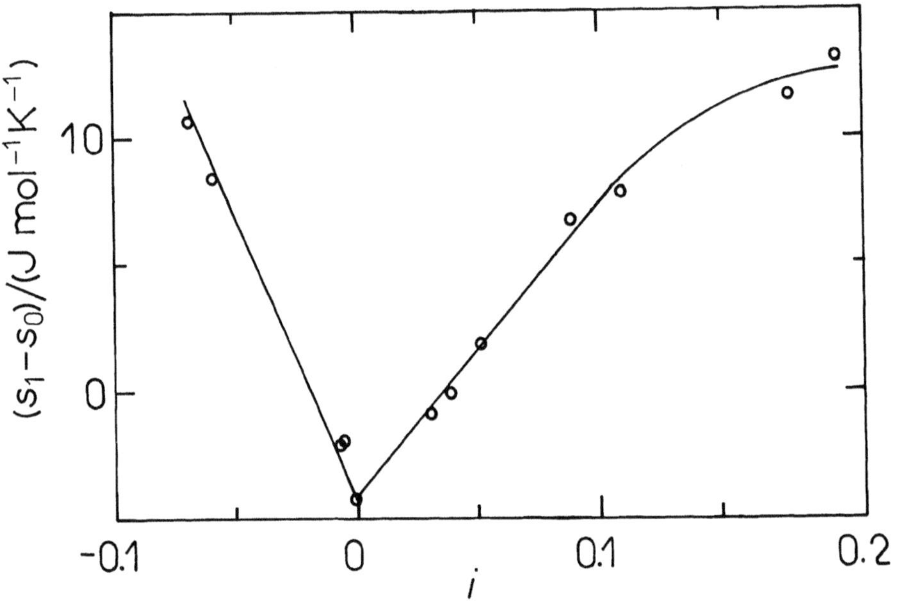

Fig. 11. Variation with size incompatibility i of the entropy difference s_1-s_0 (see table IV).

It is important, however, to emphasize the approximations implied in our analysis. First, we supposed that s_1 was the entropy of the dense monolayer formed during the 2D condensation, which is valid only at temperatures well below T_{2c}, as we stated in Section 2. But in fact, our values of s_1 are derived from the curves ln p_1 versus $1/T$ in a range of T quite close to T_{2c}. At the outset this would pose some problems because s_1 takes at T_{2c} the undetermined form 0/0. However our data do not show any measurable change in s_1 as T_{2c} is approached from below. For instance, for the Kr-$CdCl_2$ system [72,73], we were not able to detect any deviation from linearity of the ln$p$ versus $1/T$ plot over a 15 K temperature interval just below T_{2c}. This suggests some kind of cancellation in u_1 and s_1, of the contributions arising from the increasing disorder in both coexisting phases as the critical temperature is approached from below. Such a cancellation is known to be perfect for the Ising model : because of a symmetry property, the configurational entropy, the only one to be taken into account in this model, has equal areal densities in the two coexisting phases, and $s_1 = 0$. Thus, it does not appear unreasonable to admit that a few degrees below T_{2c}, the main contribution to s_1 would result from the vibrations of the adatoms in the condensed phase.

By comparing equations (19) and (20) we are tacitly suggesting that the main cause for the entropy increase with i is the decrease of the in-plane vibration frequency of the adatoms as their lateral interaction with their neighbours weakens. Of course the corrugation of the surface can also influence the lateral motion, but probably no so much as to radically change the slope of the curves s_1-s_0 versus i. The value s_1-s_0 = - 4.2 J mol^{-1} K^{-1} at i = 0 is expected to result firstly from the motion of the adatoms perpendicular to the surface of the layered halides. It is negative because the attraction of the adatoms by this surface is stronger than by the bulk crystal. However a non-negligible part might also be associated with the motion parallel to the surface, since the corrugation can increase the vibration frequency with respect to what it is in a bulk crystal. Theories such as those developed by Asada [78], which will be discussed later, are of some interest, in that they illustrate how the various degrees of freedom contribute to the entropy. We feel however that our knowledge of the interaction potentials of the adatoms with the substrate, and consequently of the corrugation, is still too bad to permit a realistic molecular description of the variation of s_1-s_0 with i.

Clearly if the monolayer would remain registered below $i = 0$, the entropy s_1-s_0 should continue to decrease along the same straight line as for $i > 0$. The observation of an abrupt increase seems to us an indication that a registered monolayer is no longer stable below i = 0.

The conjecture that the monolayer in equilibrium with the 2D gas is registered for $0 < i < 0.12$ is supported by a number of additional experimental facts : first, the results of structural investigations by neutron diffraction of two systems, ^{36}Ar-MnI_2 (i = 0.078) and CD_4 - PbI_2 (i = 0.066) [79] ; second, the observation for Kr on PbI_2 (i = 0.12) of a transition within the condensed monolayer, which suggests a commensurate-incommensurate transition.

Since similar conclusions are derived from both structural studies, we will present the results of a single one, ^{36}Ar on MnI_2. ^{36}Ar is used because the coherent scattering cross section for neutrons is about 150 times higher than for natural Ar. Figure 12 shows a diffraction spectrum at a coverage slightly below one monolayer, so that the 2D crystals are in equilibrium with their 2D vapor. A dissymmetric Bragg peak is observed, although less clearly than for CD_4 on $NiCl_2$ (see Figure 9), because the slowly decreasing right wing is overlapped by crossed contributions from both the monolayer and the substrate at the positions of the (002) and (104/3) Bragg reflections of the latter. If this non-symmetric Bragg peak is considered as the (10) reflection of a close-packed triangular structure, a parameter $a = 4.094$ Å is derived from the values of the Bragg angle at the maximum of the peak, differing by 1.3 percent from the parameter $a = 4.146$ Å of bulk MnI_2[22]. Because the non-symmetric peak results from a convolution of a symmetric Bragg peak by the function describing the orientational distribution of the 2D crystallistes, the angular position of its maximum is shifted rightwards with respect to the Bragg angle $\theta_{10} = \sin^{-1} (0.5\lambda / d_{10})$. Therefore, our estimate of the crystal parameter has to be augmented by a

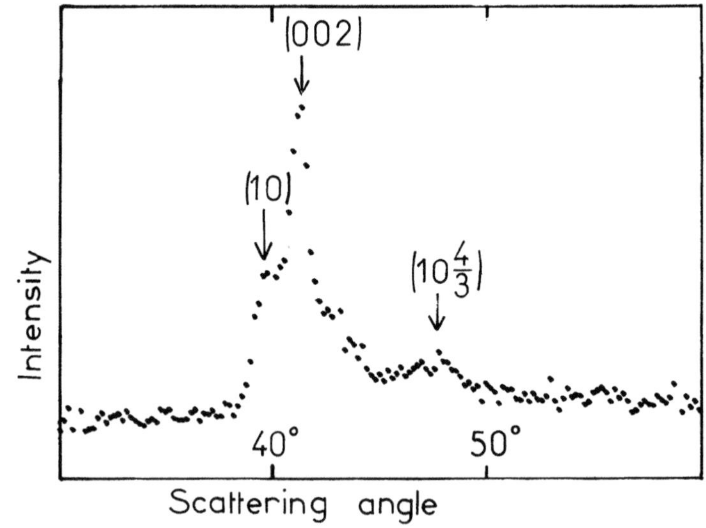

Fig. 12. Neutron diffraction spectrum of a monolayer of ^{36}Ar adsorbed on $MnI_2(0001)$(measured at ILL Grenoble, France). The wavelength is $\lambda = 2.41$ Å. This is a difference spectrum [I(monolayer + substrate) - I(substrate)] containing in addition to the (10) Bragg peak of the monolayer, crossterms at the Bragg reflections (002) and (104/3) of the substrate. The latter shows reflections from a rhombohedral structure which have been indexed in the hexagonal structure, whence the fractional index.

quantity which is of the order of 1 percent for layered halides [42,55]. And consequently we can conclude that the diffraction spectrum supports the conjecture of a registered monolayer of Ar on MnI_2. A similar conclusion is obtained for the methane monolayer adsorbed on PbI_2. For both systems, when the pressure is raised beyond the 2D condensation, the peak at the commensurate position shifts rightward, indicating a change towards a denser incommensurate monolayer.

As the size incompatibility increases, the interaction between registered adatoms becomes less and less favorable and beyond a certain value of i the commensurate adlayer is expected to become unstable in any condition. We have indications that this occurs somewhat beyond $i = 0.12$. Indeed we know from a neutron diffraction investigation of the ^{36}Ar-PbI_2 system ($i = 0.194$) [79] that the 2D crystal of Ar in equilibrium with its 2D vapor is already incommensurate. On the other hand, this does not seem to be the case for the Kr-PbI_2 system ($i = 0.12$), since beyond the 2D condensation we observe in the adsorption isotherms (Figure 13) an inflexion point most simply interpreted as corresponding to a commensurate-incommensurate transition. In the adsorption isotherm at 78.41 K, the step corresponding to the 2D condensation of the monolayer suggests a reasonably good uniformity of the substrate, since the pressure change from bottom to top is around 3.2 percent corresponding to a variation of the chemical potential $\Delta\mu = 21$ J mol^{-1} for a transition involving an energy change of 13 000 J mol^{-1} (derived from Equation (22) below). The commensurate-incommensurate transition, shown on an expanded scale in the insert, seems rather continuous. The variations of the transition pressure with temperature is expressed by the following linear regression :

$$\log_{10} (p_{ci}/\text{Pa}) = 732 \text{xK/T} + 10.68 \qquad (21)$$

with mean root square deviations on the coefficients of 2.7 and 0.035 respectively. From a previous study of the 2D condensation of Kr on PbI_2 [72] we know that the pressure of this transition obeys the following relation :

$$\log_{10} (p_1/\text{Pa}) = 679.5 \text{xK/T} + 9.414 \qquad (22)$$

We note that, although it is derived from experimental results obtained in a range of temperature between 86 and 92 K, it predicts the value of p_1 at 78.43 K with an accuracy of 0.2 percent (Figure 13).

The linear regressions (21) and (22) cross at a temperature of 41 K, below which the 2D gas will condense directly into the incommensurate monolayer. Although this temperature is not known with a good accuracy, since its determination involves an extrapolation over a large temperature interval, about 30 K, our results however clearly indicate that the incommensurate monolayer is the 2D dense phase stable at low temperatures and therefore has an energy lower than the commensurate monolayer. Conversely the latter has a higher entropy so that it can exist at high temperature and low pressure. What this means at the molecular level is that the lowering of the energy corresponding to the localization of the adatoms at the sites is insufficient to compensate the increase in lateral energy resulting from the expansion of the adlayer. We will come back to this question in the next Section.

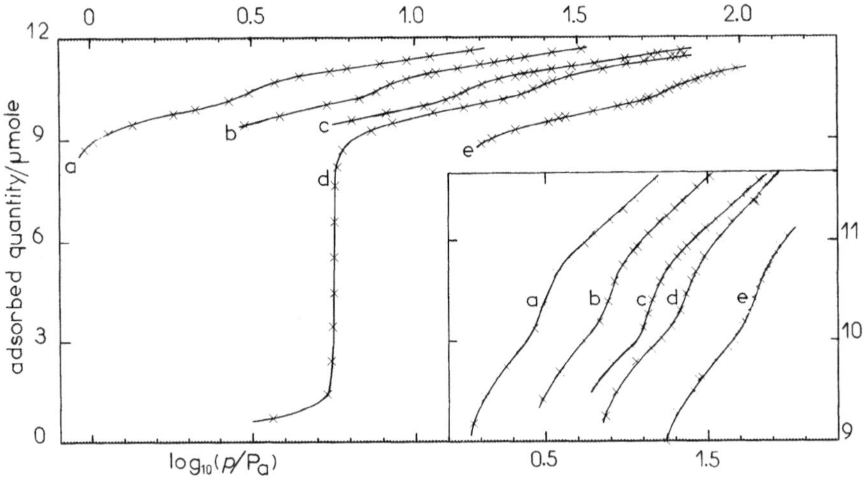

Fig. 13. Monolayer adsorption of Kr on $PbI_2(001)$. The "commensurate-incommensurate" transition appears more clearly in the insert, in which the abscissae have been contracted by a factor 2 and the ordinates expanded by a factor 3. The temperatures in K are : a,71.79 ;b,74.74 ; c,76.72 ; d,78.43 ; e, 90.09. For the isotherm e a quantity 0.8 must be added to the abscissae indicated in the graph.

5.2. CRITICAL TEMPERATURE OF 2D CONDENSATION

One might imagine that interpreting the values of the critical temperatures of 2D condensation in the range $0 \leq i \leq 0.12$ should not constitute a major problem since there has been plenty of theoretical studies of critical phenomena for lattice models and even more because a rigorous solution can be found in 2D , as shown first by Onsager in 1944 [80] for a square lattice. However, such a view happens to be much too optimistic, because such models are unrealistic in many respects.

The first problem is that interactions are restricted to first neighbours only, which means that a fraction of the attraction energy is ignored. For a complete coverage of a triangular lattice by (6-12) Lennard-Jones particles an underestimation of about 12 percent can be easily calculated which should give us a rough idea on the undervaluation of T_{2c} resulting from the neglect of neighbors beyond the first. We will see later that this cause of error is far from being the most important.

Let us now look at the prediction of the theory for three specific systems : Ar-$CdCl_2$, Ar-$CdBr_2$ and Ar-MnI_2. The interaction between nearest neighbours w is calculated by using again the effective (6-12) Lennard-Jones potential and taking for Ar $\varepsilon/k = 120.6$ [19]. For the triangular lattice $kT_{2c}/w = 0.910$ [46]. In Table V the theoretical estimates are compared to the experimental values. It appears that the statistical theory overestimates the critical temperature by at least 22 percent. Clearly, if the neighbors

TABLE V

Critical temperature of 2D condensation of Ar.

Substrate	$CdCl_2$	$CdBr_2$	CdI_2
a/Å	3.846	3.985	4.24
i	- 0.003	0.033	0.105
T_{2c}(th)/K	110	105	87
T_{2c}(exp)/K	79	83	71
T_{2c}(th)/T_{2c}(exp)	1.39	1.26	1.22

beyond the first would be taken into account, the deviation between theory and experiment would be even larger. Of course there is another source of error on the theoretical estimate of T_{2c} resulting from the use of an approximate interaction potential.

However from our discussion of the properties of the monolayers in the range of i in which the effect of the corrugation is negligible (see Section 4.1.2) it seems very unlikely that the effective (6-12) Lennard-Jones potential we are using could induce errors larger than 10 percent. Since an underestimation of the same order of magnitude results from the neglect of neighbors beyond the first, no less than a 20 percent difference between theory and experiment remains still to be explained. A reason for that may be found, we think, in the model assumption that adatoms are restricted to discrete sites, which of course cannot be correct neither in the high-density nor in the low-density monolayer.

In the high-density phase, adatoms, due to the vibrational motion,spend most of the time in relative positions where the interaction energy has a value above the minimum corresponding to the site positions. A number of theoretical studies have tried to take such vibrations into account (see for instance ref. [78,81-84]). Unfortunately, this cannot be done within a rigorous treatment of the configurational degree of freedom. The price paid for this is severe. The mean field approximation, used for instance by Asada in his first two articles [81,82] devoted to adsorption on a triangular lattice, yields, when no account is taken of vibrations, $T_{2c}/w = 3/2$ [85], which is 65 percent above the rigorous value $kT_{2c}/w = 0.91$. The quasi-chemical approximation, which is a bit less crude, yields $kT_{2c}/w = 2 \ln(3/2) = 1.233$ [85], 35 percent above the rigorous value. Another more elaborate approximate theory by Asada [83], yields $kT_{2c}/w \approx 1.10$ (estimate from the curves, in figure 3 and 5, corresponding to an infinite in-plane vibration frequency), still 20 percent above the exact value. As a consequence, agreement between theory and experiment could not be achieved but by assigning incorrect values to molecular parameters in order to compensate for the error introduced by the approximation.

A second defect of lattice models is to consider the adatoms in the low density monolayer as localized around the sites, whereas there are good reasons to assume they are already mobile well below the critical temperature of 2D condensation. Indeed Doll and Steele [86] have calculated that an isolated adatom on a corrugated surface becomes mobile at a temperature $T_{mob} \approx (V_m(SP) - V_m(S))/ 5k$. The upper bound we get from the data in Table I is $T_{mob} = 17.6$ K, for a system, Ar-$CdCl_2$, for which $T_{2c} = 79$ K. Even if the relation of Doll and Steele is not rigorous and the data of Table I are very imprecise, the estimated T_{mob} is so much lower than T_{2c} that there is little doubt that well below T_{2c} the atoms of the low density monolayer will already be mobile. This is an additional reason for considering very cautiously the predictions of lattice theories. We think however that , in spite of their serious defects such approximate theories are helpful in disclosing the order of magnitude of the variation of various thermodynamic quantities as a function of molecular parameters. We will give an illustration of this view in the next paragraphs, in which we will examine the important issue of the variations of T_{2c} as the size incompatibility varies from 0 to 0.12.

The simple lattice model tells us that for a given type of lattice T_{2c} is proportional to the absolute value of the potential energy of attraction between nearest neighbours. Within the scope of this approximation if r_0 is the value of the distance at which the effective potential we are using is minimum, then T_{2c} will be maximum at a value of the size incompatibility $i = (r_0 - d)/d$ which turns out to be zero to better than 1/500, but this extreme closeness to zero is a just a consequence of the conventions we used to define i (see Section 3.3). Given the numerous approximations involved in our reasoning, we can just predict that the maximum should occur within a range of i a few percent around $i = 0$. It actually occurs at $i \approx 0.035$. If account were taken from neighbours beyond the first, the total potential energy would achieve its minimum at $i \approx - 0.01$. As for the values of critical temperatures, taking into account all neighbours worsen the agreement between theory and experiment. It is probable that the main cause for the disagrement should be looked for again in the vibrations of the adatoms.

This intuitive point of view is supported by the approximate theory of Asada we mentioned earlier [83]. The author describes the corrugation by an in-plane vibration frequency ν_s. From his Figure 5a it can be seen that for finite values of ν_s, T_{2c} achieves its maximum at a positive value of i. In the limit case $\nu_s = \infty$, for which the adatoms are perfectly localized at the sites, the maximum is at $i = 0$, which is expected since only first neighbours are taken into account in the theory and since in the definition of i, somewhat different from ours, $i = 0$ corresponds to the minimum of the Lennard-Jones potential between adatoms. As ν_s is decreased the root mean square displacement of the adatoms increases, bringing about a decrease in the maximum value of T_{2c} and an increase of the size incompatibility at which this occurs. If Asada's theory predicts correctly a trend which seems to be intuitively obvious, we do not feel that it could pretend to give correct quantitative predictions. For instance from the same figure we can see that, in the range of ν_s of interest, multiplying this frequency by a factor 2 results in an increase of T_{2c} by 20 percent. Since Asada approximate treatment of the configurational degree of freedom overestimates T_{2c} by a comparable amount, it is clear that the assessment of ν_s from the application of the theory to the experimental results cannot be but mediocre.

It seems reasonable to conjecture that the commensurate monolayer achieves its optimum stability at the value of the size incompatibility at which T_{2c} is maximum, i.e. i = 0.035, for which we should observe a minimum of its free energy f_1 [75]. Unfortunately this seems difficult to check experimentally because the irregular and uncontrolled variations of the holding potential from one substrate to the other, although small, are sufficient to obscure the quadratic variation of f_1-f_0 with i around the mimimum of this function. This was not a problem for the entropy difference s_1-s_0 which varied linearly with i. Since the behavior of the entropy suggests that the commensurate monolayer is stable only down to $i = 0$, a value of a few percent for the optimum stability is not unexpected.

In Section 4.1.2. we noticed that, when the monolayer could be described by a continuous model, the ratio T_{2c}/T_{3c} took for simple molecules, vz CH_4, O_2, etc..., the same value as for rare gases. Surprisingly, at $i \approx 0$ this ratio is 8 percent lower for CH_4 than for Ar, as noticed by Nardon and Larher as soon as 1974 [28]. The subsequent data of Millot for Ar [26,75] indicate that a comparable difference exists at least from $i = 0$ to about 0.07. To explain this observation it could be suggested that because CH_4 has a lower mass than Ar, the surface could be less efficient in decreasing the amplitude of the lateral vibrations of the adspecies. We are not sure at all that this alone could explain the 8 percent difference.

Since our experimental results suggested that the dense monolayer in equilibrium with the low-density monolayer was registered it seemed to us wise to try to account for the parameters of the transition in using a lattice model. Our analysis shows the shortcomings of such a model. Clearly, it is less efficient in the range of size incompatibility $0 \leq i \leq 0.12$ than the continuous model was for $- 0.21 < i \leq - 0.09$. Given the potential success of the latter it could be wondered if it would not be wiser to try to use it even for $0 \leq i \leq 0.12$ in the following way : it would be considered as a reference model and the corrugation would be introduced as a perturbation. Such an approach was initiated by Steele, Monson and Henderson [87] and later used by Dominko, Patrykiejew and Sokolowski [88] to try to account for the values of our critical temperatures. The maximum increase they obtained with respect to the value for a continuous model was 6 percent, much lower than the 40 percent experimentally observed. This justifies a posteriori our use of a lattice model.

To conclude this section, we would like to say however, that, at present, the hope for a realistic and rigorous molecular theory of registered monolayers seems to us more remote than for corrugation-independent monolayers.

6. Monolayer structure and corrugation : an Overwiew

In the two previous Sections we examined two distinct ranges of size incompatibilities, vz $- 0.21 < i < - 0.09$, and $0 < i < 0.12$, in which the monolayers of Ar, Kr, Xe or CH_4 adsorbed on the basal face of the layered halides behave very differently. In the first range, the phase diagram has the same aspect as for a 3D one-component system, while in the second range it appears analogous to the phase diagram of a lattice gas or of the isomorphous Ising model. From a structural point of view, the crystal monolayer in equilibrium with its vapor appears very little influenced by the corrugation in the first case, while it is registered in the second.

Between these two extremes, the simplest which can be envisaged a priori in physisorption, there is place for more intricate structures, involving a more balanced influence between the adatom-adatom interaction on the one hand and the corrugation on the other hand. They are expected to exist at higher coverages for the monolayers mentioned above, and at any coverage for the monolayers occurring in the ranges of size incompatibilities we have not yet looked at. In the absence of any structural information we could not but suggest a vague idea of what these structures could be, basing ourselves on both the thermodynamic properties derived from our adsorption isotherms and our approximate knowledge of the corrugation. This is one of the objective of this Section but not the main one.

Indeed the two types of behaviours already discussed are expected to have some generality and we would like to know to what extent they could be predictable from an approximate knowledge of the corrugation. With this goal in mind we will try to deepen the understanding of our own results as well as of a number of significant data obtained elsewhere on uniform substrates other than the basal face of layered halides.

Let us come back first to the registered structures. On layered halides they occur over a limited range of size incompatibilities, from about 0 to 0.12. It is intuitively expected that a similar behavior should be observed for any other surface having a triangular sublattice of sites in a range of size incompatibilities which will depend upon the amplitude of the surface corrugation : the larger this amplitude, the wider this range.

It is on purpose that we included the Kr-graphite system in Table I for comparison with the rare gases-layered halides systems. With respect to the $\sqrt{3}x\sqrt{3}R30°$ sublattice of sites of graphite, Ar, Kr and Xe have respectively size incompatibilities of about : 0.12, 0.046, - 0.035. At equilibrium with its 2D vapor the crystal monolayer of Kr alone can be registered(see previous Chapter). Given that the amplitude of the corrugation, which is of the same order of magnitude for Ar, Kr and Xe on graphite, but more than five times lower than on layered halides, it is by no means surprizing that graphite (0001) turns out to be less efficient in localizing the adatoms. Monolayers of Ar and Xe on graphite have phase diagrams similar to those we observed for rare gases on layered halides at i <-0.09. Kr, although its 2D vapor condenses into a registered crystal on graphite, has a phase diagram completely different from what we observe for rare gases on layered halides at comparable size incompatibilities. We mentionned its peculiar aspect at the end of Section 4.2. It suggests a virtual critical temperature such that the ratio T_{2c}/T_{3c} is quite close to 0.39, a value differing little from that for the continuous model and consequently indicating a behavior little affected by the corrugation, completely opposite to what is observed for adsorption on layered halides at comparable size incompatibilities. Clearly graphite should be considered as an exceptionally smooth substrate for the noble gases considered in this chapter.

It seems worth discussing at this point the properties of the monolayer of Xe adsorbed on Pt(111), which has been thoroughly investigated by Comsa and coworkers [89-92], since it could be expected to show similarities with the monolayer of Kr on PbI_2(0001). Indeed, the size incompatibility of Xe with the √3x√3 R30° sublattice of sites of Pt(111) is 0.09, close to the value of 0.12 we have for Kr on PbI_2(0001). Additionally, according to Kern *et al.* [92], for Xe on Pt (111) the interaction energy difference between the average and the lowest level would be about 30 meV ≈ 350 K, which is of the same order of magnitude as what can be expected for Kr on PbI_2 and almost an order of magnitude higher than for Kr on graphite (see Table I). A first indication of the similarities of the monolayers is that in both cases a transition exists from an incommensurate to the expected commensurate monolayer at comparable reduced triple point temperatures (the 3rd phase is the 2D vapor), namely : 62/289.7 = 0.21 for Xe-Pt(111) and 41/209.4 = 0.20 for Kr-PbI_2(0001). Note that the evidence for this behavior should be considered stronger in the first case since a structural study has been done over a large temperature interval, below and above the triple point. However the existence of a vapor-liquid coexistence domain, with temperature limits of 98 K (triple point) downwards and 120 K (critical point) upwards [90], agreeing within experimental uncertainty with those of a corrugation-independent monolayer, suggests that Pt(111) is much less corrugated than the basal face of layered halides. At the size incompatibility of Xe-Pt(111), i = 0.09, we would have for Xe on layered halides, according to Figure 7, a critical temperature of 0.50

x 289.7 = 145 K, 20 percent above the value for Xe-Pt (111). We are tempted to think that the corrugation of Pt(111) for Xe is probably more comparable to the corrugation of graphite than to that of the layered halides, although it could be significantly higher than the former. In fact the estimate of the mean amplitude of the corrugation by Kern *et al.* [92] seems to us questionable.

These authors consider that the average corrugation corresponds to the drop in the isosteric heat, $\Delta q_{st} = 30$ meV, occurring at the commensurate-incommensurate transition as coverage is increased. It is not clear at all why it should be so, because q_{st} is not a simple physical quantity, and even if it were, ignoring the lateral energy change due to a crystal parameter increase expected to be of the order of magnitude of $i = 0.09$ is unwarranted. Even more embarrassing is an apparent contradiction between the sign of the q_{st} variation with coverage and the aspect of the phase diagram : indeed the incommensurate monolayer appears to be the stable phase at low temperatures so that we expect its vapor pressure to become closer and closer to that of the commensurate phase as temperature is lowered, which necessitates q_{st} (incommensurate) > q_{st} (commensurate) in contradiction with the experimental results.

It is intuitively obvious that the denser a surface appears to the adatom, the lower the amplitude of its corrugation for this atom. The atomic densities of graphite (0001) and Pt (111) are respectively about 5 times and 2 times higher than that of the basal face of the layered halides. Consequently it is not surprizing that at good size incompatibilities the corrugation of graphite affects the monolayers of Ar, Kr and Xe, much less than that of the layered halides. For the intermediate case of Pt(111) nothing can be said a priori, particularly since this surface has a holding potential significantly stronger than the layered halide and graphite. It is indeed intuitively obvious that V(A) - V(S) will increase with $|V(S)|$, and, for instance, this is the main reason why graphite appears strongly corrugated to He [18].

Considering the two above factors, period and amplitude of the corrugation, numerous surfaces of hexagonal symmetry can be expected to behave in the same way as the basal face of layered halides. Let us mention inter alia the (111) face of crystals having the blende (ZnS) or fluorite (CaF_2) arrangments, or the (0001) face of crystals having the wurzite (ZnS) arrangment, etc... Unfortunately there are no thorough adsorption studies on such surfaces. We quote however an isolated result which seems most beautiful. Indeed, on CaF_2, Davis [93] has determined a single adsorption isotherm of Ar at 77 K which clearly shows a first order monolayer condensation. This means a ratio T_{2c}/T_{3c}

above 0.51. Since the size incompatibility is small, about 0.005, this high value of T_{2c}/T_{3c} strongly suggest a lattice gas behaviour, comparable to what is observed on layered halides at favorable size compatibility.

Similar observations are possible for adsorption on surfaces of different symmetry. A most interesting example is the registered monolayer of methane adsorbed on the square MgO(100) face, which Coulomb and coworkers have studied both thermodynamically and structurally [94,95]. A lattice gas behavior could be expected since the diameter of CH_4, about 4.20 Å, is very close to the parameter of a square sublattice of sites, vz 4.21 Å. But for Kr or Xe, whose diameter are respectively 3.4 percent below or 4.5 percent above this parameter, there is no more registered monolayer, and the phase diagram looks like that for the continuous model [96]. Thus MgO(100) appears less efficient than the basal face of layered halides to impose to rare gases registered structures. A first obvious reason is that square structures have a lateral cohesion energy which is less favorable than the triangular structures, simply because the number of nearest neighbors of adatoms is 4 instead of 6. As a result the critical temperature of 2D condensation is much lower : the estimated value for CH_4 on MgO is 80 K [94] instead of 91 K for CH_4 on CdI_2 which also has a good size compatibility ($i \approx -0.007$). Additionally it is also possible that the corrugation of MgO (100) could be more favorable than that of the basal face of layered halides for the appearance of a 2D liquid.

From this short survey we can conclude that indisputable evidence for a lattice gas behavior, resembling to what we observe at favorable size compatibilities for monolayers of Ar, Kr, Xe and CH_4 on the basal face of layered halides, exists only for the CH_4-MgO system. And still our discussion suggests that such a behaviour should be observed quite frequently, in fact as soon as the conditions we stated above are fulfilled.

Let us look now at the corrugation-independent behaviour, which until now has been more frequently observed. In addition to the examples already mentioned, it seems worth quoting the adsorption studies on BN, a substrate differing very little from graphite. And indeed adsorption isotherm studies [41,97-100] suggest for the Ar, Kr, Xe and CH_4 monolayers, only minor differences with respect to graphite, which besides can be explained by a crystalline parameter *a* of BN, larger by 1.8 percent.

We get evidence for the coexistence of two dense monolayers adsorbed upon the basal face of a layered halide for the first time in 1978 while studying the adsorption of Xe on $NiCl_2$ [57]. We suggested to interpret it as a solid-solid coexistence mainly because our rough estimates of potential energies of interaction (the then unpublished results of Table

I) indicated that even at $i = -0.21$, V(A) - V(S) remained for adsorption on layered halides about 5 times higher than for adsorption on graphite. The subsequent thermodynamic [26,42,54] and structural [42,55] investigations showed we were wrong. Still, even now it seems to us astonishing that the main thermodynamic properties appear, within experimental accuracy, unaffected by the surface corrugation.

As intuitively obvious the amplitude of the corrugation diminishes as i is decreased, but no so much (see Table I). From $i = 0.15$ to -0.21, (V(A) - V(S))/3ε varies by a factor somewhat lower than 2, and remains about 5 times higher than at the surface of graphite. On the other hand V(SP) - V(S) has values comparable to those estimated for graphite. As a consequence it may be expected that in the monolayer the adatoms will stay preferentially around the honeycomb path formed by the sites (SP is midway between two S), avoiding a small space around the A positions. Suppose we take a snap-shot of a crystalline or a liquid monolayer adsorbed on a smooth surface and we superpose it above the lattice of the substrate in trying to avoid the vicinity of the unfavorable A positions for the adatoms by straining the adlayer. It seems intuitive that as soon as we are definitely below $i = 0$, say $i < -0.10$, the necessary strain will decrease as i decreases. Also the diffusive motion of atoms or molecules in the liquid layer, avoiding the A position will become easier. This is because the 2D space above which the atoms can sit becomes closer if compared to the diameter of the adatoms. Thus we can imagine that below $i < -0.09$ the liquid and solid monolayers could be comparable to what they are on a smooth surface, even though the absence of a clearcut first order melting for certain systems should be considered, as in the case of Ar on graphite, as already indicating some influence of the corrugation.

This tentative explanation of our results [54] is of course very qualitative. It receives some support from a computer simulation due to Prasad and Toxvaerd [101], who studied the behavior of a monolayer of fluid Ar adsorbed upon a (111) crystal plane of Ar which would have been expanded by 5 percent. For such a system a lattice gas behavior is expected in the range of temperature of interest to us. However, Prasad *et al.* calculations are carried out at much higher temperatures, namely at $kT/\varepsilon = 1$ or 1.5. For a liquid density of about 80 percent of its value at the 2D triple point, the density is almost the same at S and SP, but is about 80 percent lower at A. In spite of the strong tendancy of the surface to localize the adatoms, the radial distribution function differs little from the radial distribution function of a monolayer adsorbed on a smooth surface. The authors write "that the main effect of the external potential is to exclude some of the area for the particles, with the results that the system corresponds to a more compressed liquid-like system". Clearly computer simulations of both the liquid and solid monolayer in a temperature range slightly above the 2D triple point and for i <-0.9 are desirable to try to better understand the almost negligible influence of the corrugation upon the thermodynamic properties of the adlayer in these conditions.

It seems appropriate at this point to wonder how the monolayer looks like in the intermediate range of i, we place at $-0.09 < i < 0$, between the two simple regimes we discussed above. The steep variations with i of both T_{2c}/T_{3c} and $s_1 - s_0$ suggest a strong dependence of the structure of the adlayer upon the corrugation. For these particular values of i, a commensurability more involved than the 1x1 might exist. We feel that in absence of any structural information in that range of i, speculations about the possible structures would be premature at least within the scope of this paper.

Our knowledge of what happens beyond $i \approx 0.12$ is equally poor. Clearly we expect an increased influence of the corrugation. But the only significant piece of evidence we have for this is given to us by our investigation of multilayer adsorption [75,76,102,103], which will be hopefully discussed in a forthcoming review paper.

So far, we have discussed mainly the lowest-density solid monolayer in equilibrium with a fluid (gas or liquid) monolayer. Of course, as pressure is increased its structure can change more or less according to the range of i. Only in one case have we discussed such a behaviour, namely for Kr on PbI_2 at $i = 0.12$. Again, the only information we have concerning these compressed solids come from our of multilayer adsorption studies and therefore we will not discuss them here.

7. Conclusion

Our approach to the monolayer adsorption of Ar, Kr, Xe and CH_4 on layered halides is mainly based upon sets of adsorption isotherms, from which a number of significant quantities, such as the critical teemperature of 2D condensation and the adsorption entropy are derived. Their striking variations with the so-called size incompatibility (see Figure 7 and 11) allow an intuitive approach to the problems of the structure and of the properties of the monolayer, which is elementary but proves quite reliable. A limited number of structural investigations satisfactorily support the interpretation of the thermodynamic results.

Maybe the more gratifying outcome of this study is that by investigating the adsorption properties of a specific family of substrates we obtain a comprehensive view of the various situations which may be encountered in monolayer adsorption.

At one end, for $i < -0.09$, we observe what we call the corrugation-independent monolayer. The adjective is not to be taken literally. Most probably, sharp structural investigations would detect small modulations of the structure by the corrugation, but the energy and entropy changes associated to these are so small that they do not show in the thermodynamic quantities we are measuring. As an illustration supporting this view let us quote for instance the calculation by Novaco and McTague [104] of the energy associated with static distortion waves for rare gases on graphite : for Ar, which is always out of registry, this energy is lower than 1K, which is about 1/500 of the lateral interaction energy. We claim that our results, along with those obtained on a few other substrates, yield for the thermodynamic quantities, which are independent on the holding potential, (T_{2t}, T_{2c}, $s_l - s_c$, etc...) values characteristic of the phase diagram of a true 2D system of Ar, Kr and Xe. Theories using (6-12) Lennard-Jones effective potentials account well for the results, except in the vicinity of the critical temperature of 2D condensation where finite-size scaling of computer simulation results should permit a notable progress [49].

An entirely different behavior is observed at $0 < i < 0.12$. Within this range of size incompatibilities we suggest that the best description of the monolayer condensation is by a lattice gas model. Such models have been solved rigorously for atoms immobile at the sites, but as yet the efforts done to make them more realistic by taking into account the mobility, in particular the vibrations of the adatoms, involve so an important loss of rigor in the treatment of the configurational degree of freedom that one is skeptic about the chance that an analytic theory would succeed within a reasonable period of time. We feel that computer simulations are presently the only possible approach. Besides, they should be carried out over the entire range of size incompatibilities we have been investigating. Even in the range $i < -0.09$, we would like to know why the monolayer seems so little perturbed by the surface corrugation. Clearly it would be fanciful to try to use the true holding potential for each system because it is badly known. Besides, in our opinion, it would not be a judicious approach, because it would prevent us from localizing the minimum of the free energy versus i, as we explained earlier. This is an important issue about which our results do not give any information. In addition to giving additional useful information in the ranges i in which we already have a more or less precise idea of the properties and structure of the monolayer, computer simulations should bring new information in the range in which our knowledge is poor.

While reaching the end of this article, it seems difficult to escape the conclusion that the molecular theory of physisorbed monolayer is a most involved matter. The idea that adsorption isotherms could be accounted for by a limited number of analytical functions, often obtained by generalizing well-known equations, such as those of Langmuir or BET, wich was not uncommon within the community of physical chemists at the time we began our work, should clearly be dropped out. Physisorption upon well-defined surfaces, at least the densest crystal faces, now appears as a branch of the general theory of phase equilibrium. As such, it benefits from a long tradition of fascinating research which luckily has speeded up in the second half of this century. From the macroscopic point of view an adfilm can be described as a one-component system, involving only the adsorbate. But quite obviously, the molecular theory also involves the substrate. The approach to the adsorbate-substrate interaction, naturally introduces the concept of corrugation, which has dominated the analysis of our results. Its theoritical description uses the Fourier expansion [18,24]. So long as the positional order alone is involved, i.e. for rare gases or even to a certain extent for methane at sufficiently high temperatures, this concept is probably sufficient. But clearly, when orientational order comes into play, for molecules of adsorbate which are unspherical and/or possess an electrostatic moment, the problem becomes more involved. A review paper covering the experimental aspects of such a subject could at present be useful, but should not be limited to adsorption on layered halides for which the number of results is too limited. Another useful extension of the present review, which we hope to undertake later, concern multilayer adsorption. Again, it is not obvious that it should be limited to adsorption studies on layered halides, but it should certainly include them, since they significantly contributed to important advances in the applications to two interesting subjects wetting [102,103] on the one hand, the roughening transition [105,106] on the other.

Appendix: Semi-empirical estimation of the corrugation amplitude.

The data contained in Table I are obtained from numerical summations of pair-wise interaction potentials between one adatom and the atoms (Ar or Xe) or the ions of the substrate. For the layered chlorides the metal ions have been ignored, which should not affect the order of magnitude of the results, because it is the outer plane of the halide ions which is mainly responsible for the corrugation. The parameters of the (6-12) Lennard-Jones potentials between two different species are obtained by using the arithmetic mean for the distance and the geometric mean for the energy except for

$\varepsilon(Cl^-, Ar)$, or $\varepsilon(Cl^-,Xe)$. We took $\sigma(Cl^-) = 3.296$ Å which is 3.5 percent below $\sigma(Ar)$, and checked that a value of σ 10 percent lower did not change significantly the results, in fact by less than 10 percent. The values of $\varepsilon(Cl^-, Ar)$ or $\varepsilon(Cl^-, Xe)$ were chosen such that the summation at position S gives an energy equal to the sum of the condensation enthalpy of the bulk crystal, $h_0 - h_g$, and of $RT\ln(p_1/p_0)$. Since the thermal contributions to h_g and h_0 are almost equal at the temperatures we have been working, h_0-h_g is close to the potential energy of the bulk crystal, while $RT\ln(p_1/p_0)$ represents the excess due to the substrate.

ACKNOWLEDGMENTS

The subject matter of this review paper is contained in four successive theses carried out in my laboratory at the Centre d'Etudes Nucléaires de Saclay, Commissariat à l'Energie Atomique, France. They began with mine in 1970 [72], and went on with those of Yves Nardon in 1972 [74], of Francis Millot in 1975 [75] and of Christine Tessier in 1983 [42]. I would like to express my deep gratitude to these three students, who worked ardently and efficiently to advance this field of research.

REFERENCES

1. L. G. Joyner and P. H. Emmett, *J. Am. Chem. Soc.* **70**, 2353 (1948).
2. M. L. Corrin, *J. Am. Chem. Soc.* **73**, 4061 (1951).
3. L. Bonnetain, X. Duval and M. Letort, *C.R. Acad. Paris* **234**, 1363 (1952).
4. Y. Larher, in *Les Surfaces des Solides Minéraux* (ed. V. Gabis and M. Lagache), Société Française de Minéralogie et de Cristallographie, Paris, Vol.1, p.183 (1982).
5. J. H. Singleton and G. D. Halsey, J. *Phys. Chem.* **58**, 1011 (1954).
6. H. Clark, *J. Phys. Chem.* **59**, 1068 (1955).
7. C. H. Amberg, W.B. Spencer and R. A. Beebe, *Can. J. Chem.* **33**, 305, (1955).
8. S. Ross and W. Winckler, *J. Coll. Sci.* **10**, 330, (1955).
9. B. B. Fisher and W. G. McMillan, *J. Chem. Phys.* **28**, 549 (1958).
10. M. Bretz, J. G. Dash, D. C. Hickernell, E. O. McLean and O. E. Vilches, *Phys. Rev.* **A8**, 1589 (1973).
11. R. B. Griffiths, in *Phase Transitions in Surface Films* (ed. J. G. Dash and J. Ruvalds), Plenum Press, New-York, p.1 (1980).

12. Y. Larher, in *Les Surfaces des Solides Minéraux* (ed. V. Gabis and M. Lagache), Société Française de Minéralogie et de Cristallographie, Paris, Vol. 1, p. 73 (1982).
13. Y. Larher, *J. Chim.Phys.* **68**, 796 (1971).
14. A. Craig and R. McIntosh, *Can. J. Chem.* **30**, 448 (1952).
15. D. M. Young and J. A. Morrison, *J. Sci. Instr.* **31**, 90 (1954).
16. R. H. Van Dongen, *Doctoral Dissertation*, Delft, Netherlands, 1972.
17. S. Ross and J. J. Hinchen, in *Clean Surfaces* (ed. G. Goldfinger), Marcel Dekker, New-York, p.115, (1970).
18. W.A. Steele, *The Interaction of Gases with Solid Surfaces,* Pergamon Press, Oxford (1974).
19. J. S. Rowlinson, *Liquids and Liquids Mixtures*, Butterworths, London, 2nd ed., p.269 (1969).
20. D. Ambrose, *Vapor-Liquid Critical Properties,* NPL Report Chem. **107**, Teddington, U.K. (1980).
21. G. L. Pollack, *Rev. Mod. Phys.* **36,** 748 (1964)
22. Landolt-Börnstein, *Zahlenwerte und Funktionen aus Naturwissenschaften und Technik*, Springer-Verlag, Berlin, Vol. III. 7a (1973).
23. D. Billerey, C. Terrier, R. Mainard and P. Mériel, *C.R. Acad. Sc. Paris,* **B284**, 495 (1977).
24. W. A. Steele, *Surf. Sci.* **36**, 317 (1973).
25. K. E. Niebel and J. A. Venables, in *Rare Gas Solids* (ed. M.L. Klein and J.A. Venables), Academic Press, London, Vol. 1, p.558 (1976).
26. F. Millot, Y. Larher and C. Tessier, *J. Chem. Phys.* **76**, 3327 (1982).
27. Y. Larher, *Molec. Phys.* **38**, 789 (1979).
28. Y. Nardon and Y. Larher, *Surf. Sci.* **42,** 299 (1974).
29. H. K. Kim and M. H. W. Chan, *Phys. Rev. Lett..* **53**, 170 (1984).
30. H. S. Nham and G. B. Hess, *Phys. Rev.* **B37**, 9802 (1988).
31. F. Ser, Thesis, Paris XI, (1990) ; C.E.A. Rep. 5551 (1991).
32. H. Au-Yang and M. E. Fisher, *Phys. Rev.* **B11**, 3469 (1975).
33. Y. Larher, *Surf. Sci.* **65**, 365 (1977).
34. E. A. Guggenheim, *Thermodynamics*, North-Holland, Amsterdam, p.167 (1959).
35. J. S. Rowlinson and F. L. Swinton, *Liquids and Liquids Mixtures*, London, 3rd ed., p.261 (1982).
36. J. A. Barker, in *Rare Gas Solids (*ed. M.L. Klein and J.A. Venables), Academic

37. I. R. McDonald and K. Singer, *Molec. Phys.* **23**, 29 (1972).
38. F. Millot, *J. Phys. (Paris) Lett.* **40**, 9 (1979).
39. Y. Larher and B. Gilquin, *Phys. Rev.* **A 20**, 1599 (1979).
40. A. Thomy and X. Duval, *J. Chim. Phys.* **67**, 1101 (1970).
41. C. Tessier and Y. Larher, in *Adsorption at the Gas-Solid* and *Liquid-Solid Interface* (ed. J. Rouquerol and K.S.W. Sing), Elsevier, Amsterdam, p. 503 (1982).
42. C. Tessier, Thesis, Nancy, (1983) ; CEA Rep. 5250 (1984).
43. Q. M. Zhang, H.K. Kim and M. H. W. Chan, *Phys. Rev.* **B34**, 8050 (1986).
44. P. F. Sullivan and G. Seidel, *Phys. Rev.* **173**, 679 (1968).
45. M. J. Buckingham, *Ber. Bunsenges. Physik. Chem.* **76**, 287 (1972).
46. M. E. Fisher, *Rep. Progr. Phys.* **30 II**, 615 (1967).
47. R. O. Watts and I.J. McGee, *Liquid State Chemical Physics*, John Wiley, New-York (1976).
48. J. A. Barker, *Proc. Roy. Soc.* London **A377**, 425 (1981).
49. M. Rovere, D. W. Heermann and K. Binder, *J. Phys. Condens. Matter* **2**, 7009 (1990).
50. J. R. Klein and M. Cole, *Faraday Discuss. Chem. Soc.* **80**, 71 (1985).
51. K. J. Strandburg, *Rev. Mod. Phys.* **60**, 161 (1988).
52. Y. Larher, *Surf. Sci.* **134**, 469 (1983).
53. M. Nielsen, J. Als-Nielsen, J. Bohr, J.P. McTague and D.E. Moncton, *Phys. Rev.* **B35**, 1419 (1987).
54. C. Tessier and Y. Larher, in *Ordering in Two Dimensions* (ed. S.K. Sinha)North Holland, New-York, p. 163 (1980).
55. Y. Larher, P. Thorel, C. Tessier and A. Terlain, *Le Vide, Les Couches Minces* **38**, 181 (1983).
56. K. Morishige, H. Hanayama and S. Kittaka, *J. Chem. Phys.* **88**, 451 (1988).
57. P. Robert and Y.Larher, *Phys. Rev. Lett.* **40**, 1275 (1978).
58. B.J. Alder and T.E. Wainwright, *Phys. Rev.* **127**, 359 (1962).
59. F. Tsien and J.P. Valleau, *Molec. Phys.* **27**, 177 (1974).
60. S. Toxvaerd, *Molec. Phys.* **29**, 373 (1975).
61. S. Toxvaerd, *J. Chem. Phys.* **69**, 4750 (1978).
62. J. A. Barker, D. Henderson and F. F. Abraham, *Physica* **106A**, 226 (1981).
63. D. Henderson and J.A. Barker, *Molec. Phys.* **14**, 587 (1968).
64. J. P. Hansen and L. Verlet, *Phys. Rev.* **184**, 151 (1969).

65. D. M. Butler, J. A. Litzinger, G.A. Stewart and R.B. Griffiths, *Phys. Rev. Lett.* **42**, 1289 (1979).
66. Y. Larher, F. Millot and C. Tessier, J. *Chem. Phys.* **88**, 1474 (1988).
67. A. D. Migone, M.H.W. Chan, K.J. Niskanen and R.B. Griffiths, *J. Phys.* **C16**, L1115 (1983).
68. Y. P. Feng and M. H. W. Chan, *Phys. Rev. Lett.* **64**, 2148 (1990).
69. A. Terlain, Thesis, Nancy (1984); CEA Rep. 5302 (1985).
70. A. Terlain and Y. Larher, *Surf. Sci.* **125**, 304 (1983).
71. Y. Larher, *J. Phys. Chem.* **72**, 1847 (1968).
72. Y. Larher, Thesis, Université de Paris XI, Orsay (1970) ; CEA Rep. 4089 (1970).
73. Y. Larher, *J. Coll. Interface Sci.* **37**, 836 (1971).
74. Y. Nardon, Thesis, Nancy (1972); CEA Note 1574 (1973).
75. F. Millot, Thesis, Nancy (1979) ; CEA Note 1865 (1976).
76. Y. Larher and F. Millot, *J. Physique Colloq.* **38** C4, 189 (1977).
77. M. S. Anderson and C.A. Swenson, *J. Phys. Chem. Solids* **36**, 145 (1975).
78. H. Asada, *Surf. Sci.* **167**, 331 (1986).
79. Y .Larher, P. Thorel , C. Marti and B. Croset, unpublished results.
80. L. Onsager, *Phys. Rev.* **65**, 117 (1944).
81. H. Asada, *Surf. Sci.* **133**, 279 (1983).
82. H. Asada, *Surf. Sci* **137**, 412 (1984).
83. H. Asada, *Surf. Sci.* **159**, 393 (1985).
84. H. Asada, *Surf. Sci.* **167**, 331 (1986).
85. T. L. Hill, *An Introduction to Statistical Thermodynamics*, Addison-Wesley, Reading Ma, p.249 and 254 (1960).
86. J. J. Doll and W.A. Steele, *Surf. Sci.* **44**, 449 (1974).
87. P. A. Monson, W.A. Steele, and D. Henderson, *J. Chem. Phys.* **74,** 6431 (1981).
88. A. Dominko, A. Patrykiejew, and S. Sokolowski, *Thin Solid Films* **144,** 139 (1986).
89. B. Poelsema, L.K. Verheij, and G. Comsa, *Phys. Rev. Lett.* **51**, 2410 (1983).
90 B. Poelsema, L.K. Verheij, and G. Comsa, *Surf. Sci.* **152-153**, 851 (1985).
91. K. Kern, R. David, R.L. Palmer, and G. Comsa, *Phys. Rev. Lett.* **56**, 620 (1986).
92. K. Kern, R. David, P. Zeppenfeld, and G. Comsa, *Surf. Sci.* **195**, 353 (1988).
93. B. Davis, *J. Coll. Interface Sci.* **31**, 353 (1969).
94. K. Madih, Thesis, Marseille (1986).

95. J. P. Coulomb, K. Madih, B. Croset, and H. J. Lauter, *Phys. Rev. Lett.* **54,** 1536 (1985).

96. J. P. Coulomb, T.S. Sullivan, and O. E. Vilches, *Phys. Rev.* **B30**, 4753 (1984).

97 J. Régnier, A. Thomy, and X. Duval, *J. Coll. Interface Sci.* **70**, 105 (1979).

98. Y. Grillet and J. Rouquerol, *J. Coll. Interface Sci.* **77,** 580 (1980).

99. M. T. Alkhafaji and A. D. Migone, *Phys. Rev.* **B43**, 8741 (1991).

100. C. Bockel, A. Thomy and X. Duval, *Surf. Sci.* **90**, 109 (1979).

101. S. D. Prasad and S. Toxvaerd, *J. Chem. Phys.* **72**, 1689 (1980).

102. Y. Larher and D. Haranger, *Surf. Sci.* **39.** 100 (1973).

103. I. C. Bassignana and Y. Larher, *Surf. Sci.* **147**, 48 (1984).

104. A. D. Novaco and J. P. McTague, *J. Physique Colloq.* **38** C4, 116 (1977).

105. Y. Larher,in *4e Conférence Internationale de Thermodynamique Chimique,* Vol.7, p.5 (1975).

106. Y. Larher and F. Angerand, *Europhys. Lett.* **7,** 447 (1988).

LIST OF ACRONYMS

ADC	analogue-to-digital converter
AFM	atomic force microscope
AGIC	alkali-metal graphite intercalation compounds
APW	augmented plane wave
Bi(2212)	BISCCO with composition $Bi_2Sr_2CaCu_2O_8$
BISCCO	bismuth strontium calcium cuprate
C-I	commensurate-incommensurate
CDW	charge density wave
DFT	density functional theory
DOS	density of states
DWF	domain wall fluid incommensurate phase
EELS	electron energy loss spectroscopy
EPM	empyrical pseudopotential method
FLAPW	(all-electron selfconsistent) full potential APW
FS	Fermi surface
FSM	free standing mode
FT	Fourier transform
FWHM	full width at half maximum
GIC	graphite intercalation compounds
GK	Gordon-Kim (model)
HAS	helium atom scattering
HIC	hexagonally symmetric incommensurate
HOPG	highly-oriented pyrolytic graphite
HREM	high-resolution electron microscopy
I-S	incommensurate-striped
IETS	inelastic electron tunneling spectroscopy
KKR	Korringa-Kohn-Rostoker (method)
KT	Kosterlitz-Thouless (theory, phase transition)
KTHNY	Kosterlitz-Thouless-Halperin-Nelson-Young (theory)
LCAO	linear combination of atomic orbitals
LDA	local density approximation
LDF	local density functional
LDOS	local electronic density of states
LEED	low energy electron diffraction
LJ	Lennard-Jones (potential)
LSDF	local spin density functional
LUT	look up table
MBE	molecular beam epitaxy
MD	molecular dynamics
MO	molecular orbital
MOCVD	metal-organic chemical vapor deposition
N-I	normal-incommensurate
NC	nearly commensurate
NMR	nuclear magnetic resonance
NWS	Nicklow-Wakabayashi-Smith (graphite dynamics)
OPW	orthogonalized plane wave (method)
PLD	parallel lattice distortion
RAW	de Rouffignac-Alldredge-de Wette (model)
S-C	striped-commensurate
SBH	Schottky barrier height

SFM	scanning force microscopy
SP	saddle point
STM	scanning tunnel microscope
SXM	any kind of scanning microscopy
TCP	tricritical point
TLOR	topological long-range order
TOF	time of flight
XPS	X-ray photoemission
YBCO	yttrium barium cuprate
123	YBCO with composition $YBa_2Cu_3O_{7-x}$
2D	bidimensional
3D	tridimensional

GPSR Compliance

The European Union's (EU) General Product Safety Regulation (GPSR) is a set of rules that requires consumer products to be safe and our obligations to ensure this.

If you have any concerns about our products, you can contact us on ProductSafety@springernature.com

In case Publisher is established outside the EU, the EU authorized representative is:

Springer Nature Customer Service Center GmbH
Europaplatz 3
69115 Heidelberg, Germany

Batch number: 10816188

Printed by Printforce, the Netherlands